Identification of
Transcribed Sequences

Identification of
Transcribed Sequences

Edited by

Ute Hochgeschwender

National Institute of Mental Health
Bethesda, Maryland

and

Katheleen Gardiner

Eleanor Roosevelt Institute for Cancer Research
Denver, Colorado

Springer Science+Business Media, LLC

Library of Congress Cataloging-in-Publication Data

Identification of transcribed sequences / edited by Ute
  Hochgeschwender and Katheleen Gardiner.
        p.   cm.
    "Proceedings of the third International Workshop on the
  Identification of Transcribed Sequences, held October 2-4, 1993, in
  New Orleans, Louisiana"--T.p. verso.
    Includes bibliographical references and index.
    ISBN 978-0-306-44835-5   ISBN 978-1-4615-2562-2 (eBook)
    DOI 10.1007/978-1-4615-2562-2
    1. Human gene mapping--Congresses. 2. Genetic transcription-
  -Congresses.   I. Hochgeschwender, Ute.   II. Gardiner, Katheleen
  Jane.   III. International Workshop on the Identification of
  Transcribed Sequences (3rd : 1993 : New Orleans)
  QH445.2.I34 1994
  573.2'12--dc20   94-36835
                                                            CIP

Proceedings of the Third International Workshop on the Identification of Transcribed Sequences,
held October 2–4, 1993, in New Orleans, Louisiana

ISBN 978-0-306-44835-5

©1994 Springer Science+Business Media New York

Originally published by Plenum Press in 1994

PREFACE

The Human Genome Project, an endeavor to map and sequence the entire human genome, has been in existence for almost seven years. One result of this effort has been the development of increasingly detailed genetic and physical maps spanning large regions of virtually every chromosome. Paralleling this has been the increasingly high resolution mapping of many genetic diseases. Together, these developments have facilitated the isolation of specific disease genes and are now motivating the construction of comprehensive transcriptional maps. This latter endeavor represents a new facet of the genome project, and as such requires the recognition and solution of a new set of problems, with the attendant development and application of a new set of techniques.

The First International Workshop on the Identification of Transcribed Sequences in the Human Genome was held in 1991 and was attended by 23 investigators. Discussions at this meeting were largely devoted to defining the magnitude of the problem and describing the available techniques. A small number of laboratories reported the development of new techniques (at that time, for example, exon trapping, cDNA hybrid selection, direct cDNA screening, use of splice junction conserved sequences, etc.), but data were too limited to permit comparisons of their relative efficiencies.

As expected, interest in the problem of transcription mapping has grown - the third workshop, held in 1993, was attended by 58 investigators. With more laboratories having experience in transcriptional mapping, an important focus of this meeting was discussion of the relative strengths and weaknesses of the more popular techniques and potential complementaries among them. This has permitted some direct determination of the robustness and portability of particular methods, as well as some evaluation of their relative efficiencies for mapping projects of different scales. Variations and improvements on these techniques, as well as the development of alternative approaches also provided important topics for discussion.

With an increasing number of laboratories interested in transcribed sequence identification, it seemed timely to make the presentations and discussions of this latest meeting accessible to a wider audience. The purpose of these proceedings is four-fold: 1) to realistically assess the magnitude of the problem of transcribed sequence identification in the human genome; 2) to give an overview of current approaches to this problem; 3) to assist researchers in choosing method(s) appropriate to their projects and resources, through reports from other laboratories; and 4) to summarize both the general problems of transcribed sequences identification and the specific concerns about specific methods.

There are many ways to classify methods for transcriptional mapping, but for most purposes, a useful division is into two types of methods: 1) those which take genomic clones of known location and attempt to isolate transcribed sequences contained within them, and 2) those which start with transcribed sequences (cDNA clones) and attempt to characterize and map them to chromosomal regions.

This volume opens with a discussion of the scope of the problem facing a transcriptional mapper - how many genes are there and what is the complexity of the transcribed sequences in any tissue ? This is followed by descriptions of some classical approaches that include use of CpG islands, evolutionary conserved sequences and whole cosmid clones to screen cDNA libraries for transcribed sequences from a defined chromosomal region. Such approaches remain effective and, depending on the circumstances, may still be the method of choice.

Approaches based on hybridization between cDNAs and genomic sequences include cDNA hybrid selection applied to YAC clones and cosmid contigs. Several groups describe experiences with this approach, making it currently one of the most popular methods. Several variations on cDNA hybrid selection are also described which make use of RNA intermediates, metaphase chromosomes, etc. Another hybridization based method, that of direct cDNA screening of arrayed genomic clones, is discussed in two reports.

Four reports discuss exon trapping, a technique that relies on retention within mRNA of spliced exons from cloned genomic fragments. Experience with its application to cosmids, cosmid pools and YAC clones is reported.

Computer based analyses include GRAIL, which uses a neural network based approach to rate the potential exon content of genomic sequences. A discussion of the GRAIL program and use is presented elsewhere, but results of its application are included in several reports. It has been particularly effective in conjunction with direct cDNA screening. Also presented is a discussion of the efficiency of direct genomic sequencing for transcribed sequence identification.

The foregoing start with genomic clones and attempt to identify the transcribed sequences contained within them. An alternative set of methodologies approaches transcriptional mapping from the opposite direction - starting with transcribed sequences and collecting mapping and sequencing data for them. While generally less well suited for disease gene isolation, these approaches may provide data useful for large scale (e.g. whole genome, whole chromosome) transcriptional map construction. Two groups report on methods for, and results of, mapping numerous cDNAs from several random cDNA sequencing efforts. Two other groups report on the use of oligonucleotide hybridization to derive sequence information from arrayed cDNA libraries, and one group reports experiences using a sophisticated mRNA display methodology.

The last chapter summarizes formal and informal discussions addressing problems important to the transcriptional mapping field. These problems include defining a transcriptional map, and setting criteria for the inclusion of a novel DNA in a transcriptional map. The chapter also summarizes problematic aspects of particular methods.

Reports in this volume present an overview of current techniques in use for transcriptional mapping; they also illustrate the infant stage at which transcriptional mapping currently lies. Equally clear, however, is that transcriptional mapping will rapidly grow beyond this stage. This volume in no way represents the last word in transcriptional mapping. Rather, it presents a collection of first words in what will be a long and entertaining novel.

Katheleen Gardiner and Ute Hochgeschwender

ACKNOWLEDGMENTS

We wish to thank the Department of Energy, the National Center for Human Genome Research, and Amgen, Inc., whose support made this meeting possible. We also wish to thank the session chairs, and particularly, the discussion group leaders whose efforts and input made it a more valuable experience for all who attended. Special thanks are due Nan Matthews whose patience, organizational skills and attention to detail made both the meeting itself and these proceedings possible.

CONTENTS

Computer Based Approaches

FROM TRANSCRIBED SEQUENCES TO GENOMIC LOCALIZATION

Generating and Sequencing cDNAs

Mapping cDNAs

INTRODUCTION: Seven Blind Men and an Elephant

Miles B. Brennan

Unit on Genomics, NIMH
9000 Rockville Pike, Bethesda, MD 20892, USA

An oft-stated goal of the Human Genome Project is the cataloging of all human genes. To know the magnitude of this undertaking, an enumeration of the genes in man would be necessary. While the question of gene number has interested geneticists at least since Muller (1), it has also influenced thought on evolution and development. Indeed, the C-value paradox was important because of the implication that "simpler" organisms had more genetic complexity than mammals. Now the number of human genes has become a question of practical relevance, affecting how we pursue comprehensive transcriptional mapping. I present here a short, critical review of papers representing the major approaches which have been used to address this question.

The first approach derived from the theory of genetic load (2). Briefly, at steady-state, the rate of new, deleterious mutations entering a population is considered to be off-set by the rate of their loss through their phenotypic effects. It follows that no more than one recessive or 0.5 dominant "lethal" mutations can arise per haploid genome per generation. (Dominant deleterious alleles result in the genetic death of two genomes, one of which does not carry the mutation; for recessive alleles, both genomes contain the deleterious allele.) The concept of a lethal mutation here is quite different from its currently common usage. In evolutionary terms, whether a mutation leads to the immediate death of an organism or to the death of one progeny after 10 generations is irrelevant: selection will act with equal rigor in either case. There is no evolutionary reason to attribute a special significance to "essential" genes, those with fully penetrant lethal phenotypes.

Thus, the rate of new lethal mutations must equal the rate of their loss. Assuming that the upper bound for genetically determined death is 1, Muller concluded that the average rate of new lethal mutations must be less than 0.5 per haploid genome per generation. An upper limit to the number of genes would be given by the inverse of the average per locus rate of deleterious mutations. Unfortunately, little data existed on the frequency of deleterious mutations in man. Assuming that most deleterious mutations are dominant and that the average per locus rate for deleterious mutations is 1 in 50,000, Muller concluded that there are a maximum of 20,000 genes in man.

Following this line of reasoning, Ohta and Kimura (3) in 1971 attempted to define the intrinsic mutation rate for mammals as well as the fraction of mutations in mammalian cytochrome c which are deleterious using recently obtained protein sequence data. They estimated the intrinsic mutation rate in mammals at 8.3×10^{-9} per amino acid codon per year.

Based upon the genetic load theory, they then calculated the maximum number of genes, using the deleterious mutation fraction for cytochrome c. An uncertainty was introduced by the temporal denominator in the mutation rate: should it be per year or per generation? Varying the rate of deleterious mutation (0.5-1) and the mutation rate (per year or per generation), they calculated that at most between 0.6-6% of the human genome could be under the same constraint as cytochrome c. If all of this were in coding sequence, it would be sufficient to encode 10^4-10^5 proteins of 500 amino acids residues each. Included in the assumptions required for this approach are the mutation rate, the deleterious mutation rate, and the distribution of deleterious mutations between coding and non-coding sequences.

In 1981, an empirical approach was taken by Bodmer (4). Using the gene density from the well studied MHC and globin regions, and assuming that clustering of related genes was a common theme in the organization of mammalian genomes, he extrapolated over the entire genome and estimated between 50,000 and 100,000 genes. Recent study of the MHC region in man (5), has identified more genes, which would necessitate an upward revision of these estimations. The strong assumption of this approach is that of homogeneity in gene density; this assumption has been challenged by the observation that gene density varies dramatically in different isochores (6).

Most recently, Antequera and Bird (7) have used the correlation between HTF (Hpa II Tiny Fragments) islands and genes. With the assumptions that all HTF islands have associated genes and that a specific fraction of all genes are associated at their 5' ends with CpG islands, the number of genes could be determined simply by counting the number of such islands. Antequera and Bird quantitated the fraction of genomic DNA found in small Hpa II fragments, extrapolated to calculate the number of HTF islands and further to the number of genes. The variation in the fraction of genomic DNA was significant: they found between 1.01 and 1.26% of the human genomic DNA was liberated as Hpa II fragments <500 bp. Assuming an HTF island size of 1 kb, they gave approximations of the number of HTF islands ranging from 39,900 to 51,600. To estimate the total number of genes they assumed that the fraction of HTF associated genes in GenBank/EMBL database was an accurate reflection of the fraction of HTF genes in the human genome. They concluded that there are between 71,377 and 92,307 genes in the human genome.

Determining the number of genes expressed in a particular tissue was more experimentally tractable, but the results have not avoided controversy. The data come largely from two hybridization based approaches. In saturation hybridization, an excess of mRNA is hybridized to labelled unique sequence genomic DNA. The fraction of genomic DNA hybridized is a measure of the complexity of the sequences expressed in the tissue; the number of genes is calculated by assuming an average size for an mRNA. In general, these experiments showed that 1-2% of the non-repetitive fraction of the genome is expressed in mammalian cells in culture (cited in ref. 8); this is equivalent to 3×10^7 nt, or $1-2 \times 10^4$ mRNAs with an average length of 2000 nt. A notable exception is the estimation from rodent brain. In mouse brain, upwards of 3.8% of unique genomic sequences were protected by hybridization with poly A^+ mRNA, corresponding to a complexity of 1.4×10^8 nt, equivalent to 7×10^4 mRNAs with an average size of 2000 nt (9). (The 2000 nt estimation of the average mRNA size is used for historical reasons; the actual value may be much higher (see below).)

The other approach was to anneal a small amount of labelled cDNA with an excess of mRNA and follow the kinetics of hybrid formation. These data give not only the number of expressed sequences, but also their relative abundances. In general, the data from R_0t experiments agree with saturation experiments, giving values between 10,000 and 15,000 mRNA species expressed per tissue (reviewed in ref. 8). One explanation for the slightly higher complexity determined by saturation hybridization was proposed by Meyuhas and Perry (10): if the size of rare mRNAs is larger than the mass average, then the saturation hybridization will reflect the complexity and the R_0t will give the gene number. Such an inverse correlation between transcript abundance and size was reported by Milner and

2

Sutcliffe for the rat brain transcripts (ref.11, see below). More significantly, the number of genes expressed in brain estimated by $R_o t$ brain was only slightly larger than those for liver and kidney, rather than the much higher figure given by the saturation hybridization experiments (12). This discrepancy in the hybridization data for brain mRNA has not been resolved. A possible contamination of the poly A^+ mRNA with hnRNA would not affect the count of genes by $R_o t$ but would lead to anonymously high complexity measurements by saturation hybridization.

In 1983, Milner and Sutcliffe published their analysis of 191 clones picked randomly from a rat brain cDNA library (11). By approximating the expression levels of their mRNAs on Northern blots, they extrapolated to calculate the abundance and total numbers of genes expressed in brain. They estimated that there were between 5,000 and 30,000 genes expressed in the adult rat brain. The large uncertainty arose from the interpretation of a problematic class of 67 clones which produced no signal on Northern blots of brain, liver, and kidney mRNA. As low abundance transcripts contribute most of the complexity to mRNA populations, gene number calculations depend critically on whether the negative Northern blots resulted from the low abundance of the transcripts or from cloning artifacts. Thirty-one of the problematic 67 clones were tested on genomic Southern blots; twelve (39%) gave no signal.

A final question is what fraction of genes is expressed in common between two tissues. This question has been approached by $R_o t$ analysis (12). Briefly, the driver RNA is derived from one tissue and the labelled tracer cDNA is derived from another. Comparison of the rates and extents of the reaction allow an estimation of the overlap between the mRNA populations. 60-80% of the mRNA species present in mouse kidney are also expressed in mouse liver or brain (12). In the analysis by Milner and Sutcliffe (11), 70 of 154 cDNA clones detected mRNAs in brain and kidney and/or liver, 41 in brain only, and 67 failed to detect transcripts in any tissue. Again, depending on the interpretation of the problematic class of clones, they concluded that between 50 and 90% of brain mRNA species are not expressed in liver or kidney.

My purpose in reviewing the work on gene number was neither to derive a definitive number still less to use 20-20 hindsight to point out the difficulties in the approaches, but rather to illustrate that our fragmentary knowledge precludes any accurate predictions about the numbers of expressed sequences. Like the seven blind men who extrapolated from their partial knowledge of the elephant we have carried partial knowledge as far as it will go. We do not know what fraction of mutations will be lethal in the evolutionary sense. We do not know whether our sampling of genes gives an accurate value for gene density or for association with HTF islands. Arguments can be made, and not disproved, for any number of genes between 50,000 and 150,000. How we think about development and evolution will be strongly influenced by the number of genes available. Indeed, the large number of genes estimated in the rodent brain by saturation hybridization was accepted largely because it was consistent with the expectation that the mammalian brain, the most complex organ, should have an equally complex genetic endowment.

Finding an accurate value for the number of genes will be only a small part of the Human Genome Project: it will answer many basic questions about the constitution of genes and chromosomes and greatly facilitate mapping and identification of human disease genes. It is less certain that knowledge of the human genome sequence per se will lead to a "paradigm shift" (13) in human biology. Cogent arguments have been raised against this expectation (14), but millennial expectations were necessary to generate the political support to fund such an undertaking. In return for this support, research funded by the Project is directed and centralized (15). Thus, balanced against the certain, and less certain, benefits of the Genome Project are significant costs from politicizing how the scientific community determines priorities. We can look forward to science fragmenting into "Projects" competing for political favor by promising immediate gains. In time we will learn whether this was a productive model for science or a Mephistophilean bargain.

REFERENCES

1. H.J. Muller, Our load of mutations, *Am. J. Hum. Gen.* 2:111 (1950).
2. J.B.S. Haldane, The effect of variation on fitness, *Am. Natur.* 71:337 (1937).
3. T. Ohta and M. Kimura, Functional organization of genetic material as a product of molecular evolution, *Nature* 233:118 (1971).
4. W.F. Bodmer, The William Allan Memorial Award Address: Gene clusters, genome organization and complex phenotypes. When the sequence is known, what will it all mean? *Am. J. Hum. Gen.* 33:664 (1981).
5. H. Wei, W.-F. Fan, H. Xu, S. Parimoo, H. Shukla, D.D. Chaplin, and S.M. Weissman Genes in one megabase of the HLA class I region. *Proc. Natl. Acad. Sci. USA* 90:11870 (1993).
6. D. Mouchiroud, G. D'Onofirio, B. Aissani, G. Macaya, C. Gautier, and G. Bernardi, The distribution of genes in the human genome, *Gene* 100:181 (1991).
7. F. Antequera and A. Bird, Number of CpG islands and genes in human and mouse, *Proc. Natl. Acad. Sci. USA* 90:11995 (1993).
8. B.M. Lewin, "Gene Expression", volume 2, Second edition, John Wiley and Sons, New York (1980).
9. J.A. Bantle and W.E. Hahn, Complexity and characterization of polyadenylated RNA in the mouse brain, *Cell* 8:139 (1976).
10. O. Meyuhas and R.P. Perry, Relationship between size, stability and abundance of messenger RNA of mouse L cells, *Cell* 16:139 (1979).
11. R.J. Milner and J.G. Sutcliffe, Gene expression in rat brain, *Nucl. Acid. Res.* 11:5497 (1983).
12. N.D. Hastie and J.O. Bishop, The expression of three abundance classes of messenger RNA in mouse tissues, *Cell* 9:761 (1976).
13. T. S. Kuhn, "The Structure of Scientific Revolutions", University of Chicago Press, Chicago (1962).
14. R.C. Lewontin, "Biology as Ideology: The Doctrine of DNA", Harper Collins, New York (1991).
15. U.S. Dept. of Health and Human Services and U.S. Dept. of Energy, "Understanding Our Genetic Inheritance, The U.S. Human Genome Project: The First Five Years FY 1991-1995", NIH Publication Number 90-1590, April 1990.

IDENTIFICATION OF GENES AND CONSTRUCTION OF A TRANSCRIPTIONAL MAP IN Xq28

C. Tribioli[1], E. Maestrini[1], S. Bione[1], F. Tamanini[1], M. Mancini[1], C. Sala[2], G. Torri[2], S. Rivella[1] and D. Toniolo[1]

[1]Istituto di Genetica Biochimica ed Evoluzionistica, CNR, 27100 Pavia
[2]DIBIT-HSR, Milano, Italy

INTRODUCTION

The identification and sequencing of the human genes and their localization on the map of the human genome will provide an invaluable tool to study genetic disorders. It will identify candidate genes for diseases mapped to specific regions of the genome, where gross rearrangements of genomic material cannot be detected and all the genes from a large candidate region need to be isolated and analyzed. Intermediate goals are the development of methods to identify and map genes, and their application to the study of selected regions of the genome. One such region is the human X chromosome which contains many mapped disease loci. However, fewer than 100 X-linked genes have been characterized and cloned, a small proportion of the 3000 to 5000 genes expected on the chromosome (1). The distal portion of the long arm is among the most gene dense portions of the human X chromosome (1). By isolation and mapping of transcripts and CpG islands to the physical map of the region and by determination of partial nucleotide sequences and study of the pattern of expression of the transcripts identified, we are constructing a transcriptional map of Xq28 which should help in the identification of candidate genes for the several uncloned rare disorders localized to this chromosomal band.

The construction of physical and transcriptional maps of mammalian genomes, in providing new clues to a better understanding of genome organization, may also shed light on how chromosomal position can influence gene expression. In viruses and prokaryotes, the position of genes is important and often essential for regulating gene expression. In higher organisms, however, the increasing complexity of genomes makes the significance of gene order less obvious. Recent large scale sequencing projects have revealed that the genomes of lower eukaryotes are densely packed with genes (2,3). The organization of the genomes of vertebrates, which are much larger, is still poorly understood and in most instances the functional significance of gene organization along chromosomes is not known. Genes are not dispersed throughout the genome, rather they appear to be clustered in specific regions, mainly in the G positive bands or near the telomeres (4). In some instances (e.g. the homeobox or the globin genes) genes related in function are arranged in groups along the chromosome and in the same topological order in which they are

Identification of Transcribed Sequences, Edited by
U. Hochgeschwender and K. Gardiner, Plenum Press, New York, 1994

expressed (5,6,7). Gene order of apparently unrelated transcripts is also often maintained (1). Evolution via chromosomal rearrangements and genome duplication is the main factor responsible for keeping genes closely linked in different species but functional relationships may also be important in maintaining gene order. The mammalian X chromosome, whose high conservation in evolution may be related to the mechanism of gene dosage compensation via chromosome inactivation, is the best example of conserved gene association (1), and it may be a special case of the wider phenomenon of genomic imprinting (8,9).

To construct a detailed transcriptional map of Xq28, we have used different approaches including identification of CpG island rich regions, analysis of evolutionary conserved sequences, and cloning of cDNAs, coupled with the construction of a physical map of the region.

Identification of CpG Islands in Distal Xq28

A large number of Xq28 specific CpG islands were isolated from a genomic library of DNA of the hamster-human cell hybrid X3000.11 digested with the rare cutter restriction enzyme EagI and with the frequent cutter enzyme EcoRI (10). In such a library, many fragments identifying CpG islands were expected which should contain additional sites for rare cutter restriction enzymes (as in CpG islands) and should correspond to genomic DNA presenting the pattern of DNA methylation characteristic of CpG islands of the X chromosome (11): such clones have indeed been identified at high frequency (>70%) in the library (12). All the CpG island clones were regionally mapped by hybridization to a panel of DNAs from somatic cell hybrids (13). Two bands, Xq24 and Xq28, were highly enriched in CpG islands. About 1/3 of the CpG islands isolated, those in Xq28, were more precisely mapped with a specific hybrid panel and by Pulsed Field Gel Electrophoresis (13). It appeared that they were mainly clustered in the distal portion of the band, in a 2 Mb region between the DXS15 locus and the G6PD gene: of the 21 CpG islands mapped in the region, 16 were found within the 300 kb between the G6PD and the CV genes. No CpG islands were mapped in a 3-4 Mb region between DXS15 and DXS304 and only 2 were found between DXS304 and the IDS gene.

Fine Mapping of CpG Islands and Construction of the Physical Map

Our probes for CpG islands in Xq28 as well as probes for known genes in Xq28 and reference markers were hybridized to genomic libraries: two Xq28-specific and one X-chromosome specific cosmid libraries were probed. STSs were also derived to screen, by PCR, YAC libraries (the ICI (14) and the CEPH MegaYAC libraries (15)). End fragments from cosmids and YACs, as well as whole YACs, were used to reprobe the libraries and enlarge the cloned region. The clustering of the CpG islands helped to establish a cosmid contig and a partial physical map of this region of the human X chromosome (Fig. 1): in agreement with our previous data most of the CpG islands isolated from the X3000.11 library were localized in two DNA contigs of 100-200 kb each (Fig. 1). At the same time, the construction of a map for rare cutter restriction enzymes has demonstrated that all the EagI clones isolated were actually CpG islands since they all identified clusters of rare cutter enzymes. Moreover with this analysis, additional CpG islands were identified in some of the cosmids and YACs. A total of 26 CpG islands were mapped in the region analyzed (Fig. 1). Some corresponded to CpG islands at the 5' end of known genes of the region, the ALD, LICAM, P3, GdX and G6PD genes, which have been therefore precisely localized by hybridization to the cosmids. The tissue specific genes V2R, CV and FVIII, independently mapped to Xq28, were also found (Fig. 1). They did not demonstrate a CpG island at their 5' end. They were also mapped to the cosmids, interspersed with genes with CpG islands. All known genes were found in the cloned genomic DNA region, suggesting that most genes in Xq28 are clustered in the distal part of the band.

Isolation of cDNAs

CpG islands were separated by 5-20 kb DNA tracts, which appeared highly conserved in evolution, based on the hybridization of randomly chosen genomic fragments to DNA of different animal species (16).

By hybridization of fragments between CpG islands to cDNA libraries (human fetal brain, the human teratocarcinoma cell line NTERA2, skeletal muscle and others) cDNAs were isolated (Fig. 1) and used to probe the corresponding cosmid and genomic DNA from human and human-hamster hybrids carrying only the human X chromosome or portions of the X chromosome. All the cDNAs were single copy sequences and mapped exclusively to Xq28. In all instances, they hybridized to the corresponding cosmid and were localized between two CpG islands. A total of 15 new cDNAs were isolated.

Few of the fragments used to screen the cDNA libraries did not identify a cDNA (indicated as white boxes in Fig. 1): the alternative technique of direct cDNA selection (17) is being used to identify cDNAs from the same libraries. Preliminary data seem to indicate that this approach is successful.

In summary, the combined use of genomic probes to identify gene rich regions and of different procedures to isolate cDNAs allowed the identification of a large number of new genes in distal Xq28. The interspersion observed in Xq28 between tissue specific genes and genes with CpG islands indicates that CpG islands are good landmarks for gene rich regions.

Transcriptional Map

The transcriptional orientation of new and known genes in the region was determined by different methods: 1) by partial nucleotide sequence and demonstration of a PolyA tail or of a splice junction, 2) by RT-PCR using two oligonucleotides complementary to the opposite strands to separately prime reverse transcription and lastly 3) by hybridization of oligonucleotides or fragments at the 5' or 3' ends of cDNA clones to appropriate restriction digestion of the corresponding cosmids. The results are schematically shown in Fig. 1, where the orientation of transcription of each cDNA is indicated by an arrow. Our data show that transcripts in this region of the genome are not randomly oriented but that one hundred kb "transcriptional domains" may be defined where genes have the same direction of transcription (Fig. 1).

To gain information on the function of the new genes, cDNAs were also hybridised to total RNA from human cell lines, tumours and normal tissues. Most of the new cDNAs were ubiquitously expressed, some were found in higher amounts predominantly in a few tissues: in the G6PD/CV genes region some of the cDNAs were highly expressed in brain and/or muscle, in the L1CAM contig region some were highly expressed in kidney. Thus a common tissue distribution seems to be shared by many of the genes and it suggests that the transcriptional order we have defined may have a functional role. The significance of such order, in the absence of sequence similarity, will have to be searched for in common functions in some tissues or at some developmental stage: this may require some kind of temporal and/or regional control of expression to open them to transcription at the same time.

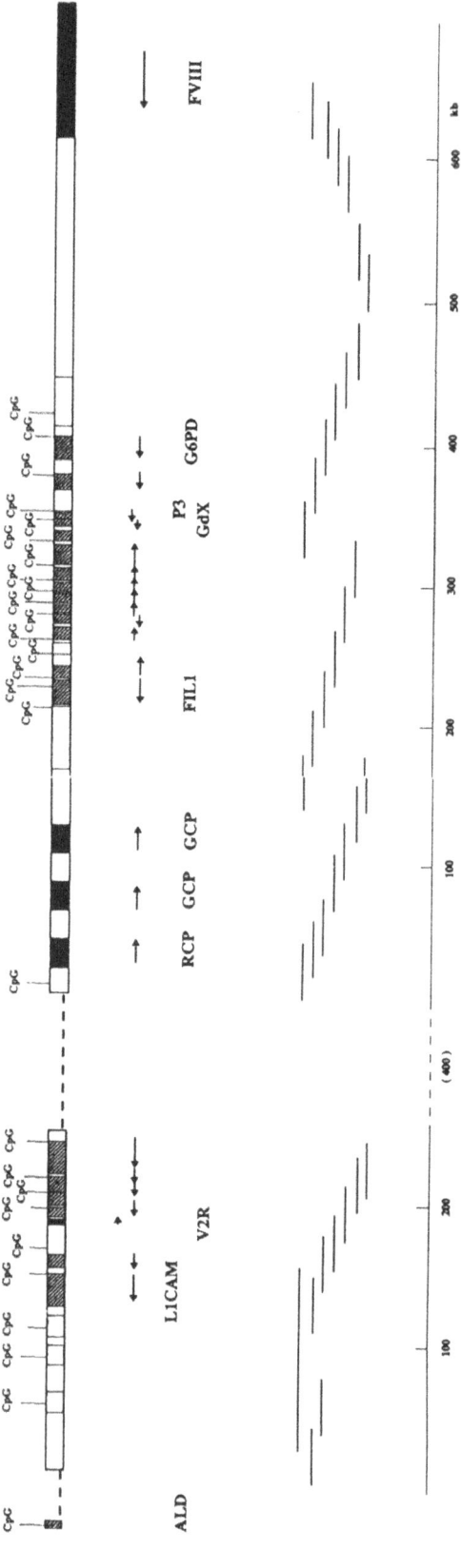

Figure 1. Schematic representation of the distal Xq28 and of the transcriptional organization of the genes. The genomic region occupied by each of the cDNAs is indicated by slashed boxes. Blackened in boxes indicate tissue specific genes. Regions where we expect to find additional transcripts are indicated by white boxes. Arrows indicate the direction of transcription of each gene. Horizontal bars indicate cosmids or YACs in the contigs.

CONCLUSIONS

Until now only the structural organization of special gene families has been studied in detail (5,6). In mammals, however, genomic regions with a concentration of genes similar to the one described in our work and not related in sequence, have been recently described (18,19). The best known is the MHC locus (20-21). The study of additional chromosomal regions with similar density of apparently unrelated genes will establish if the gene organization we have described in Xq28 is the rule in the human genome and will be the basis for functional studies on how chromosomal position may affect gene expression. This would be particularly important for X linked or imprinted genes on autosomes, which are subject to chromosomal or locus control of their expression.

Many genes responsible for inherited disorders are mapped to distal Xq28 (22). The results of our studies on the expression of the new genes in cell lines and tissues suggest that some of them may be candidates for those diseases. The muscle specific expression of some of the genes indicates that they may be responsible for the muscle disorders mapped to this region, the Emery-Dreifuss muscular dystrophy and the Barth syndrome (22). In the same way, the brain specific expression of others makes them candidates for X linked mental retardations or neurological disorders.

Studies are in progress to determine if this is the case and to complete the transcriptional map of the region.

ACKNOWLEDGEMENTS

This work was supported by grants from Progetto Finalizzato CNR "Ingegneria Genetica" and by "Telethon Italy". We thank Drs. A. Poustka, H. Lehrach and P. Vezzoni for cosmid library filters and for cosmids and M. D'Urso for the YAC 527.

REFERENCES

1. Human gene mapping 11. *Cytogenet. Cell Genet.* 58:1 (1991).
2. S.G. Oliver, Q.J. Van der Aart, M.L. Agostoni Carbone, M. Aigle, L. Alberghina, D. Alexandraki, G. Antoine, R. Anwar, J.P. Ballesta, P. Benit, *et al.*, The complete DNA sequence of yeast chromosome III, *Nature* 357:38 (1992).
3. R. Waterston, C. Martin, M. Craxton, C. Huyn, A. Coulson, L. Hillier, R. Durbin et *al.*, A survey of expressed genes in C. elegans, *Nature Genetics* 1:114 (1992).
4. W.A. Bickmore and A.T. Sumner, Mammalian chromosome banding - an expression of genome organization, *Trends Genet.* 5:144 (1989).
5. E. Boncinelli, A. Simeone, D. Acampora and F. Mavilio, HOX gene activation by retinoic acid, *Trends Genet.* 7:329 (1991).
6. F.S. Collins and S.M. Weissman, *Prog. Nucl. Acid Res. Mol. Biol.* 31:315 (1984).
7. C.G. Kim, E.M. Epner, W.C. Forrester and M. Groudine, Inactivation of the human β-globin locus control region, *Genes and Dev.* 6:928 (1992).
8. M.A. Surani, R. Kothari, N.D. Allen, P.B. Singh, R. Fundele, A.C. Ferguson-Smith and S.C. Barton, Genomic imprinting and development in the mouse, *Development (Suppl.)* pp. 89 (1990).
9. W. Reik, Genomic imprinting and genetic disorders in man, *Trends Genet.* 5:331 (1989).
10. E. Maestrini, S. Rivella, C. Tribioli, D. Purtilo, M. Rocchi, N. Archidiacono and D. Toniolo, Probes for CpG islands on the distal long arm of the human X chromosome are clustered in Xq24 and Xq28, *Genomics* 8:664 (1990).
11. A.P. Bird, CpG islands as gene markers in the vertebrate nucleus, *Trends Genet.* 3:342 (1987).
12. C. Tribioli, F. Tamanini, C. Patrosso, L. Milanesi, A. Villa, R. Pergolizzi, E. Maestrini, S. Rivella, S. Bione, M. Mancini, P. Vezzoni, and D. Toniolo, Methylation and sequence analysis around EagI sites:identification of 28 new CpG islands in Xq24-Xq28, *Nucl. Acids Res.* 20:727 (1992).
13. E. Maestrini, F. Tamanini, P. Kioschis, E. Gimbo, P. Marinelli, C. Tribioli, M. D'Urso, G. Palmieri, A. Poustka, and D. Toniolo, An archipelago of CpG islands in Xq28: identification and

fine mapping of wo new CpG islands of the human X chromosome, *Hum. Mol. Genet.* 1:275 (1992).

14. R. Anand, J.H. Riley, R. Butler, J.C. Smith, and A.F. Markham, A 3.5 genome equivalent multi access YAC library: construction, characterization, screening and storage, *Nucl. Acid Res.* 18:1951 (1990).

15. H.M. Albertsen, H. Abderrahim, H.M. Cann, J. Dausset, D. Le Paslier, and D. Cohen, Construction and characterization of a yeast artificial chromosome library containing seven haploid human genome equivalents, *Proc. Natl. Acad. Sci. USA* 87:4256 (1990).

16. S. Bione, F. Tamanini, E. Maestrini, C. Tribioli, A. Poustka, G. Torri, S. Rivella, and D. Toniolo, Transcriptional organization of a 450 kb region of the human X chromosome in Xq28, *Proc. Natl. Acad. Sci. USA* 90:10977 (1993).

17. B. Korn, Z. Sedlacek, A. Manca, P. Kioschis, D. Konecki, H. Lehrach, and A. Poustka, A strategy for the selection of transcribed sequences in the Xq28 region, *Hum. Molec. Genet.* 1:235 (1992).

18. J. Yon, T. Jones, K. Garson, D. Sheer, and M. Fried, The organization and conservation of the human surfeit gene cluster and its localization telomeric to the *c-abl* and *can* proto-oncogenes at chromosome band 9q34.1, *Hum. Molec. Genet.* 2:237 (1993).

19. F. Larsen, J. Solheim, T. Kristensen, A. Kolstoe, and H. Prydz, a tight cluster of five unrelated human genes on chromosome 16q22.1, *Hum. Molec. Genet.* 2:1589 (1993).

20. J. Trowsdale, J. Ragoussis, and D.R. Campbell, Map of the human MHC, *Immunol. Today* 12:443 (1991).

21. J. Trowsdale and S.H. Powis, The NHC: relationship between linkage and function, *Curr. Opin. Genet. Dev.* 2:492 (1992).

22. V.A. McKusick, "Mendelian Inheritance in Man", 9th ed. The Johns Hopkins University Press, Baltimore (1990).

USE OF cDNA SELECTION AND EVOLUTIONARILY CONSERVED SEQUENCES TO ISOLATE TRANSCRIBED SEQUENCES FROM REGION Xp11.21

Eric N. Burright[1], N. German Pasteris[1], Michael D. Bialecki[2], and Jerome L. Gorski[1,2]

Departments of [1]Human Genetics and [2]Pediatrics and Communicable Diseases, University of Michigan, Ann Arbor, MI 48109

ABSTRACT

Although advances in the development of positional cloning techniques have rapidly accelerated the pace of physical mapping and gene localization, the complete and efficient isolation of transcribed sequences from within large targeted genomic regions remains a significant challenge. Here, we describe two approaches used to isolate transcripts encoded within region Xp11.21. First, subcloned genomic fragments derived from regional YAC clones were used to identify evolutionarily conserved sequences (ECSs); ECSs were used to screen cDNA libraries and isolate regional transcripts. Second, YAC DNA was immobilized on a membrane and hybridized to PCR-amplified cDNA clone inserts; selectively retained inserts were subcloned to construct an enriched region-specific cDNA library. Both techniques were successfully used to isolate region-specific transcribed sequences. The advantages and limitations of each approach are discussed.

INTRODUCTION

Despite recent advances in positional cloning techniques, the implementation of efficient and comprehensive strategies for the isolation of regional transcribed sequences has remained problematic and challenging. Several contributing factors have been identified: (a) only a small fraction of the mammalian genome appears to be transcribed (1); (b) the number, size of and distance between exons varies dramatically among different regions of the genome (2,3); and (c) cell types differ in their mRNA populations (4,5). In addition to these biologic considerations, typically, to clone a region-specific disease-related gene, an investigator must opt to use one or more of the multiple, disparate, and independently developed published methods or, alternatively, develop yet another approach for gene isolation. Techniques of proven utility include the use of: (a) evolutionarily conserved sequences (6,7) and CpG islands (8,9) to screen cDNA libraries, (b) radiolabelled cDNAs to identify encoding genomic clones (10,11), (c) *in vivo* assays that detect mRNA splice sites to isolate exons (12,13), and (d) immobilized genomic DNA to construct enriched region-specific cDNA libraries (14,15). Although each of these approaches has been limitedly successful, none appears to be superior to all others. Limited resources commonly preclude a researcher's ability to try all methods; therefore, strategy selection can be critical.

In comparing and selecting among described methods, several factors can be considered. First, to facilitate technical importation and expeditious application, the method should be technically simple and not dependent upon limited or difficult to obtain resources or expertise. Second, to isolate the gene of interest, the technique should provide a sufficiently detailed transcriptional representation of the region. Third, the technique should yield regional candidate cDNA clones that are readily verifiable. Finally, to facilitate an analysis of a relatively large (1-2 Mbp) region, such as those identified in linkage analyses, the technique should be amenable to automation, thereby allowing for the characterization of larger genomic regions without considerable method redesign. Since none of the published methods completely fulfills all of these stringent criteria, a decision concerning which approach to use is often ultimately based on size of the region under study, available resources, and the individual expertise of the laboratory.

Our laboratory has an interest in characterizing and isolating genes from region Xp11.21. This region is approximately 4.5 Mbp in size (16,17), has been largely covered in a YAC clone contig (18), and contains several disease-related genetic loci including: a gene responsible for X-linked sideroblastic anemia (ALAS2) (19), a gene involved in neural crest differentiation, incontinentia pigmenti type 1 (IP1) (16,17,20), and a gene responsible for a craniofacial/skeletal dysplasia malformation complex, Aarskog syndrome (FGDY) (21). Here, we report on our experience of using (a) evolutionarily conserved sequences and (b) a modified cDNA selection strategy to isolate transcripts encoded within region Xp11.21. Our considerations used in choosing these approaches, as well as the advantages and limitations of each approach, are discussed.

METHODS

Standard Hybridization Analyses

Southern transfers were performed and radioactive probes were prepared as previously described (20). Hybridizations were performed for 16-24 h at 65°C; membranes were washed to a final stringency of 0.1XSSC, 0.1% SDS at 65°C for at least 15 min. Southern "zoo" blots containing DNA of divergent species (EcoRI, PstI, and BamHI-digested human, baboon, dog, cat, mouse, and duck DNA) were washed at a final stringency of 0.1XSSC, 0.1% SDS at 55°C; DNA fragments were considered to contain evolutionary conserved sequences if, by using these conditions, one or more fragments were reproducibly detected in non-primate DNA. A regional somatic cell hybrid mapping panel that dissects band Xp11.21 into six distinct regions was used to map DNA markers to a particular interval of Xp11.21; this hybrid mapping panel has been described (16,17,20).

cDNA library hybridization screenings were performed by standard methods (22). A human 17 week fetal brain λZAPII cDNA library (Stratagene) was used exclusively; cDNA clone inserts were recovered in plasmids by *in vivo* excision as recommended by the supplier.

Yeast Artificial Chromosomes (YACs)

By using DNA markers mapped to Xp11.21, regional YAC clones were isolated and restriction mapped as described (17). To identify non-chimeric clones, YAC clones were analyzed by fluorescence *in situ* hybridization and by characterizing isolated YAC clone-specific end fragments (23) by using a regional somatic cell hybrid mapping panel. To facilitate the fine mapping of YAC clone inserts, λ(EMBL3) or cosmid (pWE15) subclone libraries were constructed from selected YAC clones. Contig maps of YAC insert subclones were constructed by using human-specific interspersed repetitive sequence (IRS)-PCR products and YAC clone-specific end fragments as described (24). Human-specific cosmid subclones derived from YAC inserts were individually picked, grown in culture, and arrayed as 15% glycerol stocks in 96 well microtiter dishes. Replica filters, each containing 384 individual clones, were prepared and hybridized by using standard methods (22).

Hybridization-based cDNA Selection

The general strategy of this technique is illustrated schematically in Fig. 1. Yeast DNA was prepared in agarose plugs and fractionated on a contour-clamped homogeneous electric field (CHEF-DRII, BioRad) gel system as described (17). YAC DNA was excised from the gel, digested *in situ* with Sau3A, and recovered on a silica matrix (GeneClean II, Bio101); YAC DNA (100 ng) was heat denatured 5 min at 100°C, snap cooled on ice for 2 min, and dot blotted onto 6 mm^2 nylon membranes (Hybond N, Amersham). Immobilized DNA was denatured 5 min in 0.5 M NaOH, 1.5 M NaCl, neutralized 10 min in 0.5 M Tris-HCl pH 7.5, 1.5 M NaCl, washed briefly in 2XSSC, air dried, baked at 80°C for 90 min, and UV-crosslinked.

To prepare PCR amplified cDNA inserts, two pairs of primers directed against vector λZAPII were used: an external pair (GGAAACAGCTATGACCATG, TTGTAAAACGACGGCCAGT) and an internal pair (CGAAATTAACCCTCACTAAAGGG, GTAATACGACTCACTATAGGGCG). All PCRs were 100 μl in volume and contained 50 mM KCl, 10 mM Tris-HCl, pH 8.3, 1.5 mM MgCl$_2$, 0.001% gelatin, 200 μM each dNTP, 0.5 μM each primer, and 2.5 U AmpliTaq DNA polymerase (Perkin-Elmer/Cetus); all PCRs involved 30 cycles of denaturation at 94°C (1 min), annealing at 55°C (1 min), and extension at 72°C (5 min). Amplified cDNA insert DNA was purified by spin column (Millipore), resuspended in TE, and preannealed as described for 90 min (25,26).

Immobilized YAC DNA was preannealed for 90 min at 65°C in hybridization buffer with 1 mg/ml sheared total human DNA, 1.25 mg/ml sheared yeast DNA, and 1 mg/ml pBR322 plasmid DNA in a total volume of 80 μl, under mineral oil. Following preannealing, filters were rinsed in 5XSSC, 0.5% SDS and transferred to individual tubes containing 1 μg of pre-annealed cDNA inserts. Hybridizations were performed for 20-24 h at 65°C; filters were washed four times for 20 min in 2XSSC, 0.1% SDS at room temperature, and three times for 30 min in 0.1XSSC, 0.1% SDS at 65°C. Filters were rinsed in 0.1XSSC and transferred to tubes for PCR amplification using the external primer set. Resultant products were amplified using the internal primer set. Amplified selected cDNA inserts were subcloned directly into the TA cloning vector, pCRII (Invitrogen), and transformed into bacteria.

Recombinant clones were individually inoculated into a series of 96 well microtiter tubes, grown overnight in culture, and stored as 15% glycerol stocks at -80°C in microtiter plates. Replica filters for colony hybridizations were prepared and screened as described above. Prior to use as probe, radiolabeled YAC DNA was preannealed with 3 mg/ml human DNA, 1 mg/ml yeast DNA, and 100 μg/ml pBR322 DNA in 5XSSC; DNAs were mixed, denatured for 10 min at 100°C, snap cooled on ice for 2 min, and preannealed for 15 min at 65°C.

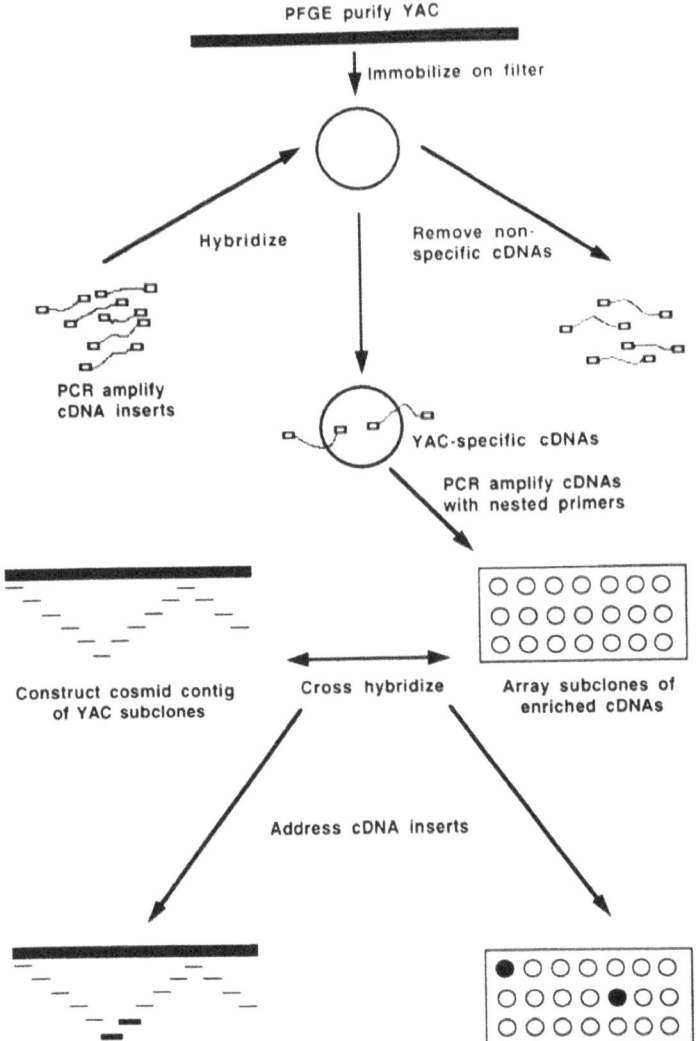

Figure 1. Schematic diagram illustrating the general strategy of cDNA selection (14,15,26). Gel purified YAC DNA was bound to a filter and hybridized to a PCR-amplified cDNA inserts. YAC-specific cDNAs were re-amplified and subcloned; recombinant clones were arrayed in microtiter dishes. Selected cDNAs were analyzed by identifying clones that cross-hybridized to either YAC or YAC subclone DNA.

RESULTS

Evolutionarily Conserved Sequences (ECSs)

YAC clones isolated from region Xp11.21 were analyzed to identify those containing non-chimeric inserts in close proximity to disease-specific X-chromosomal translocation breakpoints. Two non-chimeric independent YAC clones, y741A and y323A, were selected for further analysis (23,24); restriction maps of these YAC inserts are shown in Fig. 2. Simultaneous to the construction of YAC contig maps, cosmid (y741A) and λ (y323A) subclones were hybridized to zoo blots to identify subclones containing evolutionarily conserved sequences (ECSs); subclones containing CpG islands were prioritized for analysis. To facilitate cDNA library screening, low-copy (LC) restriction fragments containing ECSs were isolated from subclones containing ECSs. Two independent LC fragments containing ECSs were identified; DNA markers $LC_{HN}1.4$, a 1.4 kb HindIII/NotI fragment, and $LC_E5.5$, a 5.5 kb EcoRI fragment, were isolated from clones y741A and y323A, respectively (Fig. 2).

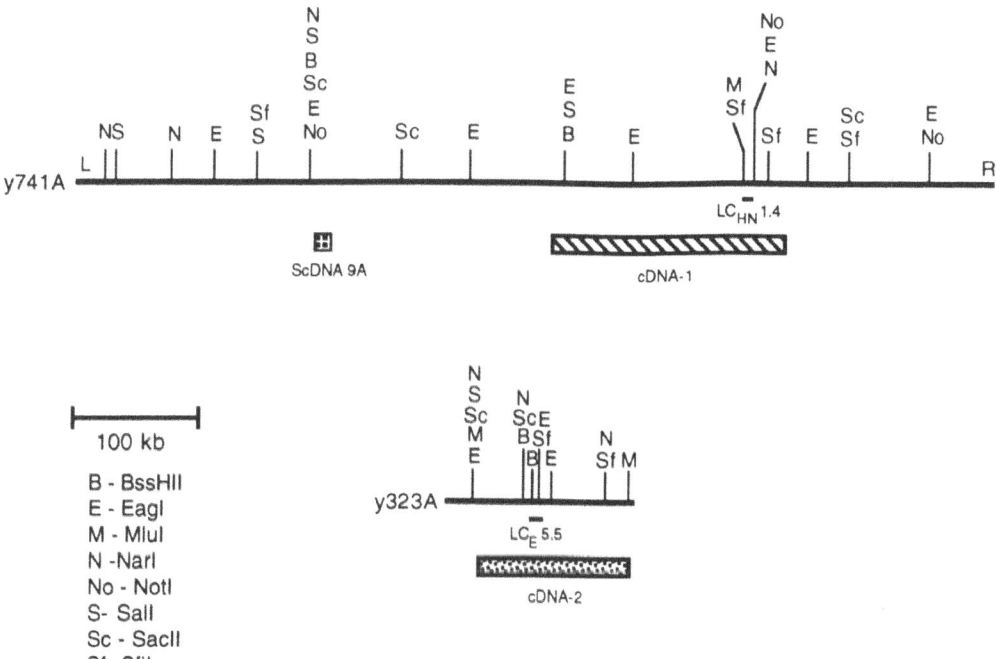

Figure 2. Schematic diagram of the derived restriction maps of YAC clones y741A and y323A showing the relative locations of low-copy fragments that detect evolutionarily conserved sequences ($LC_{HN}1.4$ and $LC_E5.5$) and isolated cDNA clones (cDNA-1 and cDNA-2). The location of selected cDNA clone ScDNA9A is also indicated.

Used as probes to screen a human 17 week fetal brain cDNA library, both LC fragments detected independent cDNA clones. $LC_{HN}1.4$ detected six partially overlapping cDNA clones representing a total of 2.4 kb of sequence (cDNA-1); $LC_E5.5$ detected two partially overlapping clones representing 3.8 kb of sequence (cDNA-2) (Fig. 2). To verify the origin of the isolated cDNAs, clones were mapped to Xp11.21 by using a regional somatic cell hybrid mapping panel (16,17,20). Both cDNA clones detected tissue-specific transcripts when hybridized to Northern blots containing human fetal tissue-specific poly(A)+mRNA; transcripts of 10.0 kb and 4.6 kb were detected by cDNA-1 and cDNA-2, respectively (data not shown). Using YAC subclones, hybridization analyses indicated that cDNA-1 spanned at least 200 kb of clone y741A; similarly, cDNA-2 spanned at least 110 kb of clone y323A (Fig. 2).

cDNA Selection

To isolate additional cDNA clones encoded by YAC clone y741A, varying quantities (2.5 x 10^9, 2.5 x 10^8, and 2.5 x 10^7 pfu) of the cDNA library were PCR amplified by using the external primers and aliquots of each reaction were hybridized to immobilized y741A DNA. Following hybridization washes, cDNA inserts selectively retained on the membranes (selected cDNAs, ScDNAs) were sequentially amplified by using the external and internal primers; to monitor hybridization efficiency, cDNA-1 served as an internal control. PCR products isolated from a typical cDNA selection experiment are shown in Fig. 3. PCR products containing portions of cDNA-1 were limited to those cDNA inserts derived from the 2.5 x 10^9 pfu cDNA library pool (Fig. 3); PCR products containing portions of cDNA-1 were subcloned and arrayed in microtiter dishes for analysis.

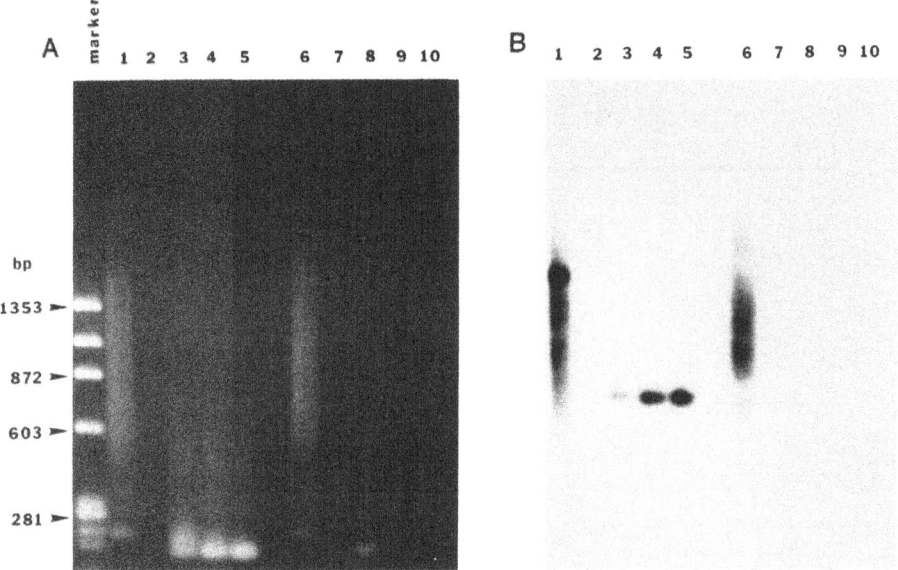

Figure 3. Comparison of PCR-amplified cDNA inserts (1/10 reaction) obtained from either 2.5 x 10^9 pfu (lanes 1-5) or 2.5 x 10^8 pfu (lanes 6-10) fractionated in a 1.5% agarose gel and stained with ethidium bromide (A) and an autoradiogram of the transferred DNA hybridized with probe cDNA-1 (B). External primers alone were used to amplify the products in lanes 1,2,6 and 7; lanes 1 and 6 contain inserts prior to selection; lanes 2 and 7 contain selected inserts following amplification with external primers. Lanes 3-5 and 8-10 contain the products of amplification using internal primers using 1/10 (lanes 3 and 8), 1/30 (lanes 4 and 9), and 1/50 (lanes 5 and 10) of the products of lane 2 (lanes 3-5) and 7 (lanes 8-10), respectively. Probe cDNA-1 detected specific PCR amplified inserts in lanes 3-5 only.

Previous investigators had observed an approximate 10^3 fold enrichment of regional cDNA clones per round of cDNA selection (14,15,26); therefore, to ensure an adequate representation of ScDNA clones, 3,000 recombinant clones were picked. To identify ScDNA clones containing repetitive inserts, membranes containing arrays of ScDNAs were

16

hybridized with radiolabeled human genomic DNA; 75 of the 3,000 (2.5%) ScDNAs cross-hybridized (data not shown). These results are in close agreement to previous observations (27,28). To identify ScDNA clones containing additional y741A transcripts, ScDNA clones were sequentially hybridized with radiolabeled probes derived from cDNA-1 and YAC y741A DNA (Fig. 4). Probe cDNA-1 detected 12 ScDNA clones and YAC y741A DNA detected 15 ScDNA clones; 4 ScDNA clones were detected by both probes. The length of these ScDNA inserts ranged from 800 to 200 basepairs (bp) with an average size of 250 bp (data not shown). Since clones containing portions of cDNA-1 had a frequency of 1 in 10^6 in the starting cDNA library, these results indicate that the selection procedure yielded a 10^3 enrichment for ScDNA clones containing a portion of cDNA-1, an finding consistent with expectations.

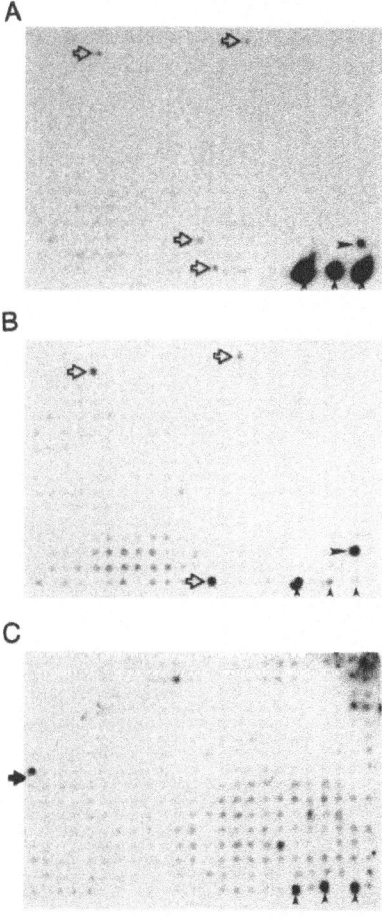

Figure 4. Autoradiograms of identical (A and B) and similar (C) membranes containing 384 arrayed selected cDNA clones sequentially hybridized with probe cDNA-1 (A), and YAC y741A insert DNA (B and C). Vertical arrowheads indicate cDNA-1 colonies (positive controls) present on each membrane; horizontal arrows indicate clones detected with cDNA-1 (A) and y741A DNA (B and C). The solid horizontal arrow shows a cDNA clone (ScDNA9A) detected only with y741A DNA (C); small horizontal arrowheads show cDNAs containing repetitive DNA (A).

Figure 4 shows one of the 11 ScDNA clones, ScDNA9A, detected with YAC y741A DNA only. The 250 bp insert of ScDNA9A was mapped to YAC y741A by PFGE analysis and by cross-hybridization studies using YAC cosmid subclone arrays (Fig. 2); ScDNA9A failed to hybridize to any of the other ScDNA clones isolated (data not shown). Characterizations of the remaining 10 ScDNA clones detected with YAC y741A DNA are in progress.

To identify additional ScDNAs encoded in YAC y741A, and identify the subcloned cosmids containing them, probes made from pools of ScDNAs were hybridized to membranes containing an array of 580 independent y741A-derived cosmid subclones; repetitive ScDNAs and ScDNA clones detected with cDNA-1 subclones and/or y741A DNA were excluded from the pools. Ten separate pools were used; each pool consisted of approximately 90 independent ScDNA clones. Seven of the 10 pools detected cosmid clones; each pool detecting a cosmid detected 6 to 16 different cosmids. A total of 54 cosmid clones were detected including 5 cosmids known to contain exons of cDNA-1, 4 cosmids found to contain ScDNA9A, and 45 cosmids which had not been identified as containing transcribed sequences; 16 (30%) of the detected cosmids were recognized by two or more different ScDNA pools. Representative results of these analyses are shown in Fig. 5. Characterizations of the 45 cosmids suspected of containing a transcribed sequence are ongoing.

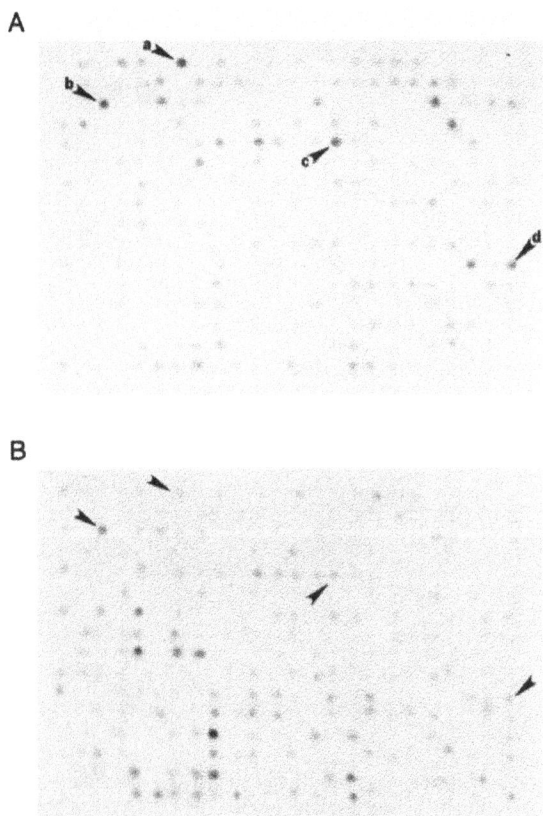

Figure 5. Autoradiograms of identical membranes containing 384 arrayed cosmid clones derived from YAC y741A sequentially hybridized with two probes (A and B) derived from two different pools of 96 independently selected cDNA clones. Arrowheads indicate the positions of a cosmid known to contain cDNA-1 exons (a), a cosmid known not to contain cDNA-1 exons (b), and a mapped (c) and an anonymous cosmid (d) not known to contain portions of a cDNA. Each pool of selected cDNA clones detected a specific pattern of cosmid colonies; results were specific, reproducible, and distinct.

DISCUSSION

Efforts to isolate disease-associated transcribed sequences have become increasingly common and, because of limits associated with human linkage analysis, relatively large areas of the genome (1-4 Mbp) are ever more frequently being analyzed to identify candidate disease genes (29). In addition, a systematic isolation of mapped human transcribed sequences has become one of the explicit immediate goals of the human genome project (30). An efficient completion of either goal will require a critical evaluation of the strengths and weaknesses of available techniques and a continued refinement and development of gene isolation technology. In this report we describe alternative approaches for isolating transcribed sequences contained in YAC clones mapped to region Xp11.21. Although these analyses are not nearly complete, in part, our results serve as a start for comparing alternative gene isolation strategies and clarifying issues to be examined.

Two methods were used to isolate regional transcribed sequences. First, YAC subclones containing CpG islands were analyzed to identify low-copy evolutionarily conserved sequences (ECSs); ECSs were used to screen a tissue-specific cDNA library and isolate transcribed sequences. Second, PCR-amplified cDNA clones were isolated by hybridization to membrane bound YAC DNA; selectively retained inserts were subcloned to construct an enriched region-specific cDNA library. These techniques were comparable in several respects. Most importantly, both methods succeeded in isolating regional transcribed sequences. In addition, both techniques were relatively simple and independent of limited resources or expertise; both techniques used the same cDNA library and clones isolated by either technique were verified and characterized with equal ease. However, despite these similarities, these techniques differed substantially in scope, yield, and operational magnitude.

Compared to the ScDNA strategy, as expected, ECSs yielded significantly larger cDNA clones. However, by the design of our experiment, the isolation of ECSs was labor intensive and focused on only small segments of the YAC clones analyzed. As described here, it would be difficult to efficiently adapt the use of ECSs to completely characterize large genomic regions. However, the hybridization of interspecies genomic DNA combined with PCR amplification may provide a ECS-based technique to analyze large genomic regions for the isolation of transcribed sequences (31).

As expected, the cDNA selection technique provided an apparently broader transcriptional representation of the region. Compared to the use of ECSs, to insure an adequate analysis, it was necessary to use relatively large aliquots of the cDNA library and utilize an internal control (cDNA-1) to ensure adequate library representation; the use of a limitedly amplified primary cDNA may decrease the quantity of cDNA required for this technique and increase the size of ScDNA inserts isolated (27). Using these hybridization conditions, we did not find the isolation of repetitive ScDNA clones to be problematic (14,15). The cDNA selection technique should be amenable to automation without considerable method redesign. In addition, by performing the selection procedure with combinations of different tissue-specific cDNAs, a more complete regional isolation of transcribed sequences may be attained.

In conclusion, both of the techniques were successful, neither was ideal. In some respects, an investigator's decision regarding the choice of gene isolation techniques may be analogous to Alice's (of Wonderland) decision regarding which road to take. This comparison may be particularly apt, since Alice was attempting to decide between visiting the March Hare and the Mad Hatter (32).

"Cheshire-Puss," she began, rather timidly, as she did not at all know whether it would like the name: however, it only grinned a little wider. "Come, it's pleased so far," thought Alice, and she went on. "Would you tell me, please, which way I ought to go from here?"

"That depends a good deal on where you want to get to," said the Cat.
"I don't much care where - " said Alice
"Then it doesn't matter which way you go," said the Cat.
" - so long as I get *somewhere*," Alice added as an explanation.
"Oh, you're sure to do that," said the Cat, "if you only walk long enough." (32).

In the absence of a clearly superior technique, the selection of a gene isolation strategy should still be guided by the specific problem at hand, the region under study, available resources, and the individual expertise of the laboratory.

ACKNOWLEDGMENTS

We wish to thank D. Bentley for helpful discussions, D. Voss for assisting in the preparing the manuscript, and L. Carroll for inspiration. This work was supported, in part, by March of Dimes - Birth Defects Foundation Basic Science Grant 1-91-176 and National Institutes of Health (NIH) Grants HD-23768 and NS-30771 to J.L.G. and NIH Training Grant NIH-5-T32-GM7544-11 to E.N.B.

REFERENCES

1. T. Ohta, and M. Kinura, Functional organization of genetic material as a product of molecular evolution, *Nature* 233:118 (1971).
2. W.A. Bickmore, and A.T. Sumner, Mammalian chromosome banding - an expression of genome organization, *Trends Genet.* 5:144 (1989).
3. E.M. Southern, Genome mapping: cDNA approaches, *Current Opinion Genet. Devel.* 2:412 (1992).
4. J.O. Bishop, The gene numbers game, *Cell* 2:81 (1974).
5. N.D. Hastie and J.O. Bishop, The expression of three abundance classes of messenger RNA in mouse tissues, *Cell* 9:761 (1976).
6. A.P. Monaco, R.L. Neve, F.C. Colletti, C.J. Bertelson, D.M. Kurnit, and L.M. Kunkel, Isolation of candidate cDNAs for portions of the Duchenne muscular dystrophy gene, *Nature* 323:646 (1986).
7. J.M. Rommens, M.C. Iannuzzi, B. Kerem, M.L. Drumm, G. Melmer, M. Dean, R. *et al.*, Identification of the cystic fibrosis gene: chromosome walking and jumping, *Science* 245:1059 (1989).
8. A.P. Bird, CpG-rich islands and the function of DNA methylation, *Nature* 321:209 (1986).
9. L. Bonetta, S.E. Kuehn, A. Huang, D.J. Law, L.M. Kalikin, M. Koi, *et al.*, Wilms tumor locus on 11p13 defined by multiple CpG island-associated transcripts, *Science* 250:994 (1990).
10. T. Yokoi, M. Lovett, Z.Y. Cheng, and C.J. Epstein, Isolation of transcribed DNA sequences from chromosome 21 using mouse fetal cDNA, *Hum. Genet.* 74:137 (1986).
11. U. Hochgeschwender, J.G. Sutcliffe, and M.B. Brennan, Construction and screening of a genomic library specific for mouse chromosome 16, *Proc. Natl. Acad. Sci. USA* 86:8482 (1989).
12. G.M. Duyk, S.W. Kim, R.M. Myers, and D.R. Cox, Exon trapping: a genetic screen to identify candidate transcribed sequences in cloned mammalian genomic DNA, *Proc. Natl. Acad. Sci. USA* 87:8995 (1990).
13. A.J. Buckler, D.D. Chang, S.L. Graw, J.D. Brook, D.A. Haber, P.A. Sharp, and D.E. Housman, Exon amplification: a strategy to isolate mammalian genes based on RNA splicing, *Proc. Natl. Acad. Sci. USA* 88:4005 (1991).
14. M. Lovett, J. Kere, and L.M. Hinton, Direct selection: a method for the isolation of cDNAs encoded by large genomic regions, *Proc. Natl. Acad. Sci. USA* 88:9628 (1991).
15. S. Parimoo, S.R. Patanjali, H. Shukla, D.D. Chaplin, and S.M. Weissman, cDNA selection: efficient PCR approach for the selection of cDNAs encoded in large chromosomal DNA fragments, *Proc. Natl. Acad. Sci. USA* 88:9623 (1991).
16. J.L. Gorski, M. Boehnke, E.L. Reyner, and E.N. Burright, A radiation hybrid map of the

proximal short arm of the human X chromosome spanning incontinentia pigmenti 1 (IP1) translocation breakpoints, *Genomics* 14:657 (1992).

17. J.L. Gorski, E.N. Burright, E.L. Reyner, P.N. Goodfellow, and D.L. Burgess, Isolation of DNA markers from a region between incontinentia pigmenti 1 (IP1) X-chromosomal translocation breakpoints by a comparative PCR analysis of a radiation hybrid subclone mapping panel, *Genomics* 14:649 (1992).

18. E.N. Burright, N.G. Pasteris, and J.L. Gorski, (unpublished results).

19. P.D. Cotter, H.F. Willard, J.L. Gorski, and D.F. Bishop, Assignment of human erythroid delta-aminolevulinate synthase (ALAS2) to a distal subregion of band Xp11.21 by PCR analysis of somatic cell hybrids containing X;autosome translocations, *Genomics* 13:211 (1992).

20. J.L. Gorski, E.N. Burright, C.E. Harnden, C.K. Stein, T.W. Glover, and E.L. Reyner, Localization of DNA sequences to a region within Xp11.21 between incontinentia pigmenti type 1 (IP1) X-chromosomal translocation breakpoints, *Am. J. Hum. Genet.* 48:53 (1991).

21. T.W. Glover, V. Verga, J. Rafael, J.L. Gorski, E. Bawle, and J.V. Higgins, Translocation breakpoint in Aarskog syndrome maps to Xp11.21 between ALAS2 and DXS323, *Hum. Mol. Genet.* 2:1717 (1993).

22. T. Maniatis, E.F. Fritsch, and J. Sambrook, "Molecular Cloning: A Laboratory Manual", Cold Spring Harbor Laboratory, Cold Spring Harbor, (1982).

23. E.N. Burright, B.J. Trask, and J.L. Gorski, Incontinentia pigmenti type 1 (IP1) gene cloned in yeast artificial chromosomes: (in preparation).

24. N.G. Pasteris, M. Bilecki, and J.L. Gorski, YAC subclone contig assembly by serial interspersed repetitive sequence (IRS)-PCR product hybridizations, *Nucl. Acids Res.* (in press).

25. P.G. Sealey, P.A. Whittaker, and E.M. Southern, Removal of repeated sequences from hybridization probes, *Nucl. Acids Res.* 13:1905 (1985).

26. D. Vetrie, I. Vorechovsky, P. Sideras, J. Holland, A. Davies, F. Flinter, L. Hammarstrom, *et al.*, The gene involved in X-linked agammaglobulinaemia is a member of the src family of protein-tyrosine kinases, *Nature* 361:226 (1993).

27. J.M. Rommens, B. Lin, G.B. Hutchinson, S.E. Andrew, Y.P. Goldberg, M.L. Glaves, *et al.*, A transcription map of the region containing the Huntington disease gene, *Hum. Mol. Genet.* 2:901 (1993).

28. W-F. Fan, X. Wei, H. Shukla, S. Parimoo, H. Xu, P. Sankhavaram, Z. Li, S.M. Weissman, Application of cDNA selection techniques to regions of the human MHC, *Genomics* 17:575 (1993).

29. F.S. Collins, Positional cloning: let's not call it reverse anymore, *Nature Genetics* 1:3 (1992).

30. F.S. Collins, and D. Galas, A new five-year plan for the U.S. Human Genome Project, *Science* 262:43 (1993).

31. Z. Sedlacek, D.S. Konecki, R. Siebenhaar, P. Kioschis, and A. Poustka, Direct selection of DNA sequences conserved between species, *Nucl. Acids Res.* 21:3419 (1993).

32. L. Carroll, "Alice's Adventures in Wonderland & Through the Looking-Glass," Ninth Printing, The New American Library, New York (1960).

IDENTIFICATION OF cDNAS BY DIRECT HYBRIDIZATION USING COSMID PROBES

Gregory G. Lennon and Kimberly Lieuallen

Human Genome Center, L-452
Lawrence Livermore National Laboratory
Livermore, California 94550

ABSTRACT

The goal of this effort is to quickly obtain as many chromosome-specific cDNAs with as much map and sequence detail as possible. Many techniques have been proposed to isolate and identify genes within defined genomic regions; the technique discussed here is direct hybridization of a relatively complex genomic probe, an entire cosmid clone, to cDNA libraries. This method continues to be a straightforward technique with a fair number of successes.

INTRODUCTION

Many techniques are available to find cDNAs, and new techniques are continually being developed and streamlined. Methods include (1) expression-dependent (northern blot analysis, cDNA library screening with complex probes, hybrid selection), expression-independent (cross-species sequence homology, CpG island selection, exon-trapping, splice-site screening), sequence-based methods, subtractive cDNA hybridization, coincident sequence cloning, and hnRNA library screening.

Our goal is the isolation and characterization of genes (and their cognate cDNAs) on human chromosome 19. The euchromatic portion of human chromosome 19 is estimated to be 45 megabases, and to contain about 2000 genes. Therefore, 1 gene is estimated to be present every 22.5 kb, or over 1 gene per 40 kb cosmid on average.

One of the most direct methods of obtaining these cDNAs, and the focus of this chapter, is the use of hybridization to directly identify cDNAs encoded within the genomic probes used. Cosmids generated from flow-sorted chromosome 19 libraries are used as probes against arrayed cDNA libraries (2), and thus the cDNAs identified should be chromosome-specific. The aspects of this method to be discussed are: probes; targets; hybridization; analysis; and, verification. The discussion will conclude with an overview of the pro's and con's of this technique. Although this effort is focused on chromosome 19, it is just as applicable to other chromosomes, and even to non-chromosome-specific efforts.

Identification of Transcribed Sequences, Edited by
U. Hochgeschwender and K. Gardiner, Plenum Press, New York, 1994

23

MATERIALS AND METHODS

Probes

The chromosome 19 specific cosmid libraries were made at Lawrence Livermore National Laboratory (3). Currently, we have three libraries in Lawrist vectors. The average insert size of the cosmids is 40 kb. In an attempt to increase the chances of identifying a gene we screened cosmids for SacII sites, since SacII sites include the dinucleotide CpG which may be indicative of nearby genes. Typically, cosmid DNA is purified using QIAGEN columns. Approximately 100 ng of cosmid DNA is incubated with 4 units of Sac II enzyme at 37°C overnight. After electrophoretic analysis to determine the number of fragments, those cosmids with 3 or more Sac II sites are singled out for use as probes.

Approximately 25 ng of cosmid DNA is random-labelled by nonamer priming including 32-P (Prime-It, Stratagene). The probe is purified from unincorporated 32-P using a G-50 column. After this purification process, we typically get 1×10^5 cpm per microliter of probe. We used 3 μg of pBluescript KS- DNA, 5 μg of poly A, 10 μg of human Cot-1 DNA and 50 μg of salmon sperm DNA to block the probe. The probe is annealed with the blocking agents for approximately 30 min at 65°C.

Targets

The normalized human infant brain cDNA library created by Dr. Bento Soares (unpublished) was the target in our hybridizations with cosmid probes. We arrayed approximately 40,000 cDNA clones from this library into 96-well dishes. From this arrayed library high density colony filters were generated. These filters are arranged such that there are 6,912 independent clones present in duplicate on each filter, for a total of 13,824 clones per filter. The ORCA (Hewlett-Packard) robot was used to spot the high density colony filters, spotting at a rate of approximately 40,000 clones per hour. After spotting, the colonies were grown overnight and the DNA fixed to the filter the next day.

Hybridizations and Analyses

The filters were typically pre-hybridized for 2-12 hours in a 2% SDS hybridization solution containing 0.6M NaCl, 10 mM EDTA, 50 mM Tris (pH 7), 0.1% Na-pyrophosphate, 10% dextran sulfate (of a 50% stock solution). We blocked the filters during this pre-hybridization period with 500 μg of salmon sperm DNA, 3 μg of pBluescript DNA, and 200 μg of poly U. Both the pre-hybridization and hybridization are done at 65°C. After the pre-hybridization period, the probe is added and allowed to hybridize overnight. Typically, two low stringency washes and two high stringency washes are performed for 30 min each at 65°C. The low stringency solution contains 2XSSC and 1% SDS, while the high stringency wash contains 0.2XSSC and 1% SDS.

After the washes, the filters are exposed to PhosphorImager screens (Molecular Dynamics) overnight. After exposure the positives (present in duplicate) on the images are evaluated and called manually. Clone array position (plate and well) is calculated, and compared with other information available for those array positions. The filters themselves can be re-used approximately ten times. Typically, the filters are stripped for about 8 hours at room temperature with a solution of 2 mM EDTA and 0.1% SDS.

Verification

The verification procedures (Fig. 1) begin with the selection of colonies which are positive. Glycerol stocks of these clones are used to inoculate mini-preps. Plasmid DNA is harvested using a Promega Wizard mini prep kit and then the DNA is sequenced.

Sequences are sent off to be searched against Genbank sequences using the programs Blastn and Blastx (4). Most cDNAs have primers developed using the "Primer" program available from the Whitehead Institute. The primers are then tested by PCR against total genomic DNA and against a hamster/human somatic cell hybrid line containing human chromosome 19. Those cDNAs that show a product in the somatic cell hybrid line (but not in the hamster only parental cell line) are then used in a hybridization against high density colony filters containing clones from the Lawrence Livermore National Laboratory chromosome 19 cosmid library (as previously mentioned) and one northern blot with mRNA from human heart, brain, placenta, lung, liver, skeletal muscle, kidney, and pancreas. The cosmids identified as positives are then verified by digesting them with EcoRI and blotting to a nylon membrane whereby this membrane is probed with the cDNA. Alternatively, DNA from identified cosmids is used as template in a PCR reaction using the primers generated from the cDNA.

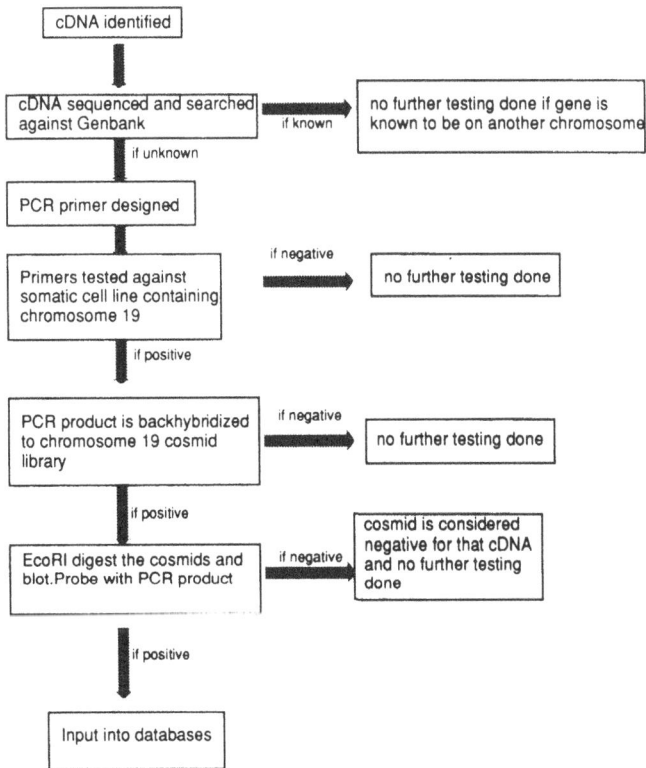

Figure 1. Flowchart of verification procedures.

A cDNA is only considered to be verified as originating from chromosome 19 if all three conditions are met, i.e. it maps by PCR to chromosome 19, cosmids from the chromosome-specific cosmid library are isolated, and these cosmids are confirmed by PCR or hybridization to contain homologous sequences. Once a cDNA has been verified, further work on other cDNAs identified in the same original hybridization is suspended unless there is independent evidence supporting the existence of another gene encoded within the same cosmid.

RESULTS

34 hybridizations have been performed using cosmids as probes on the normalized infant brain cDNA library ; a reasonably typical example of a good hybridization is shown in Fig. 2. These 34 hybridizations have resulted in 330 positives, for an average of 9.7 clones per probe. About half of the hybridizations yield no positives, so the average number of positive clones per positive probe is about 20.

Figure 2. Hybridization result. Total cosmid DNA from cosmid f29297 was used as probe; a portion of one filter, containing 3,360 independent cDNA clones (duplicated, upper and lower), is shown. In addition to the one positive, the reference clones present in the upper left corner of each 6x6 array unit are clearly visible.

Sequence has been derived from approximately 67% of the positive cDNAs. Sequence analysis reveals that even with blocking of both the probe and the filters during hybridization, approximately 5% of the positives contain an Alu element. These clones are not analyzed further. Another 1% of the positives are known genes from other chromosomes. The remaining 94% of sequenced positives lack significant homologies with known sequences.

Currently, primer pairs for 33 cDNAs await testing, and others are in various stages of the verification process. On a per cosmid basis (and thus per hybridization) results for clones that have been completely analyzed indicate that about 50% of hybridizations giving at least one positive yield a cDNA that can be verified to be from chromosome 19. Thus far, twelve new chromosome 19 cDNAs have been identified. It is difficult to estimate the success rate on a per cDNA clone basis (rather than per hybridization) since analysis stops on most cDNA clones once one clone from the same hybridization has been verified as from chromosome 19.

Over 800 cosmids have been digested with SacII, and of these, 29 had 3 or more SacII sites. This is in keeping with studies on SacII site frequency for other chromosomes (5). Surprisingly, we have not seen a significant difference between 17 SacII-rich and 17 randomly chosen cosmids in the percentage of hybridizations that succeed in identifying a cDNA eventually verified to be from chromosome 19.

DISCUSSION

With this method we have identified and fine-mapped 12 independent anonymous cDNAs on chromosome 19 in approximately one person-year. Interestingly, this number (12) is also the total number of chromosome 19 cDNAs assigned over the last year by four groups (6,7,8,9) mapping cDNAs (from EST sequencing projects) using somatic cell hybrids. These latter cDNAs are initially mapped to chromosome resolution. By contrast, the cDNAs identified by the direct hybridization approach are mapped to a single restriction band of usually less than 10 kb. As the physical map becomes more precise, the precision of the location of these cDNAs also increases since the cDNA is linked to an element of the physical map.

The positive aspects of the direct hybridization technique therefore include fine resolution mapping, and the use of straightforward and robust laboratory techniques. In contrast to methods involving PCR amplification, cDNA clones isolated are of the average length of the cDNA library, and thus significantly over 1 kb in length. The two main disadvantages involve the yield of cDNAs. First, about half of all hybridizations result in zero positives. Second, although 50% of the remaining hybridizations will yield a verifiable chromosome 19 cDNA, the amount of work required to sift the false positives out from the true positives is high, limiting the throughput of this approach.

We have found that the number of SacII sites within a cosmid is not necessarily indicative of increased success with the direct hybridization approach; we have not found a higher true positive rate for the SacII rich cosmids compared with randomly selected cosmids. Because of this and the time it takes to screen cosmids for SacII, we no longer choose cosmid probes based on their SacII-site richness. Note, however, that chromosome 19 is a GC rich (and gene-rich) chromosome, and that screening for CpG-rich cosmids may be of greater aid in identifying candidate probe regions on chromosomes that are GC poorer in general than chromosome 19.

In the future, we hope to apply this technique to screening arrayed full-length cDNA libraries. Currently, in order to isolate full-length versions of our new chromosome 19 cDNAs we usually screen a number of cDNA libraries, and check clone insert sizes against the mRNA size as estimated by the Northern blots. For numerous reasons, both for our and other groups, a great savings in time would be achieved if arrayed full-length libraries were available. Since many genes are not ubiquitously expressed, we are also planning on arraying high quality cDNA libraries from other tissues. The use of arrayed DNA spots (from either PCR or automated DNA mini-preps) for these or our current libraries would also give better results. Other improvements for this technique include the promise of automated image analysis programs to reliably call positives from the digitized images.

Although cDNAs have been isolated from phage cDNA libraries using probes as complex as YACs of size 270 kb (10), and thus the equivalent of about 6 cosmids, it is likely that the practical limits on complexity are somewhere in between 40 and 250 kb for routine screening of cDNA colony filters. The use of pooled genomic probes that are of this range of complexity may therefore be appropriate. These include microdissected material, pooled exon trap products, hybrid selection products, CpG islands isolated from flow-sorted chromosomes, or PCR products from chromosome specific material using consensus motif primers (such as splice site sequences). For example, if there are 2,000 genes on chromosome 19, every gene could contribute up to approximately 100 bp of sequence on average to a "whole chromosome" probe and the estimated complexity would still be within an acceptable range for routine screening of colony cDNA libraries.

CONCLUSIONS

The identification of cDNAs by direct hybridization of complex genomic probes continues to be of great utility. Using the arrayed normalized human infant brain cDNA library, we have achieved a high rate of discovery of novel chromosome 19 cDNAs. The keys to our success to date rely on at least two factors: first, a high quality cDNA library that is arrayed and thus available for reproducible screening by filter hybridization, and second, a rigorous verification process. Disadvantages of the technique include a high false positive rate, and thus a labor-intensive verification process. Future improvements aimed at reducing the false positive rate and increasing throughput include the use of full-length cDNA libraries, the spotting of DNA rather than colonies, and the use of pooled genomic probes and automated image analysis.

ACKNOWLEDGEMENTS

This work was performed under the auspices of the U.S. Department of Energy by Lawrence Livermore National Laboratory under contract no. W-7405-Eng-48.

REFERENCES

1. J.E. Parrish and D.L. Nelson, Methods for finding genes: a major rate-limiting step in positional cloning, *Gene Anal. and Tech.* 10:29 (1993).
2. G.G. Lennon and H. Lehrach, Hybridization analyses of arrayed cDNA libraries, *Trends Genet.* 10: 314 (1991).
3. A.V. Carrano, P.J. de Jong, E. Branscomb, *et al.*, Constructing chromosome- and region-specific cosmid maps of the human genome, *Genome* 31:1059 (1989).
4. S.F. Altschul, W. Gish, W. Miller, et al., Basic local alignment search tool, *J. Mol. Biol.* 215:403 (1990).
5. W.C. Golembieski, S.C. Smith, F. Recchia, *et al.*, Isolation of large numbers of chromosome-3 specific cosmids containing clusters of rare restriction-endonuclease sites, *Am. J. Hum. Gen.* 49:581 (1991).
6. K. Okubo, N. Hori, R. Matoba *et al.*, Large scale cDNA sequencing for analysis of quantitative and qualitative aspects of gene expression, *Nature Genetics* 2:173 (1992).
7. A.S. Durkin, D.R. Maglott, and W.C. Nierman, Chromosomal assignment of 38 human brain expressed sequence tags (ESTs) by analyzing fluorescently labeled PCR products from hybrid cell lines, *Genomics* 14:808 (1992).
8. M.H. Polymeropoulos, H. Xiao, A. Glodek, *et al.*, Chromosomal assignment of 46 brain cDNAs, *Genomics* 12:492 (1992).
9. L. Gieser and A. Swaroop, Expressed sequence tags and chromosomal localization of cDNA clones from a subtracted retinal pigment epithelium library, *Genomics* 13:873 (1992).
10. M.R. Wallace, D.A. Marchuk, L.B. Andersen, *et al.*, Type 1 neurofibromatosis: identification of a large transcript disrupted in three NF1 patients, *Science* 249:181 (1990).

LOCUS SPECIFIC IDENTIFICATION OF TRANSCRIBED SEQUENCES USING YACS AND WHOLE YEAST GENOMIC DNA

Sankhavaram R. Patanjali, Hong Xia Xu, Satish Parimoo and Sherman M. Weissman

Department of Genetics, Yale University School of Medicine, New Haven, CT 06510

INTRODUCTION

A large fraction of mammalian genomes contain DNA which appears to be very poor in biologic information. Unless or until nucleic acid sequencing procedures emerge that are capable of a much higher throughput of good quality sequence at a lower cost than the present protocols, it appears attractive to approach characterization of these genomes by focussing attention on transcribed regions and immediately adjacent sequences. The advantages gained by such an approach are that nearly all the genes of man will be identified and probes for these genes made available long before the first human genome is sequenced. Our laboratory has been concentrating on hybridization selection approaches (1) to identify transcribed sequences in yeast artificial chromosomes (YACs) or other cloned genomic DNA, and on affinity selection procedures to identify cDNAs that embed particular motifs.

In hybridization selection YAC DNA is hybridized to inserts from cDNA libraries, and the cDNA inserts are recovered from immobilized YAC DNA and cloned. Although the procedure was developed initially using PFGE purified YAC DNA, we later modified the method and made use of whole yeast genomic DNA containing YACs as target. This approach allows us to perform a number of selections simultaneously that lead to a high throughput of selected transcribed sequences, thereby facilitating comprehensive mapping of the target chromosomal loci in a short time. In this paper, we present some of the recent studies performed in our laboratory that include construction of normalized cDNA library pools, effective use of quencher to reduce non-specific hybridization and the development of a modified selection protocol.

METHODS AND MATERIALS

YACs

Yeast colonies containing YACs 746B4 and 865G6 were provided to us by Dr.

Schellenberg of University of Washington, Seattle, and Dr. R. Spritz of University of Madison, Wisconsin, respectively. Both 746B4 (750 kb) and 865G6 (950 kb) are non chimeric and encode c-fos and c-kit receptor, respectively. AB1380, the host yeast strain for the preparation of YAC libraries, was grown in YPD medium and YAC clones were grown in AHC medium (2).

cDNA Libraries

Short-fragment cDNA libraries were prepared by priming Poly (A)+ RNA isolated from 6 different human tissues with random hexanucleotide primers (3). Human thymus, spleen, liver, and adult testes were obtained from the National Disease Research Interchange (NDRI, Philadelphia). Spontaneously aborted total fetus (9 weeks) and fetal brain (11 weeks) were received from the Yale-New Haven Hospital. Total fetus and fetal brain libraries were amplified from about 12×10^6 plaque forming units (PFUs), and all the other libraries were amplified from 3 to 7×10^6 PFUs. The insert sizes range from 300 to 2000 bp. All the short fragment cDNA libraries were prepared in λgt10 vector with C600hfl as the host strain, and cDNA inserts were recovered by PCR with the primer set C (1) for the cDNA selection. Full-length cDNA libraries were primed with oligo (dT) primer and prepared with Poly (A)+ RNA from human brain, spleen, thymus and liver. They were cloned into Charon BS vector (4). The cDNA inserts were amplified by PCR with the SK and KS primers (Stratagene).

Normalization of cDNA Libraries

Normalization of cDNA libraries was performed essentially as described earlier (3,5). cDNA library inserts (spleen, thymus, total fetus, fetal brain, adult brain and testes) were prepared by PCR amplification using Set C primers, and purified over Chromaspin-100 column as per the instructions of the manufacturer. The range of cDNA fragments was estimated to be 400-2000 bp. A pool of six libraries was prepared by mixing 333 ng of each of these libraries in 0.3 M sodium phosphate buffer (pH 7.2) containing 1 mM EDTA and 0.1% SDS in a total of 50 μl volume. This solution was overlaid with mineral oil, denatured at 98°C for 3 min and incubated at 65°C for 2 h. The sample was diluted to 1 ml on 0.01 M sodium phosphate buffer, single stranded DNA was recovered and cDNA libraries were constructed from the normalized DNA in gt 10 vector as described earlier (5).

cDNA Selection

We have modified the cDNA selection method by changing the hybridization buffer reagent and reducing the number of quenching reagents. However some of the selections were performed exactly as described earlier (5) and are mentioned in the text appropriately. Preparation of the yeast genomic DNA, spotting on the membrane and washing conditions of the nylon membrane are unchanged from our earlier report (5). A modified selection procedure is described below briefly.

Yeast DNA

Total yeast genomic DNA was prepared as described earlier (5). 10 μg of total genomic DNA was digested with 100 units of Hind III in the presence of 100 units of RNase One (Promega) and purified by phenol:chloroform: isoamyl alcohol (25:24:1) (v/v) extraction followed by ethanol precipitation. 200 ng of the DNA was denatured by heating to 98°C, quick chilled on ice and spotted on a piece of nylon membrane (2.5 mm x 2.5 mm) in two 0.5 μl aliquots in the presence of 10XSSC. The concentration of the digested genomic DNA was estimated by comparing the ethidium bromide staining patterns on agarose gels of commercially available poly dIC (Pharmacia) at predetermined concentrations. Since the concentration estimation of genomic DNA is critical, it is recommended that more than three dilutions of DNA solution should be used in order to get accurate estimates. The membranes are air dried and processed as described earlier (5).

Hybridization Buffer

Equal volumes of 17.5% SDS and 1.0 M sodium phosphate buffer (pH 7.3) were mixed and the solution was incubated at 65°C for 15 min and mixed thoroughly to ensure proper mixing. This mixture can be stored at room temperature at least for six months. The hybridization buffer should be incubated at 65°C for at least 15 min, and mixed thoroughly every time before use.

Quenchers and Hybridization

Ribosomal DNA quenchers from human, yeast and *E.coli*, poly dIC, Cot-1 DNA, Line-1 and AB1380 genomic DNA (wild type yeast) were prepared as described earlier (5,6). 8 μg each of human and yeast ribosomal quenchers and 5 μg of all the other quenchers were pooled in a total volume of 40 μl, heat denatured, quick chilled for 1 min and added to 200 μl of hybridization buffer maintained at 65°C. 200 μl of the buffer containing the quenchers was immediately transferred to 0.5 ml tubes containing the nylon membranes. In case of multiple selections, the volumes were scaled up appropriately by mixing the quenchers and hybridization buffer into a single tube, the mixture was incubated at 65°C and 240 μl of the cocktail was added to each of the membrane discs containing the target YAC. These tubes were overlaid with mineral oil and incubated at 65°C for 24 h. 10 μl of cDNA pool containing 3 μg of DNA was heat denatured at 98°C for 5 min quick chilled on ice for 1 min and added to 240 μl of hybridization solution making up the final volume to 250 μl. The resulting concentrations of phosphate buffer and SDS in the hybridization solution were 0.5 M and 7%, respectively. Hybridization was performed for up to 36 h. The filters were washed and the selected fragments were recovered as described earlier (5). Two rounds of selection were performed using the first round selected cDNA as the input library for the second round of selection.

cDNA selection with c-fos YACs was performed as described by Parimoo *et al.* (5,6). Selection with the YACs containing c-kit receptor was performed as described in this paper.

Characterization of Selected cDNA Libraries

cDNA selections were performed with a mixture of the 6 short-fragment cDNA libraries described above. After two rounds of cDNA selection the cDNA inserts were recloned into λgt10. The cDNA inserts from about 100 clones were amplified by PCR with the primer set C. Dot blots of 96 clones gridded in 12 x 8 arrays were prepared with PCR products and DNA from a pool of twelve clones was hybridized to the array. Clones that hybridized to this mixed probe were identified and in a second round of probing twelve of the negative clones were chosen to hybridize to the remaining clones. The non-overlapping cDNA clones were sequenced with a cycle sequencing kit (GIBCO BRL), and the cDNA sequences were checked for homologies in the GENEMBL databases by the program FASTA (version 73). Selected cDNA clones from the YAC containing c-fos gene were plated at a density of 1000 clones per plate, probed with c-fos probe and the positives were eliminated for further analysis. A set of 96 negative clones were amplified and the matrix analysis was performed as described in the previous paragraph. Short-fragment cDNA clones were then used to screen oligo (dT)-primed cDNA libraries.

Dot Blots and Southern Blots

Dot blots and Southern blots were performed by standard methods (7). Probes were labeled with [α-^{32}P]dCTP by random hexanucleotide priming (8). The hybridization buffer was the same as described by Church and Gilbert (9). Hybridization was performed at 65°C for 8 to 16 h.

RESULTS AND DISCUSSION

As initially developed (1), hybridization selection required purification of YAC DNA by pulsed field gel electrophoresis (PFGE) before selection. This is inconvenient when large numbers of samples are being processed and is often difficult when the size of a YAC is close to that of a yeast chromosome or when the YAC is very large. To investigate the use of whole yeast DNA for selection, we first tried a selection with a mixture of six cDNA libraries and DNA from the host YAC strain AB1380 as target. One hundred randomly chosen selected fragments were sequenced and all proved to be ribosomal DNA fragments, indicating that the principle region of long conserved sequence between human and *S. cerevisiae* is the ribosomal DNA. Ribosomal cDNA inserts are relatively abundant in our short fragment libraries because (a) the libraries were prepared by random hexanucleotide priming rather than oligo dT priming and (b) conventional affinity purification of poly A+ RNA on oligo dT cellulose column does not completely eliminate ribosomal sequences even after two rounds of purification.

As recently published (6), we were able to adjust the mixture of quenching reagents and the ratio of quenching reagents to target and cDNA libraries so that cDNA selection could be routinely performed on whole DNA extracted from yeasts carrying YACs of interest. Ribosomal insert contamination in these libraries was generally under 15% of the total number of inserts, and commonly much less than this. To further improve selections we are currently experimenting with the use of prenormalized cDNA mixtures in the selection process.

Normalization of a single cDNA library was described earlier by us and others (3, 10). In our experience, we observed that the ribosomal DNA clones are the first to reassociate because of their very large abundance in the random primed cDNA libraries. Although prolonged reassociation results in the equalization of almost all the cDNA messages, some of the medium abundance messages may be lost during the process. In order to overcome this problem, we limited the normalization process to a Cot value of 5 mol.sec^{-1}.lit^{-1}.

Single stranded DNA from such incomplete normalization showed considerable (70%) depletion of ribosomal DNA while retaining medium abundance, cDNAs like MHC class cDNAs (Fig. 1). The partially normalized library was constructed from about 5 million clones and 95% of the clones showed the presence of inserts in the range of 400-1600 bp. We intend to use the partially normalized libraries in our modified selection approach.

cDNA selection approaches cannot detect genes that are not expressed at some level in at least one of the tissues used for cDNA library preparation. However a large fraction of genes appear to be expressed in either brain or testes. The method is very sensitive in that genes expressed at very low levels can be readily detected. For example selections with a megabase YAC containing the c-fos gene resulted in a library in which more than ten per

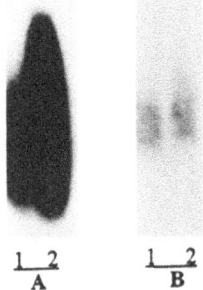

Figure 1. Single (1) and double (2) strand cDNA fractions of the 6 cDNA pool were amplified and probed with ribosomal (a) and MHC class I (B) probes.

cent of the inserts were derived from c-fos. A representative Southern blot of the c-fos selection is shown in Fig. 2.

Similarly we have detected the other early response genes, a number of transcription factors, at least one cytokine gene (TGF-ß) and one cytokine receptor (c-kit receptor) (Fig. 3) in various selections (data not shown). cDNA selection of the c-kit receptor were

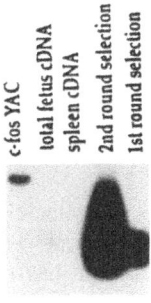

Figure 2. cDNA selection of the c-fos cDNA using a 6 cDNA library pool. Southern blot was prepared using 2 μg of EcoR I digested c-fos YAC (746B4), 1 μg of total fetus cDNA, Spleen cDNA and 1st round selected material amplified by using Set C primer; and twice selected cDNA (2nd round selection) amplified by Set A primer. The blot was probed with c-fos cDNA and exposed for 12 h.

performed with the modification described in this paper. As shown in Fig. 2, the modifications retained the same specificity and the selectivity of the method originally reported (6). Since the new method involves a fewer number of reagents and minimal manipulations, we feel that these modifications help to quicken the method and facilitate multiple selections with no loss in the specificity and the selectivity of the procedure.

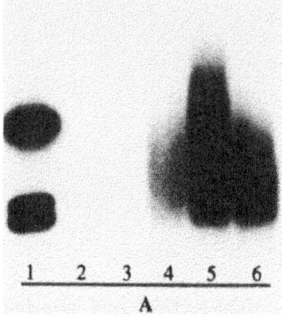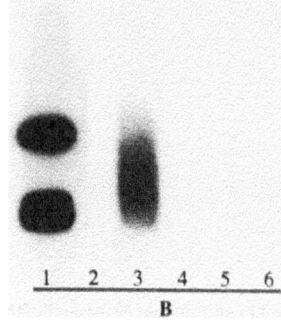

Figure 3. Panel A and B were probed with glutathione reductase and c-kit receptor cDNAs, respectively. Lanes (1) 1 kb ladder (2) 6 cDNA pool (3) selected cDNA from c-Kit containing YAC and (4,5,6) are selected cDNA from glutathione reductase containing YACs.

In a recently published analysis (11), the abundance of cDNAs selected from MHC YACs was estimated by using the cDNAs to screen an oligo dT primed library. The selected cDNAs were confirmed to come from different genes. More than half of these cDNAs were present at an abundance of less than one per hundred thousand in any of the three oligo dT primed cDNA libraries, and this low abundance didn't seem to be a consequence of the use of fragments from the 5' end of longer mRNAs. The implication of this result, if it holds up over a larger region of the genome, is that random sequencing of cDNA fragments as seen in other approaches would be likely to miss a substantial fraction, perhaps more than half, of all genes, even if over one hundred thousand inserts were sequenced.

A laborious step in the analysis of selected cDNA libraries is determination of when different selected fragments represent different genes, and when they are derived from different portions of a single longer transcript, or from differently spliced transcripts of the same gene. In our work to date we have relied on identification of oligo dT primed cDNA clones that embed a short fragment and hybridization to identify other fragments contained in the same cDNA. In principle, this problem could be decreased if short fragments libraries were prepared that represented either the 3' or 5' end of the original mRNAs. Preparation of 3' end short fragments would be accomplished readily by oligo-dT priming. They have the disadvantage that a number of mRNA have alternate 3' ends, and a significant fraction, up to 5-10% of mRNAs, have repetitive sequences in their 3' untranslated region. The presence of these repeats would tend to lead to loss of these short fragments and failure to detect the particular gene. In addition the 3' untranslated regions of mRNA are relatively long so that many of the short fragments would give no hint about the coding capacity of the gene.

An alternative procedure that would be highly desirable would be the preparation of 5' end libraries. Methods to prepare cDNA inserts from the 3' end of first strand cDNA exist, and a protocol for preparing libraries of such inserts has just been published (12). A difficulty in principle of such an approach is that many reverse transcripts arise from fragmented mRNAs or incomplete transcription of mRNAs. As a consequence, many overlapping fragments might be present from a single cDNA and the true 5' end might be missing in the case of long mRNAs or mRNAs that contain internal GC rich sequences that are difficult to reverse transcribe. For these reasons and for the purposes of generating full length cDNAs and potentially amplifying full length mRNA for translation, it would be desirable to prepare 5' fragment libraries in which the fragments were present precisely because they extend up to or at least close to the cap structure on mRNA.

Direct cDNA selection may be effectively employed to prepare chromosomal region-specific cDNA libraries. Currently, cDNA selection has been evaluated for the efficiency in enriching cDNAs complementary to inserts from cosmids or yeast artificial chromosomes (YAC) and used to prepare minilibraries of cDNAs representative of the coding regions in such clones. We have been concerned with adapting the mechanics of these procedures so that they can be conveniently applied to multiple samples in parallel and with a high output, and with optimizing their sensitivity both with respect to the amounts of cDNA required and the amount of homologous genomic DNA sufficient for the enrichment. We intend to perform multiple cDNA selections on gridded arrays of well characterized YACs and individually isolate and analyze the selected material from each of the YAC targets. Since such attempts would offer large surface area for the hybridizing cDNAs we believe that reduction of the ribosomal DNA and repetitive sequences as seen in the normalized libraries would be of considerable importance in improving the specificity of the method. In addition, modifications of the selection process leading to the simplification of the method considerably facilitate scaling up of the method at least by an order of magnitude. We feel that a high throughput of locus specific cDNA clones is feasible by the combination of normalized cDNA libraries and modified cDNA selection.

REFERENCES

1. S. Parimoo, S.R. Patanjali, H. Shukla, D.D. Chaplin, and S.M. Weissman, cDNA selection: Efficient PCR approach for the selection of cDNAs encoded in large chromosomal DNA fragments, *Proc. Natl. Acad. Sci. USA* 88:9623 (1991).
2. F. Sherman Fink and J.B. Hicks, "Methods in Yeast Genetics: A Laboratory Manual," (1986).
3. S.R. Patanjali, S. Parimoo, and S.M. Weissman, Construction of a uniform abundance (normalized cDNA library), *Proc. Natl. Acad. Sci. USA* 88:1943 (1991).
4. A. Swaroop and S.M. Weissman, Charon BS (+) and (-) versatile λ phage, *Nucl. Acids Res.* 16:8739 (1988).
5. S. Parimoo, S.R. Patanjali, and S.M. Weissman, Normalization and selection with short-fragment cDNAs, *in* "Methods in Molecular Genetics," (Academic Press) 23:50 (1993).
6. S. Parimoo, R. Kolluri, and S.M. Weissman, cDNA selection from total yeast containing YACs, *Nucl. Acids Res.* 21:4422 (1993).
7. T. Sambrook, E.F. Fritsch, and T. Maniatis, "Molecular Cloning: A Laboratory Manual," 1:7.37 and 9.31 (1989).
8. B. Feinberg and Vogelstein, A technique for radiolabeling DNA restriction endonuclease fragments of high specific activity, *Anal. Biochem.* 132:6 (1983).
9. G. Church and W. Gilbert, Genome sequencing, *Proc. Natl. Acad. Sci. USA* 81:1991 (1984).
10. M.S.H. Ko, An 'equalized cDNA library' by the reassociation of short double-stranded cDNAs, *Nucl. Acids Res.* 18:5705 (1991).
11. W. Fan, X. Wei, H. Shukla, S. Parimoo, H. Xu, S.R. Pantajali, Z. Li, and S.M. Weissman, Application of cDNa selection techniques to regions of the human MHC, *Genomics* 17:575 (1993).
12. J.B.D.M. Edwards, J. Delrot, and J. Mallet, Anchoring a defined sequence to the 5' ends of mRNAs, *Meth. in Mol. Biol.* 15:365 (1993).

TOWARDS A TRANSCRIPTIONAL MAP OF HUMAN CHROMOSOME 21

K. Gardiner[1], H. Xu[2], W. Bonds[2], F. Tassone[1], S. Parimoo[2], R. Sivakamasundari[2], F. Hisama[2], A. Rynditch[3] and S. Weissman[2]

[1]Eleanor Roosevelt Institute, Denver, CO 80206
[2]Yale University School of Medicine, New Haven, CT 06536
[3]Institute of Molecular Biology and Genetics, Kiev, Ukraine

ABSTRACT

Chromosome 21 is the smallest and one of the best mapped of the human chromosomes. It, therefore, represents a good model system for transcriptional mapping efforts. To construct a comprehensive transcriptional map, the technique of cDNA hybrid selection is being applied to a minimal contig of YAC clones spanning the long arm. Presented here are preliminary results for four YACs representing ≈ 2 Mb of non-overlapping DNA. While the cDNA hybrid selection approach is rapid and robust, several difficulties remain to be solved in actual map construction. Some of these are associated with YAC chimerism questions, pseudogenes and members of gene families; others involve verification of exonic material and rapid generation of a non-redundant gene set.

INTRODUCTION

Chromosome 21 is an excellent choice for concerted transcriptional map construction. It is the smallest of the human chromosomes, and is also one of the best mapped extended regions of the human genome. Physical mapping data includes an extensive NotI restriction map and detailed pulsed field analysis for many regions (1-5). As a result, for most of the 40 Mb comprising the long arm, any probe can be accurately positioned to within a few hundred kb of defined markers. Patterns in gene density have been assessed by localizing known genes and by mapping CpG islands in YAC clones (1,6). This has identified the distal 1/3 of the long arm as being significantly gene rich as compared to the proximal 2/3. Combining these data with the base compositional map gives a physical picture that correlates GC level, CpG island frequency, gene distribution and Giemsa band patterns (7,8). Thus, new genes can be integrated into existing maps, and the added detail will contribute greatly to our understanding of real gene density and gene size, and human genome organizational features in general.

The approach to transcriptional map construction taken here is cDNA hybrid selection from YAC clones (9). Extensive resources exist for making such an undertaking feasible. The chromosome 21 joint YAC screening effort has screened the St. Louis human

YAC library for over 75 probes, identifying over 400 clones (10). From this collection, >8 Mb (20%) of nonchimeric, non-overlapping DNA has been restriction mapped in detail (6). A YAC contig, consisting of 810 clones, including mega YACs, has been constructed from the CEPH libraries (11). This contig is currently under analysis to identify a minimum tiling path of non-chimeric, stable, apparently non-deleted clones that spans the long arm. Currently, approximately 50% is covered (12). To construct a transcriptional map, cDNA hybrid selection will be applied to each YAC in the final, verified, minimal contig.

While contig verification is underway, selected library construction has begun. To be discussed here are the results of a preliminary effort with 4 non-overlapping YACs which total 2 Mb. YACs map to different kinds of regions of the chromosome (Fig.1): GC-rich/gene-rich, AT-rich/gene rich, and AT-rich/gene-poor. The aim of the current work is the identification of a non-redundant gene set for each YAC. Additional characterization would include tissue and developmental timing of expression, and identification of alternatively processed mRNA forms. Ultimately, it is hoped that regional characteristics may translate into predictions regarding gene size and gene density.

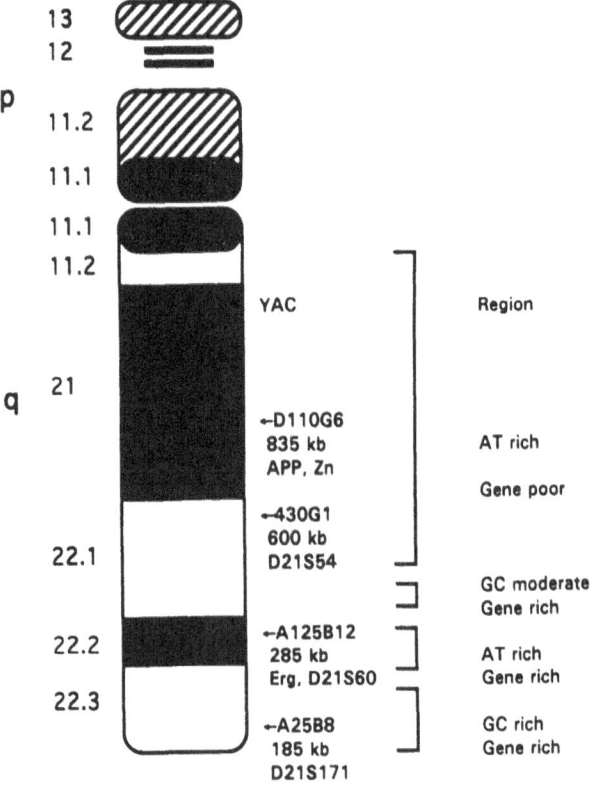

Figure 1. YAC clones used in cDNA hybrid selection. Localization of each YAC clone along chromosome 21 is shown, together with size (kb) and marker content. To the right are given characteristics of the region surrounding the YAC (6,8).

METHODS

YAC Clones

YAC clones D110G6, A125B12 and A25B8 are from the St. Louis total human library, and were obtained under the auspices of the International Chromosome 21 Joint YAC Screening Effort. D110G6 and A125B12 have been described in detail, as has A25B8 (6, and manuscript in preparation). YAC 430G1 was obtained from the CEPH complete contig of chromosome 21. 430G1 and A25B8 are non-chimeric by FISH; D110G6 and A125B12, by end probe analysis. Sizes, marker content and regional localization are shown in Fig.1.

cDNA Sources and Preparation

For selection from D110G6, polyA+ RNA was obtained by standard procedures (13) from the following cell lines: HT1080, CGM-1 (lymphoblastoid), GM00010 (fetal fibroblast), HTB10 (neuroblastoma) and HTB148 (neuroglioma). CGM-1 was obtained from M.V. Olsen; all others, from ATCC. PolyA+ RNA was reverse transcribed using random primers, ligated to linkers, and stored as cDNA without cloning, essentially as described (14). Prior to selection, cDNAs were separately amplified using primers to the linkers, and pooled. For selection, from the remaining YACs, short insert cDNA libraries (15) from fetal brain, whole fetus, adult brain, liver, spleen, thymus, testes and thyroid were amplified and pooled (9).

Selection Procedures

For selection from D110G6, intact YAC was prepared by pulsed field electrophoresis, digested in agarose with Sau3A, ligated to linkers and amplified using a biotinylated primer specific to the linker (14). Blocking reagents and conditions were as described in Parimoo *et al.* (9). Hybridization conditions, and binding and elution from magnetic beads (Dynal, Inc.) were as described (16). Eluted selected cDNAs were digested with BamHI and cloned into Bluescript (Stratagene). For the remaining YACs, selection was carried out with YAC DNA immobilized on nylon membranes; procedures were essentially as described by Parimoo *et al.* (9), except that total yeast YAC clone DNA was used, without purification of the YAC. Selected cDNAs were digested with EcoRI and cloned into λgt10.

Colony/Plaque Screening

For D110G6, the selected cDNAs cloned in Bluescript were electroporated into *E. coli* DH10B and 500 colonies picked from the transformation plates into microtitre plates. For screening, colonies were stamped onto nylon membranes (Hybond N+), and grown overnight on LB+Amp plates. DNA was denatured and fixed to the membrane simply by directly autoclaving the membranes (1 minute, fast exhaust) after growth, followed by uv-cross linking (Stratalinker). Hybridizations were in aqueous buffer at 65^0, followed by washes at 62^0C, in 0.1XSSC/0.1% SDS. For selected libraries in λgt10, 2000 plaques were plated at 1000 pfu/150 mm plate. Approximately 50-100 plaques/100 kb of YAC were picked to microtitre plates. For plaque screening, phage were stamped from the microtitre plates onto LB plates, grown for four hours, and transferred to nylon membranes, and the membranes autoclaved as in colony screening.

Probe Preparation

Probes for APP, ERG and D21S60 have been described previously (1,6). To identify Zinc-finger containing clones, the mouse Kruppel-like gene, mkr2 (17,18), was obtained from M. Aubry (McGill University). To identify ribosomal RNA clones, plasmids pA and pB, spanning the 45S ribosomal RNA precursor, were used (gift of J. Sylvester, Hahnemann University). Inserts were labelled by random oligo priming (19). For individual selected cDNA clones, inserts were amplified by PCR (using T3/T7 primers for D110G6 inserts, and A1/A2 or C1/C2 primers for λgt10 inserts), digested with the appropriate enzyme to release the insert, purified on low melt agarose, and labelled as above.

Hybrid Cell Lines

To determine if clones mapped to chromosome 21, a reduced chromosome 21 panel was used, consisting of Chinese hamster/human hybrid cell lines 153E7b (normal chromosome 21, deleted for ribosomal RNA genes on 21p), 21q+ and 8q- (carrying separate products of a reciprocal translocation: t(8;21)(q22;q22.2) (1). As a total human control, DNA from the human lymphoblastoid cell line CGM-1 was used; as the hamster control, the hybrid parent cell line, CHO adeC- was used.

Sequencing

Selected cDNA inserts were sequenced directly from the PCR amplifications, using the cycle sequencing method (BRL), and the appropriate T3/T7 or C1/C2 primers end labelled with ^{32}P. Clones from D110G6 were also sequenced by J. Sikela (University of Colorado Health Sciences Center) using the ABI automated sequencer.

Grail/BlastX Analysis

The Grail sequence analysis program (Oakridge National Laboratory) (20) and Genbank BlastX homology searches (21) have been described.

Northern Analysis

Total RNA was isolated from whole fetus (8 weeks), and HTB10 and HTB158 cell lines using standard procedures (13). Fifteen to twenty micrograms were electrophoresed in standard formaldehyde agarose gels, transferred by capillary action to Hybond N+ membranes, and uv-crosslinked.

cDNA Library Screening

One million clones from large insert cDNA libraries from fetal brain (Stratagene), testes (Clontech) and/or placenta (gift of M. Lovett, University of Texas, Dallas) were plated at 5 X 10⁵ pfu/150 mm plate, and grown for six hours. Plaques were transferred to Hybond N+, autoclaved and crosslinked. Hybridizations were by standard 50% formamide procedures. Filters were exposed for 4-5 days. Positive plaques were taken through second and third screens to obtain pure clones.

RESULTS

Scheme for Analysis of Selected cDNA Libraries

Figure 2 shows the general plan for isolation and characterization of a non-redundant set of genes from each YAC. Actual library construction (Step 1) is described in the references (9,14,16) and will not be discussed here. Step 2 covers the preliminary characterization of the libraries, and generally consists of determining the quality of the selection based on: the % contamination of rRNA and repetitive clones, control gene representation, and the % of clones that map back to the parent YAC and the correct chromosomal region, plus their % representation in the library. Step 3 characterizes individual clones by sequence and Northern analysis, determines overlaps and identifies the non-redundant gene set. It is important to bear in mind that clones from the selected library represent cDNA fragments (often small fragments, 100-500 bp), not complete cDNAs. Thus, two clones sharing no sequence overlap may derive from different regions of the same message, a circumstance that may be determined by Northern analysis (if patterns of alternative processing are seen) or by isolation of a "full length" cDNA.

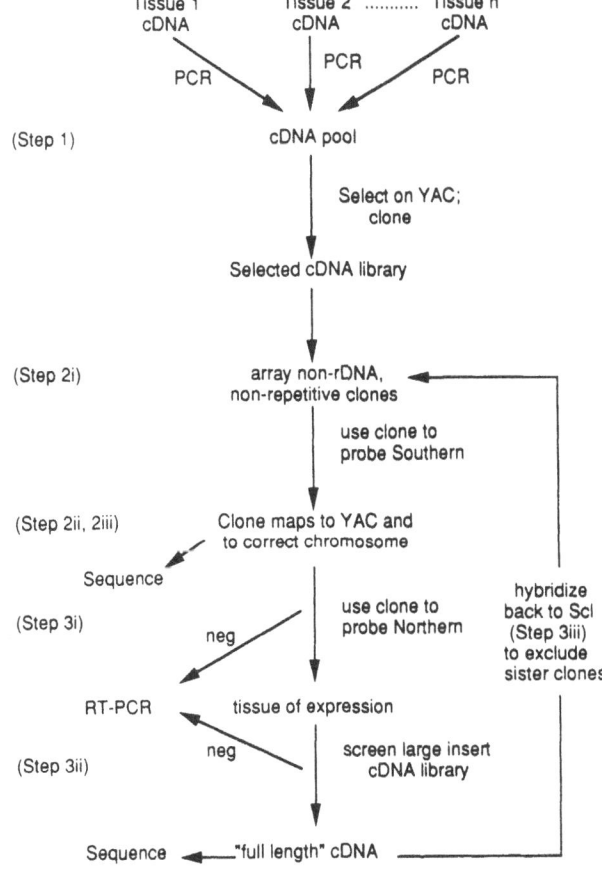

Figure 2. Schematic representation of the steps in selected cDNA library analysis. Discussion in the text primarily involves Step 2.

Preliminary Analysis

The quality of each library was assessed using three criteria: i) low representation of rRNA and repetitive clones, and adequate representation of control genes (if applicable); ii) high proportion of clones mapping to the parent YAC and to the correct chromosomal region; and iii) determining the proportion of the library represented by each clone from (ii).

i) Each selected library was screened with probes for total human DNA (to identify repetitive clones), ribosomal RNA (to determine the % contamination), and control genes, if any (genes known to be present within the YAC). For YACs cloned into λgt10, 1000 pfu were plated on two plates and directly screened. For the library in Bluescript (D110G6), the 500 colonies picked from the transformation plates were screened by colony hybridization. Table 1 summarizes the results of these screenings. These libraries were considered to be of good quality because they showed low proportions of repetitive and rDNA clones, and contained some proportion of the known genes. The expected proportion of these latter was estimated from their expression levels in the tissues used. Failure to detect the control genes, or unexpectedly low levels, can be an easy way to detect problems with the selection itself.

Table 1. Preliminary characterization of selected cDNA libraries.

YAC	Size (kb)	rDNA	% Repeat	% Control
D110G6	835	8%	4%	12% APP
				3% Zn fingers
				NA
430G1	600	12%	N	25% ERG
A125B12	285	4%	3%	8% D21S60
A25B8	185	2%	4%	NA

500 colonies were screened for the D110G6 YAC; 2000 pfu were screened for the other YACs. % rDNA was determined by hybridization with pA + pB ribosomal clones; % repeat, with total human DNA. N, none hybridized; NA, not applicable. Zn fingers were detected by hybridization with a mouse Kruppel cDNA (18).

In the D110G6 selected library, the APP gene was represented at 12%, an enrichment of approximately 500 over what is found in fetal brain libraries (unpublished observations), and Zn finger sequences were represented at 3% (enrichments cannot be determined because selection was done at high stringency, while screenings are done at low stringencies). One problem was noted, however. Prior to cloning, when the selected and amplified cDNA population (ranging from 200-600 bp) was digested with BamHI, approximately 70% of the mass of the cDNA was reduced to < 100 bp (data not shown). It must be anticipated that much of this material is lost in cloning, and with it possibly additional genes. It is unknown if this is a peculiarity of cDNAs from this YAC (for example, the APP cDNA does contain a few BamHI sites), or if it will be encountered frequently. BamHI is not predicted to cut particularly frequently, nevertheless, EcoRI or HindIII may be better choices for cloning enzymes. An alternative is use of the UDG system for direct cloning of PCR products (22).

The ERG gene represented 25% of the clones from the A125B12 library; the probe D21S60, mapping within the same YAC, detected 8% of clones in screenings of 2000 pfus.

D21S60 is a unique genomic fragment assumed to contain exons, based on the results of Northern blot analysis.

ii) Several clones from each library that were non-ribosomal, non-repetitive and non-control genes were assessed for hybridization to the parent YAC, and to chromosome 21. These data are shown in Table 2. Generally, >75% of the clones mapped back to the parent YAC; greater than this when repetitive clones not detected by the total human hybridization are ignored. Of these, 66%-90% also mapped to chromosome 21. Some failure to map to chromosome 21 is not unexpected with the use of non-chimeric YACs, and YACs deemed non-chimeric by FISH analysis (which may fail to detect some regions of chimerism). This latter is the case with the 430G1 YAC. However, several clones mapping to the D110G6 YAC do not map to chromosome 21, even though this YAC was determined to be non-chimeric by end probe analysis, generally considered stringent criteria for non-chimerism. We have no reassuring explanation for these results. Selecting a gene family member based on homology, but not identity, to sequences within the YAC should still result in some cross hybridization to chromosome 21. Postulating an internally chimeric YAC is unpleasant without further data. Because a detailed restriction map of the D110G6 has been constructed (6), several selected cDNAs were positioned within it (Fig. 3). It is noteworthy that several of these, lacking APP homology by sequence, mapped within the APP gene (which spans >300 kb (6)), presumably within introns. The majority of those mapping 3' to the APP gene did not map to chromosome 21.

Table 2. Mapping of anonymous selected clones.

YAC	Total Clones	Repeat	-YAC	+YAC	+21	% to YAC	% to 21
D110G6	20	1	2	17	13	85%	65/76%
430G1	12	2	1	9	6	75%	50/66%
A125B12	4	-	-	4	4	100%	100%
A25B8	11	1	-	10	9	90%	80/90%

Clones negative in screening with ribosomal, total human and control genes were checked for YAC and chromosomal location. Repeat: apparently repetitive clones, not detected with total human hybridization. -YAC: failed to hybridize to parent YAC; +YAC and +21, mapped to parent YAC and correct region of chromosome 21. % to 21: % of total mapping to chromosome 21/ % of clones mapping to YAC which map to 21.

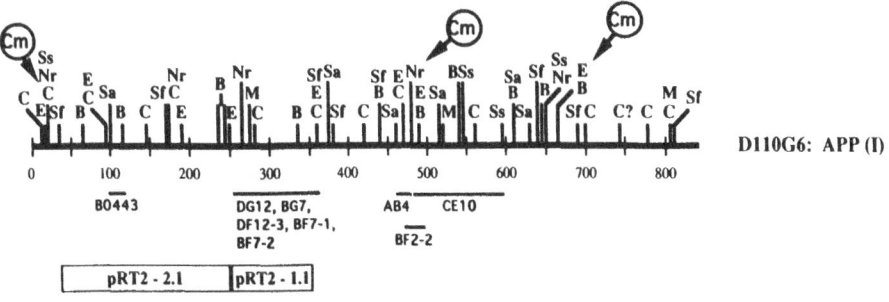

Figure 3. Position of selected cDNA clones within D110G6. The rare restriction site map is from Ref. 6. RT1-2.1 and 1.1 indicate the regions spanned by the 5' and 3' portions, respectively, of the APP gene. Short horizontal bars beneath the map indicate approximate positions of nine cDNA fragments. Distances are given in kb from the left end of the YAC clone. B, BssHII; M, MluI; Nr, NruI; E, EagI; Ss, SstII; Sf, SfiI; Sa, SalI; and C, ClaI. Cm, clusters of sites indicating potential CpG islands.

iii) The complexity of each selected library was assessed using cross hybridization to determine the % representation of various clones. These data are shown in Table 3. Representations range from <1% for some clones from D110G6 and A125B12, to 50% for clones from A25B8. In no case has the whole library so far been accounted for. A note of caution is in order when using this approach. As a rule of thumb, 50 to 100 clones per 100 kb of YAC are analyzed. While this is likely to be adequate for an overview of the library, it cannot be considered to give a complete picture. Certainly, it would be likely to miss a gene represented at 1/1000 in the selected library, a possibility for a clone originally present at $1/10^6$ and enriched 1000 fold. As a case in point, sequences for D21S60 were found to represent 8% of the A125B12 selected library when 2000 pfu were screened. This sequence ended up at only 1.5% when 400 clones were picked for analysis, and was missing entirely in the first 200 clones picked at random. Thus it is probably good policy to return periodically to the complete selected library and screen for clones that are negative in the accumulating non-redundant set. These may represent additional genes.

Table 3. Complexity of selected cDNAs.

D110G6	500 Clones	430G1	800 Clones	A125B12	400 Clones	A25B8	400 Clones
B04	0.4%	*1A5	27%	D21560	1.5%	+1A2	50%
DF12	15%	1A7	13%	3D2	1%	1A3	2%
BA3-4	4%	*1H3	13%	2D11	26%	+1A7	16%
		1H4	13%			+1A8	37%
		*1H1/ 1H9	30%	2D3	15%	+1A9	15%
		*1H12	11%			1B9	22%
						+1F9	2%
						1G8	4%

# of clones screened for each YAC are shown with the % cross-hybridizing. No library has been completely accounted for; * contained within same full length cDNA; + some cross hybridization and contained within the full length cDNA for 1B9.

Sequence Analysis

Confirmation that selected cDNA library clones indeed represent expressed sequences is important information. The more quickly this can be obtained, the fewer artifacts will be pursued. Use of the Grail exon identification program and Genbank sequence homology searches are two methods that, if positive results are obtained, rapidly generate support for the identification of expressed sequences.

Table 4 shows the results of Grail and Genbank analyses with ten fragments from these selected libraries. Of the six sequences showing no homologies in Genbank, three received good-excellent scores with Grail, and, considering the sequence lengths (>150 bp), this implies up to 90% confidence that bona fide exonic material is present. For two of the remaining clones (SW5 and 1H12), sequences were too short (approximately 100 bp) to obtain better than marginal ratings, and thus Grail results cannot be adequately judged in these cases. The GC values have been included in Table 4 as a point of interest. Current data suggests that the majority of genes are located within the GC rich (>45% G+C) regions of the genome (23), a fact not supported by the admittedly limited data of Table 4. However, a PCR bias towards less excessively GC-rich sequences may contribute to observing a lower average GC value, a value that will be corrected with either full length sequences, or more representation of 5' ends. Correlating GC values with the presence of expressed sequences and chromosomal location remains an avenue towards eventual increased understanding of genome organization.

Table 4. Sequence analysis.

YAC	cDNA	Size	% GC	Grail	Genbank
430G1	1H1	183	56	Excellent	NH
	1H4	150+170	48	Marginal	NH
	1H8	135	50	Good	NH
	1H9	168	56	Excellent	NH
	1H12	103+104	40	NE	NH
D110G6	*BG7	356	37	NE	99% APP Blast N
	*DG12	296	39	Excellent	98% APP Blast N
	B0443	260	43	Marginal	>60% Zn fingers Blast X
A125B12	SW5	118	55	Marginal	NH
	SW8	103	63	Marginal	ERG

* Sequenced by T3/T7 cycle sequencing and by automated method. NE: no exons detected; NH: no homology found; Size: total # nucleotides sequenced for each clone.

Table 5 summarizes Genbank search results. Of 32 segments, ten showed homologies, of which six corresponded to genes (APP and ERG) known to lie within the YACs. Of the "new" genes, of potential interest are the novel zinc finger sequence mapping within the D110G6 YAC, and the ferritin sequences, possibly representing a pseudogene previously mapped to chromosome 21. In A25B8, two clones showed homology to B-glucuronidase, not known to map to chromosome 21. From the same YAC, another fragment showed homology to an LFA sequence. The homology was low, but might increase with cleaner sequencing data. This is an intriguing possibility, because this YAC maps close to the CD18 gene, LFA-1, potentially indicating the presence of a clustered gene family.

Table 5. Genbank search results.

YAC	# Fragments	# Matches	# Novel
D110G6	7	2APP, 1 Zn finger	4
430G1	11	-	11
A125B12	9	4 ERG, 1 ferritin	4
A25B8	5	2 β-glucuronidase, 1 LFA-like	2

Several fragments from each library were sequenced and analyzed using BlastX. Criteria for a match were: >55% similarity, >24 amino acids, p ≤ 0.01.

Expression Studies

Expression studies are valuable ultimately for confirmation of transcribed sequence identity, but in preliminary stages, they are also useful for developing a non-redundant gene set (as opposed to a non-redundant fragment set) for each YAC. Both Northern analysis and cDNA library screening have been used here.

For cDNA library screening, fetal brain, placenta and testes were chosen, because of these tissues' reputations for expressing a more complex collection of genes. Standard procedures stipulate screening 10^6 clones to have reasonable chances of detecting sequences expressed at low levels. Such an effort has been made much less laborious by plating 5 X 10^5 phage per 150 mm plate, rather than the conventional 50,000. Perhaps because these fragments are entirely exonic, they appear to make remarkably good probes in such high density screenings, and give clean, clear signals in 4-5 days (Fig.4). Nine fragments were tested in this manner in pools of two; six gave positive results. Frequencies were highly variable. Both DF12 and EF7-1 (from the D110G6 YAC) detected only one clone in 10^6 from the fetal brain library, whereas 1A3 detected approximately 1 in 10^4 in both the fetal brain and the testes libraries (Fig. 4). Two pairs of fragments detected identical cDNAs, and thus, although members of a pair showed no sequence overlap, they likely derive from a single gene. Thus, from nine fragments, and six positives, four "full length" cDNAs were obtained. One of the three fragments giving no positive cDNAs was D21S60, and from Northern analysis, it appears that this sequence may be T-cell specific and thus not present in the libraries screened.

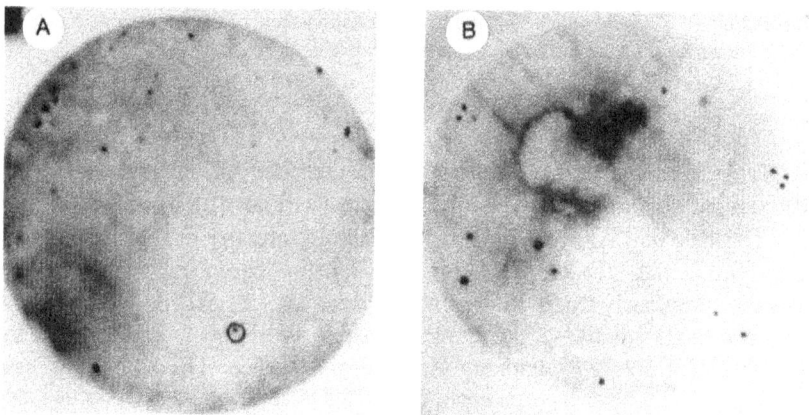

Figure 4. High density screening of cDNA libraries. cDNA libraries were plated at 5 X 10^5 pfu per 150 mm plate. After plaque transfer to nylon, filters were hybridized with probes from selected cDNA libraries. A, fetal brain library (Stratagene) screened with DF12 and EF7-1 from the D110G6 selected library. The single positive clone is shown circled; other signals are those remaining from the previous screen with 1B9 and 1A3 from the A25B8 selected library. B, testes library (Clontech) screened with 1H1 and 1H9 from the 430G1 selected library. 1H1 and 1H9 were shown to be parts of the same cDNA.

The consequences of obtaining a large insert cDNA were immediate in the case of two fragments from A25B8 YAC. Upon rescreening a microtitre plate of the selected library with the 1B9 large cDNA, 75/96 clones hybridized (this is not likely to be due to repetitive sequences). Only two are hybridized with the 1A3 cDNA. That the parent YAC maps to a GC-rich, gene-rich region of chromosome 21 suggests that more than two genes are likely to be present within the YAC, even though it is only 185 kb in size.

In a second case, the large cDNA obtained with two fragments from the 430G1 YAC failed to map to chromosome 21. This may indicate that either a gene family is detected by these fragments, and the chromosome 21 gene was not examined (5-6 positives were detected in each of testes and fetal brain libraries, but only one clone was purified to homogeneity and used in mapping), or that the chromosome 21 fragments represent a pseudogene. How frequently such situations will arise in using selected libraries is not yet clear.

Northern analysis was also successful in identifying non-overlapping fragments of a single gene. Fragments 1B9 and 1F6 from the A25B8 YAC both recognize transcripts of 2.4 and 2.8 kb in RNA from HT1080 and a whole embryo cell line (data not shown). This observation was confirmed in analysis of the large cDNA clone.

CONCLUSIONS

Analysis of the selected cDNA libraries from these 4 YACs is clearly not complete, but even this preliminary work indicates that new genes have been obtained. Clearly, the technique of cDNA hybrid selection from YAC clones generates large numbers of clones likely to contain expressed sequences. The important tasks that remain are the rapid identification of a non-redundant gene set from the collection of fragments, and the rapid verification that the fragments indeed represent exonic material. Neither of these tasks are trivial, in particular if the objective is construction of a detailed transcriptional map, and not simply identification of some new genes.

Particular problems that have been encountered so far include: i) apparently non-chimeric YACs yielding clones mapping to the YAC but not mapping to chromosome 21. This is likely to be an important consideration when using YACs non-chimeric based solely on FISH analysis; ii) the presence of a gene family member driving selection of cDNA fragments not mapping to chromosome 21; and iii) the presence of a pseudogene selecting cognate gene fragments.

The first of these is easy to detect, and such clones can simply be discarded. However, the experience with selection from the D110G6 YAC that is apparently non-chimeric by end analysis, yet yielded fragments mapping to the YAC but not to chromosome 21, is not easy to explain. This phenomenon has not been observed in selections from numerous other YACs (S. Weissman, unpublished), and forces the conclusion of an internally chimeric YAC. Fortunately, this can be further investigated using overlapping YACs. On the other hand, how to detect and deal with gene families and pseudogenes remains an issue. Obtaining 5' ends of cDNAs or analyzing genomic sequences for comparison with cDNA sequences offer solutions, but are not attractive solutions when mapping large genomic regions or dealing with large numbers of potential genes. Nevertheless, such problems will need to be addressed.

To define a non-redundant gene set, complex patterns on Northern blots can help to identify members of the same mRNA. For fragments that see only a single band on Northerns, however, concluding a common origin from the same sized mRNA is problematic given the resolution typical of RNA gels. Two fragments mapping to the same full length or long parent cDNA is a good criterion for their derivation from the same gene, and also allows the rapid identification of additional overlapping fragments by rescreening of the whole selected library. High density screenings of large insert cDNA libraries makes this task less laborious, but will be limited in success by knowing the correct tissue of expression.

As is discussed elsewhere in this volume, verification that a selected cDNA clone indeed derives from exonic material can be subject to strict criteria, in particular where expression levels are low or restricted in tissue or development timing. Nevertheless, cDNA hybrid selection from YACs presents the advantages of rapidly generating large numbers of potentially new genes, whose map location is known and who may not contain introns. As physical analysis of chromosome 21 advances, and as YAC clones themselves are analyzed in more detail, new facets of human genome organization can be addressed. These will include determination of the sizes of genes, identification of genes within introns, real gene densities and associations with CpG islands. For chromosome 21, this presents the potential for a rapid isolation of many new candidate genes for phenotypic features of Down Syndrome.

ACKNOWLEDGEMENTS

The authors wish to thank Gregory Dolganov (Genelabs, Redwood, CA) for helpful discussions concerning selection using magnetic beads and Heather Wade for technical assistance. This work was supported in part by grants from the National Institutes of Health HD17449 and HG00001 to KG.

REFERENCES

1. K. Gardiner, M. Horisberger, J. Kraus, U. Tantravahi, J. Korenberg, V. Rao, S. Reddy, and D. Patterson, Analysis of human chromosome 21: correlation of physical and cytogenetic maps; gene and CpG island distributions, *EMBO J.* 9:25 (1990).
2. H. Ichikawa, F. Hosoda, Y. Arai, K. Shimizu, M. Ohira, and M. Ohki, A complete NOTI restriction map of the entire long arm of human chromosome 21, *Nature Genetics* 4:361 (1993).
3. M.J. Owen, L.A. James, J.A. Hardy, R. Williamson, and A.M. Goate, Physical mapping around the Alzheimer disease locus on the proximal long arm of chromosome 21, *Am. J. Hum. Genet.* 46:316 (1990).
4. M. Burmeister, S. Kim, E.R. Price, T. de Lange, U. Tantravahi, R.M. Myers, and D.R. Cox, A map of the distal region of the long arm of human chromosome 21 constructed by radiation hybrid mapping and pulsed-field gel electrophoresis, *Genomics* 9:19 (1991).
5. D.R. Cox, M. Burmeister, E.R. Price, S. Kim, and R.M. Meyer, Radiation hybrid mapping: A somatic cell genetic method for constructuring high-resolution maps of mammalian chromosomes, *Science* 250:245 (1990).
6. F. Tassone, S. Cheng, and K. Gardiner, Analysis of chromosome 21 Yeast Artificial Chromosome (YAC) clones, *Am. J. Hum. Genet.* 51:1251 (1992).
7. K. Gardiner, Physical mapping of the long arm of chromosome 21, *in*: "Molecular Genetics of Chromosome 21 and Down Syndrome," D. Patterson and C.J. Epstein eds., Wiley-Liss, Inc., New York (1990).
8. K. Gardiner, B. Aissani, and G. Bernardi, A compositional map of human chromosome 21, *EMBO J.* 9:1853 (1990).
9. S. Parimoo, S.R. Patanjhali, G. Shukla, D.D. Chaplin, and S.M. Weissman, cDNA selections: Efficient PCR approach for the selection of cDNAs encoded in large chromosomal DNA fragments, *Proc. Natl. Acad. Sci. USA.* 88:9623 (1991).
10. D. Patterson, Report of the Second International Workshop on Human Chromosome 21, *Cytogenet. Cell. Genet.* 57:167 (1991).
11. I. Chumakov, P. Rigault, S. Guillou, P. Ougen, A. Billaut, G. Guasconi, P. Gervy, I. LeGall, P. Soularue, L. Grinas, L. Bougueleret, C. Bellanne-Chantelot, B. Lacroix, E. Barillot, P. Gesnouin, S. Pook, G. Vaysseix, G. Frelat, A. Shmitz, J.-L. Sambucy, A. Bosch, X. Estivill, J. Weissenbach, A. Vignal, H. Riethman, D. Cox, D. Patterson, K. Gardiner, M. Hattori, Y. Sakaki, H. Ichikawa, M. Ohki, D. Le Paslier, R. Hellig, S. Antonarakis, and D. Cohen, A continuum of overlapping clones spanning the entire human chromosome 21q, *Nature* 359:380 (1992).
12. S. Graw, D. Patterson, and K. Gardiner, Chromosome 21 YAC contigs, *Am. J. Hum. Genet.* 53:1297 (1993).
13. B. Chomzynski and N. Sacchi, Rapid RNA isolation, *Anal. Biochem.* 162:156 (1987).
14. J.G. Morgan, G.M. Dolganov, S.E. Robbins, L.M. Hinton, and M. Lovett, The selective isolation of novel cDNAs encoded by the regions surrounding the human interleukin 4 and 5 genes, *Nucl. Acids Res.* 20:5173 (1992).
15. S.R. Patanjali, S. Parimoo, and S.M. Weissman, Construction of a uniform-abundance (normalized) cDNA library, *Proc. Natl. Acad. Sci. USA* 88:1943 (1991).

16. B. Korn, Z. Sedlacek, A. Manca, P. Kioschis, D. Konecki, H. Lehrach, and A. Poustka, A strategy for the selection of transcribed sequences in the Xq28 region, *Hum. Molec. Genet.* 1:235 (1992).

17. K. Chowdhury, G. Dressler, G. Breier, U. Deutsch, and P. Gruss, The primary structure of the murine multifinger gene mKr2 and its specific expression in developing and adult neurons, *EMBO J.* 7:1345 (1988).

18. M. Aubry, C. Marineau, F.R. Zhang, L. Zahed, D. Figlewicz, O. DeLattre, G. Thomas, P.J. DeJong, J.P. Julien, and G.A. Rouleau, Cloning of six new genes with zinc finger motifs mapping to short and long arms of human acrocentric chromosome 22 (p and q11.2), *Genomics* 13:641 (1992).

19. A. Feinberg and B. Vogelstein, Addendum to "A technique for radiolabelling DNA to high specific activity," *Anal. Biochem.* 137:266 (1984).

20. E.C. Uberbacher and R.J. Mural, Locating protein-coding regions in human DNA sequences by a multiple sensor-neural network approach, *Proc. Natl. Acad. Sci. USA.* 88:11261 (1991). 21. W. Gish and D.J. Status, Identification of protein coding regions by database similarity search, *Nature Genetics* 3:266 (1993).

22. P.E. Nisson, A. Rashtchian, and P.C. Watkins, Rapid and efficient cloning of Alu-PCR products using uracil DNA glycosylase, *in:* "PCR Methods and Applications," Cold Spring Harbor Laboratory Press (1991).

23. D. Mouchiroud, G. D'Onofrio, B. Aissani, G. Macaya, C. Gautier, and G. Bernardi, The distribution of genes in the human genome, *Gene* 100:181 (1991).

ISOLATION OF EXPRESSED SEQUENCES FROM THE CHROMOSOME 17q21 BRCA1 REGION BY MAGNETIC BEAD CAPTURE

Fergus J. Couch[1], Barbara L. Weber[1], Francis S. Collins[1,2,3] and Danilo A. Tagle[3]

Departments of [1]Internal Medicine and [2]Human Genetics, University of Michigan Medical School, Ann Arbor, Michigan 48109
[3]Laboratory of Gene Transfer, National Center for Human Genome Research, National Institutes of Health, Bethesda, Maryland 20892

ABSTRACT

Magnetic bead capture of cDNAs utilizes biotin-streptavidin magnetic bead technology to isolate expressed sequences from large genomic regions, resulting in several thousand-fold enrichment of selected cDNAs. The technique allows parallel analysis of several genomic segments of varying complexity. Expressed sequences from a variety of tissue sources can also be identified simultaneously. To evaluate this approach, we have applied it to pools of cosmid clones from the interval on chromosome 17q21 which contains the familial early onset breast cancer gene (BRCA1). We describe the characterization of 9 potentially unique cDNAs which were isolated from one pool of 7 minimally overlapping cosmids representing a subset of the BRCA1 candidate region. Overall, the method is shown to detect a large fraction of coding sequences in cosmid clones. Advantages and limitations of the approach are discussed.

INTRODUCTION

Rapid isolation and identification of transcribed sequences from specific chromosomal regions is a central problem of positional cloning. Conventional methods of transcript isolation, such as using conserved sequences between divergent species and CpG rich island fragments as hybridization probes, are often labor intensive and cumbersome when applied to genomic intervals of several thousand to a few million base pairs in length. Recently several new approaches to transcript isolation from large genomic clones have been described. These techniques include the direct cloning of human transcripts from human-rodent somatic cell hybrids (1), trapping of internal exons based on the presence of splice junctions (2,3), trapping of 3' terminal exons (4), direct screening of cDNA libraries with total yeast artificial chromosomes (YACs) (5), and enrichment of cDNAs using immobilized YACs (6,7). Each methodology has certain advantages and disadvantages which dictate the applicability of the technique. The isolation of human transcripts from somatic cell hybrid cells has the advantage of screening whole chromosomes or sub-chromosomal regions, however, generally only constitutively expressed genes are isolated. Exon amplification captures exons directly from genomic clones thereby removing dependency on availability of tissue specific cDNA libraries. However, the method requires small genomic clones such as cosmids for trapping, fails to isolate 1 or 2 exon genes and tends to generate false positives by cryptic splicing at sites in Alu repeats. Direct cDNA screening using YACs and/or cosmids scans several hundred thousand base pairs simultaneously but due to the complexity of the probe, the technique suffers from a low signal to noise ratio, resulting in low reproducibility. In contrast, cDNA enrichment approaches can be powerful and straightforward. Several different approaches have been taken beginning with immobilization of YACs on membrane filters (6,7) and progressing to immobilization of biotinylated genomic DNA on streptavidin coated magnetic beads (8,9,10).

In this report we present a cDNA enrichment strategy which utilizes magnetic bead capture of PCR amplified cDNAs (8). This technique has been utilized to isolate cDNAs from both the Huntington disease region on chromosome 4p16.3 (11), and the familial early onset breast cancer region (BRCA1) on chromosome 17q21. We describe here a study of pools of cosmids from the BRCA1 region and demonstrate that the method allows amplification of a known gene from one of the cosmid pools.

METHODS

General Strategy

In this procedure cloned genomic DNA is digested with a frequently cutting enzyme and ligated to compatible linkers. The linkered genomic segments are PCR amplified using a biotinylated primer and hybridized at high stringency to PCR-amplified inserts from the desired cDNA libraries. The biotinylated genomic DNA/cDNA complexes are captured using streptavidin coated paramagnetic beads and washed to remove unbound and non-specific cDNAs. The captured cDNAs, enriched for sequences from the genomic clones, are eluted by boiling and PCR-amplified. The captured fragments are then subcloned as a region and tissue specific sublibrary. An outline of the procedure is presented in Fig. 1.

Preparation of Genomic DNA

Cosmid DNA was prepared by alkaline lysis from specific clones isolated from the human chromosome 17 specific cosmid library (Los Alamos National Laboratory). Pools of 7 minimally overlapping cosmids were used to isolate cDNAs from the BRCA1 region. Approximately 1 mg of each of 7 cosmids was digested separately with Sau3AI and AluI restriction endonucleases. A 10-100 fold molar excess of Sau3AI and AluI linkers were ligated to the complete digest.

Sau3AI linker: 5' GATCTCGACGAATTCGTGAGACCA 3'
                                AGCTGCTTAAGCACTCTGGT

AluI linker: 5' GATCTCGACGAATTCGTGAGACCA 3'
                            CTAGAGCTGCTTAAGCACTC

The ligation reaction was incubated at room temperature for 16 h. One tenth of the ligation reaction was added to a 100 ml PCR reaction containing 200 pmoles of a single

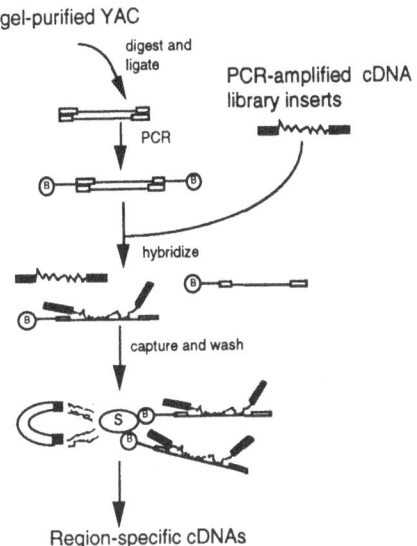

gel-purified YAC

PCR-amplified cDNA
library inserts

Figure 1. A schematic diagram for magnetic bead capture of region-specific cDNAs.

5' biotinylated primer (5' Biotin-14-TGGTCTCACGAATTCGTCGA-3'). Once resuspended, biotinylated primers were stored at 4°C. Repeated freeze-thaw may result in dissociation of the biotin group from the primer. Thermal cycling was performed for 35 cycles at 94°C denaturing for 30 sec, 60°C annealing for 30 sec and 72°C extension for 1 min 30 sec plus an additional 5 sec per cycle. An aliquot of each PCR reaction was run on a 1.5% agarose gel to assess the quality of the linkered material. Further assessment of the random representation of genomic fragments was carried out by blotting and hybridization with known probes from the genomic region.

Amplification and Preparation of cDNA Inserts

Inserts from two cDNA libraries, 1) an oligo dT primed breast λgt11 library and 2) a random primed breast λgt10 library, were PCR amplified using their respective vector primers. Approximately 2 X 10⁶ recombinants were added per 100 ml PCR reaction. Three 100 ml reactions were prepared for each magnetic bead capture reaction using the PCR conditions described above. PCR reactions were pooled and an aliquot was run on a 1.5% agarose gel to assess the quality of the amplified material. A generalized smear from 200-1500 bp suggested good amplification. Excess primers, dNTPs and Taq polymerase were removed using Magic PCR prep minicolumns (Promega). The cDNA inserts were then ethanol precipitated and resuspended in TE at 1 mg/ml.

Vector primers:
lgt10   forward: 5'-TTGAGCAAGTTCAGCCTGGTTAAG-3'
        reverse: 5'-CTTATGAGTATTTCTTCCAGGGTA-3'

lgt11   forward: 5'-GGTGGCGACGACTCCTGGAGCCCG-3'
        reverse: 5'-TTGACACCAGACCAACTGGTAATG-3'

Transcribed Sequence Selection

Approximately 5 ml of biotinylated genomic DNA fragments from each of 14 PCR reactions (7 from Sau3AI linkered material and 7 from AluI linkered material) were pooled and alcohol precipitated. In order to prevent hybridization of cDNAs non-specifically to repetitive and vector based sequences, the DNA was pre-annealed with 10 mg of human Cot-1 DNA (Gibco-BRL) and 2 mg each of Sau3AI and AluI digested sCos1 cosmid vector in a final volume of 25 ml of 0.1 M sodium phosphate at 65°C. Approximately 2 mg of denatured cDNA was added to the pre-blocked DNA and the mixture was hybridized at 65°C for 20-24 h. 50 ml of streptavidin coated magnetic beads (Dynal) were added to the reaction and incubated at room temperature for 20-30 min. Washings were carried out as specified by the company in 1X Taq polymerase buffer (Promega) and consisted of two washes at room temperature for 10-15 min followed by four washes at 65°C for 10-15 min. Finally, tightly bound cDNAs resulting from specific hybridization were removed by boiling for 10 min. All eluted fractions were used directly for PCR amplification. Amplified material was dot blotted and hybridized with 1) a known cDNA fragment from the region, to demonstrate specific capture, 2) Cot-1 DNA, to demonstrate removal of repetitive sequences from the genomic DNA/cDNA complex, and 3) a cDNA probe from a chromosomal region not represented in the original cosmids (β-actin or glucose 3-phosphate dehydrogenase (G3PD)) to demonstrate removal of non-specific cDNAs from the DNA/cDNA complex by sequential washing leading to low levels of non-specific cDNAs in the final cDNA fraction. Amplified material was cloned by two methods: 1) The material was cleaned with Magic PCR prep columns (Promega), concentrated and digested with EcoRI restriction enzyme. The products were separated on a 1.25% low melt agarose gel, fragments larger than 0.2 kb were removed from the gel in agarose and cloned into the EcoRI site of pBluescript (Stratagene). 2) The material was concentrated and separated on a 1.25% agarose gel. The agarose was melted and the product directly cloned into the pAMPI vector using the uracil DNA glycosylase (UDG) cloning kit (Gibco-BRL). In this case the vector primers used for amplification were tailed with CUA and CAU as specified by the manufacturer.

Characterization of Captured cDNAs

Individual transformants were picked into four 96 well plates containing 100 ml of LB medium with 100 mg/ ml ampicillin and grown overnight. Transformants were stamped from the 96 well plates onto Hybond-N nylon membranes (Amersham) in 4 X 96 formation and grown overnight on ampicillin plates. Grown colonies were denatured, neutralized and baked in a vacuum oven (12). Cot-1 DNA was radio-labelled using the random primer method with [α-32P] dCTP to a specific activity of 10^8 cpm/mg (13) and used to screen the gridded cDNA clones for the presence of repetitive sequences. Non-repetitive clones were used directly as template in PCR reactions with 1) plasmid vector primers for EcoRI cloned fragments or 2) phage vector primers for UDG cloned fragments. PCR products were assessed for size on 1% agarose gels and fragments greater than 200 bp in size were excised from the gel and stored at -20°C.
Isolated cDNA fragments were random primer labelled and used to screen the previously described cDNA grids for detection of redundant cDNA clones. Grids of 400 cosmids from the BRCA1 region of chromosome 17q21 were prepared as described for the cDNA grids and were screened to verify that each cDNA clone mapped back to the original genomic clones. Southern blots of total human genomic and pooled BRCA1 cosmid DNA cut with PstI restriction enzyme were also screened to verify that the cDNAs were single copy human DNA from the correct region of chromosome 17q21. An outline of the strategy for characterization of the cDNA clones is shown in Fig. 2.

Sequencing and Analysis

Plasmid DNA containing subcloned captured cDNAs or subcloned cDNAs derived

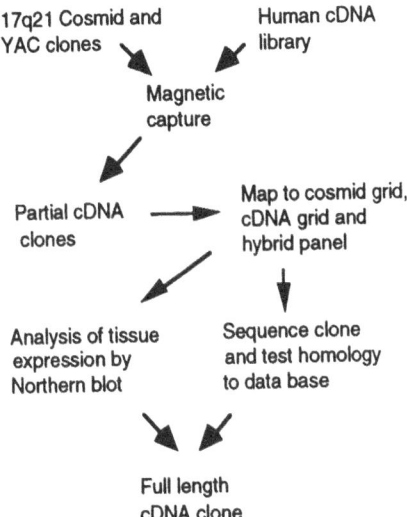

Figure 2. Strategy for isolation and characterization of magnetic bead captured cDNA clones from the BRCA1 region of chromosome 17q21.

from cDNA library screening was prepared with Magic miniprep DNA purification columns (Promega) according to the manufacturer's instructions. Automated (ABI 373A) sequence data for each cDNA fragment was generated by the University of Michigan Human Genome Center Sequence Core. Sequence data was sent to the e-mail server at the National Center for Biotechnology Information (NCBI) and compared with the non-redundant GenBank, dbEST and protein databases using the BLAST programs (14). Searches for open reading frames, translations and alignments of cDNA sequences were performed using MacVector software on a Macintosh IIsi.

RNA Hybridization Analysis

Single copy, non-redundant cDNA fragments which mapped back to the parent cosmid clones were hybridized to Northern blots to identify 1) size of the parent transcript, 2) tissue expression of the parent mRNA, and 3) variations in abundance of the transcript in normal breast tissue RNA and breast cancer cell line RNA. Each blot contained total RNA prepared from breast, ovary, benign epithelial ovarian cyst, and cultured breast cancer cell lines MCF7 and T-47D (ATCC) total RNA. RNA was prepared by the method of MacDonald (15), was fractionated on agarose gels (1%) containing 0.6M formaldehyde and transferred to Hybond membrane (Amersham). Following cross-linking with UV irradiation, blots were repeatedly hybridized with radio-labelled cDNA fragments. All hybridization conditions for genomic DNA, cDNA and RNA consisted of overnight incubation at 65°C in 6XSSC, 0.5% SDS, 1X Denhardt's, 100 mg/ml denatured salmon sperm DNA, and 125 mg/ml sonicated human placental DNA. Membranes were washed in 3XSSC, 0.1% SDS at 65°C and exposed to Kodak XOMAT AR X-ray film at -70°C. Labelled probes were removed from membranes between hybridizations by heating to 80°C with 0.1XSSC, 0.1% SDS.

Extension of cDNA Clones

The cDNAs which had been mapped back to the region of interest were used to screen the two breast cDNA libraries by standard methods to isolate larger fragments of each transcript. Each library screen was performed with 5 pooled cDNA probes labelled with both [α-32P] dCTP and dGTP. Tertiary clones were sorted into overlapping groups by dot-blotting and hybridization with the original cDNA probes. Tertiary purified clones were picked into 500 ml of SM and stored for 6 h. 2 ml of the stock was then used in a PCR amplification reaction with phage vector primers. Products were subcloned by two methods: 1) concentration, EcoRI digestion and ligation to the EcoRI site of pBluescript, or 2) direct UDG cloning of PCR product into pAMP1 vector. Subcloned large cDNA fragments were then sequenced, analyzed for homology to known genes by searching through databases and finally used for BRCA1 mutation identification studies as follows: 1) Southern blot hybridization of DNA from breast cancer patients to detect any large genomic DNA rearrangements, and 2) design of PCR primers for single strand conformational polymorphism (SSCP) (16), studies of breast cancer patient RNA and DNA.

RESULTS

Efficiency of Magnetic Bead Capture

We have used the magnetic bead capture technique to isolate cDNAs from groups of seven minimally overlapping cosmids from the BRCA1 region on chromosome 17q21 (17). The single cosmid group described contained the estradiol 17-β dehydrogenase 2 gene (EDH17B2) which was used as a positive selection control in the magnetic bead capture reaction. A total of 325 colonies were chosen at random from the captured fragment sublibrary which had been generated by cloning into the EcoRI site of pBluescript. cDNA grids were prepared and all 325 clones were PCR amplified to determine size. Fragments greater than 200 bp in size were selected for mapping studies as detailed above.

32 cDNA fragments were selected and 23/32 were identified as cDNAs which mapped back to the pooled cosmids. Figure 3 shows an example of mapping and redundancy analysis of a cDNA fragment. During the course of analyzing the cDNA clones, 2 cDNA fragments which mapped to the genomic clones were identified as fragments of the γ-tubulin gene (TUBG) (18). One of the γ-tubulin cDNA fragments, MTO-101, was mapped to the cosmid pool by hybridization to the cosmid grids. MTO-101 identified 19 cosmids from the BRCA1 region which span 4 different minimally overlapping cosmids suggesting that the γ-tubulin gene is a large gene containing large intronic regions. MTO-101 hybridized to 3 cDNA fragments on the cDNA grids. The 2 redundant clones were not further analyzed. All clones were sequenced and tested for homologies as described. Only the γ-tubulin gene was identified by database search. Figure 4 shows the comparison between the γ-tubulin cDNA and the γ-tubulin fragments identified by magnetic bead capture. 100% homology over 145 bp was observed between clone MTO-101 and positions 715-859 bp of the 1568 bp human γ-tubulin cDNA (Fig. 4). cDNAs with no significant sequence homology were hybridized to Northern blots to determine if the fragment was from a true transcript.

As many as 11 genes expressed in breast tissue may be present in this genomic interval of approximately 150 kb. This estimate represents: 1) Northern blotting indicating different sizes of transcripts, 2) Sequence homology between cDNA clones which were isolated from the breast cDNA library by the sublibrary cDNA fragments, 3) Overlap of large cDNA clones as detected by hybridization, 4) Mapping of cDNA clones to different regions of the 7 cosmid contig, 5) Gene Recognition Analysis Internet Link (GRAIL) assessment of protein coding potential, 6) Known genes in the region, EDH17B2 and TUBG, and 7) Comparison with the exon amplification (3), and direct screening techniques currently being used in our laboratory to isolate cDNA clones from the BRCA1 candidate interval. Nine of these 11 potential transcripts were isolated from the breast cDNA library by magnetic bead capture. We are currently in the process of obtaining full length cDNA clones and sequences for each of these eleven putatively unique cDNAs. As more sequence becomes available the true number of unique genes in the region, particularly those expressed in breast tissue, will become apparent. Analysis of the region by magnetic bead capture with other cDNA libraries may yield still more unique transcripts.

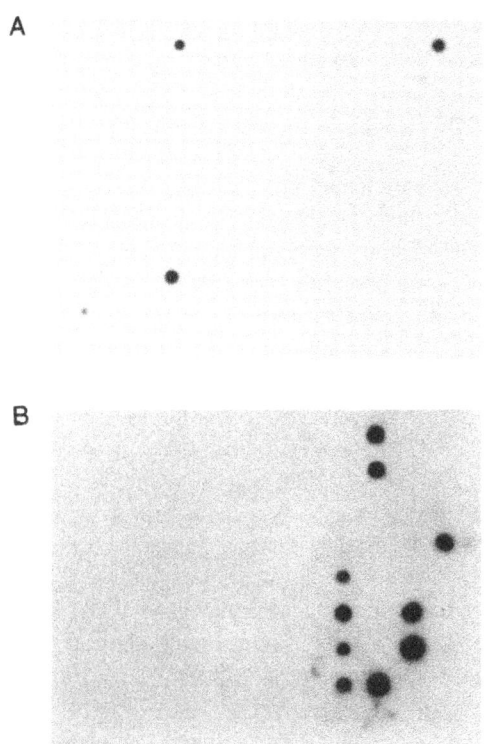

Figure 3. Hybridization of cosmid and cDNA grids with captured cDNA fragments. A Hybridization of a grid of 325 cDNA clones with MTO-101, a γ-tubulin cDNA fragment. Three clones were observed to contain γ-tubulin sequence which was present in MTO-101. The positive clone at bottom left of the grid was MTO-101. B Hybridization of a grid of 380 cosmid clones from the BRCA1 region with MTO-29. The 200 bp MTO-29 cDNA fragment mapped back to the EDH17B2 genomic region, but did not detect any homology with sequences present in a variety of databases. The MTO-29 probe detected 10 cosmids, 2 of which were from the pool of 7 minimally overlapping cosmids used in the capture experiment.

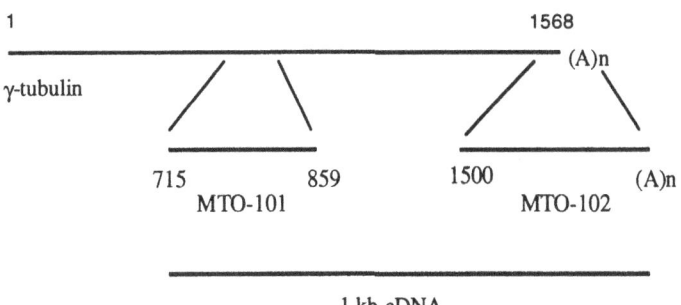

Figure 4. Position of the γ-tubulin cDNA fragments, MTO-101 and MTO-102, relative to the γ-tubulin cDNA sequence. A 1 kb γ-tubulin cDNA probe which was isolated with MTO-101 is also shown.

Magnetic bead capture compared favorably with exon amplification and direct screening as a cDNA isolation procedure for the EDH17B2 genomic region. Direct screening identified 2 unique cDNAs which cross hybridized with captured clones. This quickly led to the extension of the original cDNA clones. The exon amplification technique which is described in the accompanying manuscript by Abel *et al.* identified 7 potentially unique transcripts in the breast cDNA library, 2 of which were not isolated by magnetic bead capture. However, only a subset of both the magnetic bead capture and exon amplification sublibraries were tested, suggesting that further analysis might reveal cDNA fragments or exons from all the genes in the region.

Specificity and Enrichment

The specificity of the enrichment process was assessed by examining the nature of the captured cDNAs as described. In the example of the BRCA1 region, PCR amplified inserts from the breast cDNA library were dot blotted along with amplified fractions from magnetic capture of a 7 cosmid pool from the BRCA1 candidate interval. An EDH17B2 cDNA probe, known to map to chromosome 17q21, hybridized with greatest intensity to the final specific enriched cDNA fraction from these 7 cosmids. This demonstrates the high specificity of the captured cDNAs despite the complexity of the target DNA. As a direct comparison, a G3PD cDNA probe was used to test the level of non specific transcripts in the enriched material. The G3PD probe hybridized with decreasing intensity to each sequential fraction of the capture experiment suggesting that very little non-specific material was captured. Cot-1 DNA was also used as a probe to test for the depletion of repetitive elements. The Cot-1 probe hybridized with equal intensity to the original library fraction and to the specifically captured fraction suggesting that substantial levels of repetitive clones would be present in the capture sublibrary (Fig. 5). Cot-1 hybridizations of fractions from capture experiments of other cosmid pools showed decreasing intensity of hybridization in sequential fractions suggesting that repetitive elements were removed in the washing steps.

To further analyze the enriched cDNAs following cloning and random selection, the 325 cDNA clones on grids were assessed for the presence of the positive control EDH17B2 gene. A 1 kb EDH17B2 cDNA fragment was used as a hybridization probe on the cDNA grids. No EDH17B2 clones were detected. The EDH17B2 fragment was detected in the breast oligo dT λgt11 cDNA library at a level of 1 in 106 clones. Two explanations may account for the lack of identification of an EDH17B2 clone: 1) An EDH17B2 clone may have been present in the captured clone sublibrary, but was not randomly selected in these 325 clones. This is the most likely explanation, because dot blots indicated enrichment of EDH17B2 in the final fraction of the capture experiment. 2) EDH17B2 was enriched in the selection procedure but was not subcloned into pBluescript.

The magnetic bead capture technique has previously been shown to isolate cDNAs expressed at a frequency of only 1 in 2×10^6 clones in the parent cDNA library. That experiment involved a 2,000 fold enrichment and capture of the MBC6 cDNA from the Huntington disease region of chromosome 4p16.3 (10). To determine the enrichment level of the BRCA1 experiment, a 1 kb γ-tubulin cDNA, isolated from the breast cDNA library by a MTO-101 probe, was utilized to make a quantitative assessment of the degree of enrichment observed with magnetic bead capture. The γ-tubulin cDNA fragment was used as a probe on the cDNA grids and the breast oligo dT primed λgt11 cDNA library. Given that the cDNA clones on the grids were randomly picked, the γ-tubulin cDNA hybridization of the grids should be representative of the entire sublibrary. Approximately 325 clones from the sublibrary and 5×10^5 clones from the original library were screened. Four γ-tubulin clones were detected in the sublibrary suggesting a frequency of 1 in 81

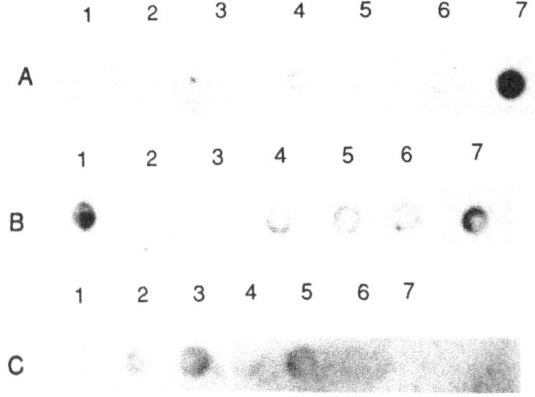

Figure 5. Dot blots of fractions from a magnetic bead capture experiment using 7 cosmids from the EDH17B2 region of chromosome 17q21. Dots 1 to 7 represent hybridizations of PCR amplified sequential washes of the genomic DNA/cDNA complex. 1 represents the supernatant from the original hybridization of cDNA to biotinylated genomic DNA. 2 and 3 represent two washes in 1X Taq polymerase buffer at room temperature for 15 min each. 4-6 represent washes in 1X buffer at 65°C for 15 min each. 7 represents the final elution of cDNA fragments from the genomic DNA by boiling in 1X Taq buffer for 10 min. 5 ml of each PCR amplified fraction was spotted manually onto Nytran nylon membrane (Schleicher and Schuell), denatured, neutralized and baked. A indicates the hybridization of the dotted fractions with a 1 kb EDH17B2 probe. A high level of enrichment for the EDH17B2 cDNA was observed in the final fraction. B shows fraction hybridization with Cot-1 DNA. No increase or decrease in intensity was observed suggesting non-specific capture of Alu repeat sequences in the final fraction. C indicates fraction hybridization with G3PD, a cDNA not present in the original genomic clones. Little or no G3PD was detected in the final cDNA fraction suggesting a low level of non-specific cDNA capture.

clones. Three γ-tubulin clones were detected in the original library suggesting a frequency of 1 in 167,000 clones, and indicating a 2,000 fold enrichment of this cDNA clone. Similar levels of enrichment of cDNAs have been reported from a 425 kb YAC from chromosome 5 (8), and from two cosmid contigs (550 kb and 350 kb) from the X chromosome (9). Further analysis of the EDH17B2 probe revealed that the clone was present in the breast cDNA library at a frequency of 1 in 106 clones. A 2000 fold enrichment of this cDNA would result in a frequency of 1 in 500 clones in the captured sublibrary. However, only 325 clones from the sublibrary were analyzed, suggesting that the inability to detect an EDH17B2 clone was not so surprising.

Artifacts

In total, 32 non-redundant cDNA fragments from the EDH17B2 cosmid group were analyzed. The results of these experiments are shown in Table 1. 72% of the clones mapped back to the original genomic clones, 61% of these clones contained some repetitive sequence in association with single copy sequence. This result was not unexpected because the library of origin was oligo dT primed and therefore was composed mainly of 3' exons of genes which are known to frequently contain Alu repeats. 12.5% of the 32 clones or 44% of the clones which did not map specifically to the parent genomic clones were also shown to contain only Alu repeat sequence. The high level of Alu associated clones in the sublibrary, 56% in total, may have been a result of using only small amounts of Cot-1

DNA for blocking purposes in the original complex hybridization. Another possibility is that Cot-1 DNA contains low levels of some Alu type repeat sequences, suggesting that these sequences would not be blocked in the hybridization reaction and would be non-specifically captured from the cDNA library. Addition of human placental DNA to the hybridization reaction was shown to reduce the levels of Alu repeat clones which were captured in subsequent reactions.

Table 1. A summary of the characterization of the captured cDNA fragments

32	100%	captured fragments were studied
23	72%	cDNA fragments mapped back uniquely to the BRCA1 region
9	28%	fragments did not map back uniquely to the BRCA1 region
4	12%	fragments contained only Alu sequence
3	9%	fragments consisted of sCos1 cosmid vector sequence
2	6%	cDNAs were single copy but did not map to the BRCA1 region
14	44%	cDNAs contained single copy and Alu sequence

Nine of 32 or 28% of clones analyzed did not map back specifically to the genomic clones. As previously stated, 4/9 clones contained only Alu repeat sequence as determined by sequence homology and non-specific hybridization to all clones on the cosmid grid except a cosmid vector only control. Three of 9 clones were determined to be cosmid vector artifacts following hybridization to cosmid vector bands on the pooled cosmid southern blot and to a cosmid vector only control on the cosmid grid. This vector only control allowed rapid discrimination between Alu repeats and cosmid vector fragments. Only 2/9 or 2/32 fragments (6%) appeared to be cDNA fragments from other genomic regions as determined by single band hybridization on a human genomic DNA southern blot. These cDNAs might have been captured non-specifically or by high homology to sequences in the BRCA1 genomic region. The low background of cDNA clones from other regions suggested that prescreening of gridded cDNA clones from the sublibrary with cosmid vector and Cot-1 plus cloned Alu or human placental DNA would eliminate the majority of the artifactual fragments from further study. A high percentage of captured fragments, perhaps greater than 90%, would then prove to be region specific cDNAs. Subsequent capture experiments on other cosmid pools have validated this approach.

DISCUSSION

Magnetic bead capture constitutes a powerful tool for the isolation of cDNAs encoded from within large genomic regions. We have shown that the technique is highly efficient and can be applied to many sources of cDNA and genomic clone pools in parallel (10). A central issue of transcript identification techniques is the degree to which any capture process is specific for gene sequences. We addressed this question by examining the magnetic bead capture products from a pool of 7 minimally overlapping cosmids from the BRCA1 region on chromosome 17q21. We tested the efficiency of the system in isolating cDNA fragments from genes known to be in this region and determined the type and frequency of artifacts. Putative genic sequences from all 7 cosmids were isolated from a breast cDNA library to a total of 9 potentially unique transcripts. 72% or 23/32 fragments mapped back to the original genomic clones and appeared to be true partial cDNAs. The system can be regarded as a success if it results in the isolation of one cDNA fragment from each gene present in a complex genomic sample and if these products are of approximately equal abundance. This was partially true of the experiment described above. The technique failed to isolate a fragment from the EDH17B2 gene known to be present in the genomic clones. However, many putative genes were identified, a number which compared very favorably with the exon amplification method of transcript identification. All of these transcripts appeared to be in almost equal abundance on the selected cDNA grids. It should be noted that these results are derived from a subpool of captured and exon trapped fragments from this region and may not represent all possible unique cDNAs which were isolated by either technique.The results presented above illustrate the ability of magnetic bead capture to isolate partial transcripts from genomic DNA clones. However, it is clear that verifying the origin of each clone is critical. Failure to do so will result in wasted effort in Northern blotting, sequencing and cDNA library screening of several non-specific cDNA fragments.

Problems have been encountered with reproducibility in magnetic bead capture. Several experiments resulted in less than 50% of clones mapping to the original genomic region. Careful analysis of these experiments suggested the following adaptations: 1) Storage of all biotinylated fragments at 4°C. Freeze-thawing of the biotin labelled PCR primer and/or the biotinylated amplified genomic fragments caused the biotin group to dissociate from the primer. 2) Use of highly specific PCR conditions for amplification of the captured fragments. Non-specific conditions may result in amplification of predominantly smaller products and/or artifactual products such as vector fragments. 3) Size selection of amplified material. Failure to carefully size select amplified cDNAs resulted in a large number of clones with insert size less 200 bp which proved difficult for use as probes on Northern blots and cDNA libraries.

We have no evidence that artifacts, particularly Alu repeats, are preferentially captured or cloned in comparison with true cDNAs. Improved blocking of the original biotinylated DNA in the complex hybridization or passage of the isolated fragments through a second round of magnetic bead capture has been shown to reduce the level of non-specific Alu contamination. However, prescreening the captured clones with repetitive probes has also been useful in greatly reducing the number of artifactual clones being fully evaluated.

The advantages of this approach to isolating cDNAs include the speed with which large region specific DNA segments can be screened for the presence of transcribed genes. As both genomic and cDNA clones are PCR amplified, a singular advantage of this technique is that only small quantities of starting materials are required. Magnetic bead capture is also relatively insensitive to the size of the genomic clones to be screened. However, enrichment with cosmid DNA seems to be more efficient than with YAC DNA. YACs are frequently difficult to isolate from yeast genomic DNA and this has led to enrichment artifacts when used as genomic template clones. It has been observed that 30-60 % of the clones from YAC sublibraries contain yeast ribosomal sequences (10). In general, the technique is highly efficient, and may be applied to several sources of cDNA simultaneously or in parallel reactions. cDNA fragments obtained may be randomly distributed over each mRNA, depending on the cDNA libraries used, and may result in rapid generation of full length cDNA clones. Furthermore, the method tends to normalize transcript levels, with a greater enrichment of rare messages.

Several limitations are also associated with this method. First, the small size of the captured subclones necessitates re-screening of cDNA libraries to obtain full length cDNA. The use of several libraries simultaneously in the capture reaction or random primed libraries for screening can substantially cut the time involved in generation of full length clones. Second, genomic clones which contain expressed repeat sequences can result in capture of a large number of enriched cDNA's containing these repeats. However, once identified these repeat sequences can be used to pre-screen the sublibrary of captured clones or can be added as a blocking agent to any future capture experiments in the region. Finally, because the technique is based on hybridization, pseudogenes and multigene family members may also be enriched. These clones can in some instances be excluded at an early stage by the more stringent hybridization conditions associated with mapping the cDNA back to the original parent genomic clones.

Magnetic bead capture provides a simple and effective method for isolation of expressed sequences from large genomic regions. It should be possible to use the technique to generate transcription maps of entire chromosomal regions of any organism for which genomic and cDNA libraries are available. As the methodology is PCR based, it may be possible to utilize chromosome microdissected material either as whole chromosome or as region specific chromosomal preparations. This approach will be particularly powerful if the transcriptional information generated can be integrated with detailed physical and genetic maps. Development of a transcription map will greatly assist in the search for any disease gene. Furthermore, once the gene is identified and isolated, the development of a detailed transcription map of a particular region will allow further assessment of the possible regulatory relationships between genes in that region. As this and other cDNA isolation techniques improve, identification of candidate genes may no longer be the rate limiting step of positional cloning.

REFERENCES

1. P. Liu, R. Legerski, and M.J. Siciliano, Isolation of human transcribed sequences from human-rodent somatic cell hybrids, *Science* 246:813 (1989).
2. G.M. Duyk, S.W. Kim, R.M. Myers, and D.R. Cox, Exon trapping: a genetic screen to identify candidate transcribed sequences in cloned mammalian genomic DNA, *Proc. Natl. Acad. Sci. USA* 87:8995 (1990).
3. A.J. Buckler, D.D. Chang, S.L. Graw, J.D. Brook, D.A. Haber, P.A. Sharp, and D.E. Housman, Exon amplification: A strategy to isolate mammalian genes based on RNA splicing, *Proc. Natl. Acad. Sci. USA* 88:4005 (1991).
4. D.B. Krizman and S.M. Berget, 3'-Terminal exon trapping: Identification of genes from vertebrate DNA, *Focus* 15:106.
5. P. Elvin, G. Slynn, D. Black, A. Graham, R. Butler, J. Riley, R. Anand and A.F. Markham, Isolation of cDNA clones using yeast artificial chromosome probes, *Nucl. Acids Res.* 18:3913 (1990).
6. M. Lovett, J. Kere and L.M. Hinton, Direct selection: A method for the isolation of cDNAs encoded by large genomic regions, *Proc. Natl. Acad. Sci. USA* 88:9628 (1991).
7. S. Parimoo, S.R. Patanjali, H. Shukla, D.D. Chaplin, and S.M. Weissman, cDNA selection: Efficient PCR approach for the selection of cDNAs encoded in large chromosomal DNA fragments, *Proc. Natl. Acad. Sci. USA* 88:9623 (1991).
8. J.G. Morgan, G.M.Dolganov, S.E. Robbins, L.M. Hinton, and M. Lovett, The selective isolation of novel cDNAs encoded by the regions surrounding the human interleukin 4 and 5 genes, *Nucl. Acids Res.* 20:5173 (1992).
9. B. Korn, Z. Sedlacek, A. Manca, P. Kioschis, D. Konecki, H. Lehrach, and A. Poustka, A strategy for the selection of transcribed sequences in the Xq28 region, *Hum. Mol. Genet.* 1:235 (1992).
10. D.A. Tagle, M. Swaroop, M. Lovett, and F.S. Collins, Magnetic bead capture of expressed sequences within large genomic segments, *Nature* 161:751 (1993).
11. D.A. Tagle, M. Swaroop, L. Elmer, J. Valdes, K. Blanchard-McQuate, M. Allard, G. Bates, S. Baxendale, R. Snell, M. MacDonald, J. Gusella, H. Lehrach, and F.S. Collins, Magnetic bead capture of cDNAs: A strategy for isolating expressed sequences encoded within large genomic segments, *in* " Magnetic Separation in Molecular and Cellular Biology ", M. Uhlen, O. Olsvik, and J. Elingboe, ed., Eaton Publishing Co., In Press.
12. T. Maniatis, E.F. Fritsch, and J. Sambrook, "Molecular Cloning: A Laboratory Manual", Cold Spring Harbor Laboratory Press, Cold Spring Harbor (1982).
13. A.P. Feinberg, and B. Vogelstein, A technique for radiolabelling DNA restriction endonucleasefragments to high specific activity, *Anal. Biochem.* 132:6 (1984).
14. S. Altschul, W. Gish, W. Miller, E. Myers, and D.J. Lipman, Basic local alignment search tool, *J. Mol. Biol.* 215:403 (1990).
15. R.J. MacDonald, G.H. Smith, A.E. Przybyla, and J.M. Chirgwin, Isolation of RNA using guanidinium salts, *Meth. Enzymol.* 152:219 (1987).
16. M. Orita, Y. Suziki, T. Sekiya, and K. Hayashi, Rapid and sensitive detection of point mutations and DNA polymorphisms using the polymerase chain reaction, *Genomics* 5(4):874 (1989).

17. J. Simard, J. Feunteun, G. Lenoir, P. Tonin, T. Normand, V. Luu The, A. Vivier, D. Lasko, K. Morgan, G.A. Rouleau, H. Lynch, F. Labrie, and S.A. Narod, Genetic mapping of the breast-ovarian cancer syndrome to a small interval on chromosome 17q12-21: exclusion of candidate genes EDH17B2 and RARA, *Hum. Mol. Genet.* 2:1193 (1993).
18. Y. Zheng, M.K. Jung, and B.R. Oakley, γ-tubulin is present in Drosophila melanogaster and Homo sapiens and is associated with the centrosome, *Cell* 65:817 (1991).

TOWARDS A TRANSCRIPTIONAL MAP OF THE q21-q22 REGION OF CHROMOSOME 7

J. M. Rommens[1,2], L. Mar[1], J. McArthur[1], L.-C. Tsui[1,2], and S. W. Scherer[1,2]

[1]Department of Genetics, Research Institute, The Hospital for Sick Children, 555 University Ave. [2]Department of Molecular and Medical Genetics, University of Toronto, Toronto, M5G 1X8, Canada

ABSTRACT

Procedures that directly select cDNAs from genomic DNA facilitate the rapid isolation of candidate genes from genomic DNA for the identification of genetic defects of known chromosome location. In such procedures, cDNA fragments are retrieved following the hybridization of cDNA pools to immobilized cosmid or YAC clones. This methodology is also applicable for the systematic identification of transcribed sequences over large regions of the genome. We propose to use such methodology to build a transcription map of the q21-q22 region of chromosome 7. This model region spans approximately 30 Megabases, corresponds to a light band with Giemsa staining and is therefore predicted to be relatively rich in genes. Extensive overlapping contigs of YAC clones from the q21-q22 region have been aligned and are being used as a starting resource. In total, over 120 YAC clones are being immobilized onto nylon filters and subjected to two rounds of hybridization with pooled cDNAs of combinations of fetal and adult tissues. Primary sources of cDNA are being generated with ends of known sequence such that each cDNA retains identification of tissue source and also instructs subsequent expansion as well as amplification and cloning steps. While the hybridization steps and washing conditions are critical, the nature of the sequences present in the starting genomic clones also influence the percentage of retrieved clones that map appropriately to the original YAC clones. We observe that rRNA or repetitive sequences are present in 4-30% of the retrieved clones, with 60-100% of those remaining originating from the correct YAC. Contributing to the population of clones that do not map appropriately are low copy repeat elements that are transcribed, pseudogenes or genes that are members of gene families. Through a combination of characterizations including physical mapping and RNA hybridization, the selected cDNAs will be arranged into tentative transcription units to provide the preliminary framework for a detailed transcription map of the region. This map will provide insight into the organization and function of this chromosome region. Further, the genes that are identified will provide candidates for the diseases and conditions that map to 7q21-q22.

INTRODUCTION

The value of a fully integrated physical, transcription and genetic map to understand functional, organizational and structural aspects of the genome cannot be understated. Although considerable progress has been achieved toward obtaining physical (1,2) and genetic maps (3) of the human genome, extensive transcriptional maps are not common. However, they are needed to aid in the studies of medical or genetic conditions. The past trials of positional cloning efforts to obtain candidate disease genes have revealed that the identification of transcribed sequences in genomic DNA is not always straightforward even when detailed physical analysis has been completed.

Systematic and reliable identification of coding regions within extensive regions of the genome is difficult since genes comprise as little as 5 - 10% of genomic DNA, are irregularly dispersed and are interrupted by introns. Further, restricted patterns of expression occur in different tissues or throughout stages of development. Ideally, the detection of genes over selected regions should: 1) be rapid, 2) be thorough for capture of all genes and 3) readily allow assembly of retrieved cDNAs or cDNA fragments into transcription units. At least two methodologies, that of exon trapping (4,5) and of direct cDNA selection (6,7), have recently been proposed that appear amenable to the generation of transcription maps over large regions of the genome (8,9). Both are geared toward the direct retrieval and identification of coding segments in genomic DNA. Further, retrieved segments provide single copy probes suitable for alignment onto the physical map.

We are pursuing the identification and mapping of transcribed sequences over the q21-q22 band of chromosome 7 with a modified direct selection methodology. The initial subcloning of yeast artificial chromosome (YAC) genomic clones and detailed mapping information is not required to initiate our strategy. We describe the preparation of primary cDNA pools from RNA, details of the selection procedure with mixtures of cDNAs from several tissues, cDNA clone characterization, and our current strategy for assembling the retrieved cDNAs into a tentative transcription map.

MATERIALS AND METHODS

Scheme to Capture cDNA Fragments

To carry out the selection and retrieval of cDNA fragments, two rounds of hybridization were carried out for three day durations with primary cDNA pools and nylon membranes containing immobilized genomic DNA. The basic procedure to retrieve cDNAs is illustrated in Fig. 1 and was similar to that described previously (6).

cDNA Selection

Preparation of cDNA pools. Primary cDNA pools were prepared as summarized in Fig. 2A from total and poly A+ RNA of tissues including brain, fetal brain, liver, fetal liver, testis, pancreas, kidney, placenta, fibroblast and from an intestinal cell line, Caco-2 (ATCC HTB 37). Poly A+ samples were purchased (Clonetech) or prepared from total RNA (10) with the PolyATract mRNA Isolation System (Promega).

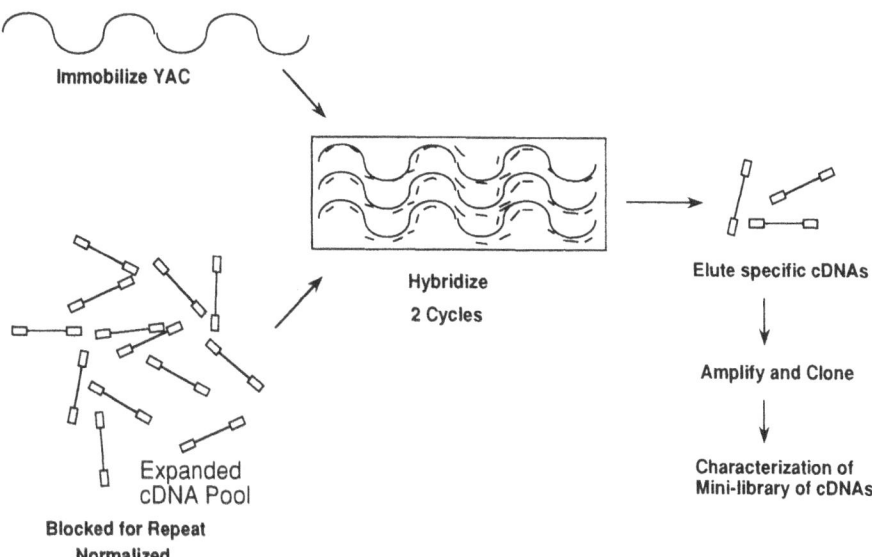

Immobilize YAC

Hybridize
2 Cycles

Elute specific cDNAs

Amplify and Clone

Characterization of
Mini-library of cDNAs

Expanded
cDNA Pool

Blocked for Repeat
Normalized

Figure 1. The basic strategy to retrieve cDNA fragments from immobilized genomic DNA. An expanded cDNA pool or mixtures of cDNA pools were pre-annealed with human placental DNA and then hybridized with immobilized genomic DNA. After extensive washing of the membrane, the hybridized material was eluted, re-blocked with excess human placental DNA and re-hybridized to the membrane. After re-washing, the hybridized material was finally eluted and amplified by polymerase chain reaction (PCR). The PCR products were then digested with restriction enzyme and separated on an agarose gel. The products were extracted with NaI and glass milk and cloned into plasmid vector.

The first strand of cDNA for each RNA was prepared by random priming (with M-MLV Reverse Transcriptase, Bethesda Research Labs) with an extended 5' end oligonucleotide, RXGA'B'C'N$_6$, and purified on agarose columns (A-5 medium, BioRad Research Labs) as described previously (11). The second strand of cDNA was then prepared from selected column fractions following tailing with terminal transferase (Bethesda Research Labs) and dATP by two rounds of extension with Taq Polymerase and the oligonucleotide, RXG(T)12, see Fig. 2B (11). The reaction buffer contained 10mM Tris-Cl, 50 mM KCl and 1.5 mM MgCl$_2$, 0.001% gelatin, pH 8.3, and reaction conditions included an initial extended denaturation step at 92°C for 4 min followed by two cycles of denaturation at 94°C for 45 sec, annealing at 37°C for 45 sec and extension at 72°C for 2.5 min.

Expansion of each pool was then achieved by PCR with 15 cycles of denaturation at 94°C for 45 sec, annealing at 58°C for 45 sec and extension at 72°C for 2.5 min. The second strand synthesis and expansion reactions were performed by consecutive linked programs. Both oligonucleotides (RXG and RXG(T)$_{12}$) were present from the start at a 15:1 molar ratio. Relatively large amounts of enzyme and the RXG oligonucleotide were used (0.8 μg RXG oligo, 0.08 μg RXG(T)12 per 100 μl reaction with 5U Taq polymerase).

Each step of the cDNA syntheses; including the first strand synthesis, purification and expansion were monitored by the specific amplification of the β-glucuronidase gene (*GUSB*) (12). Amplification with the oligonucleotides GUSB3 (5'ACTATCGCCATCAACAACACACTCACC) and GUSB4 (5'GCTCTGAATAATGGGCTTCTG) required the presence of 1.7 mM Mg++ and was carried out for 32 cycles with denaturation at 94°C for 45 sec, annealing at 63°C for 45 sec and extension at 72°C for 120 sec to two products of 1.0 and 1.1 kilobases (kb). This gene was selected as it is expressed at extremely low levels in all cell types (12, 13).

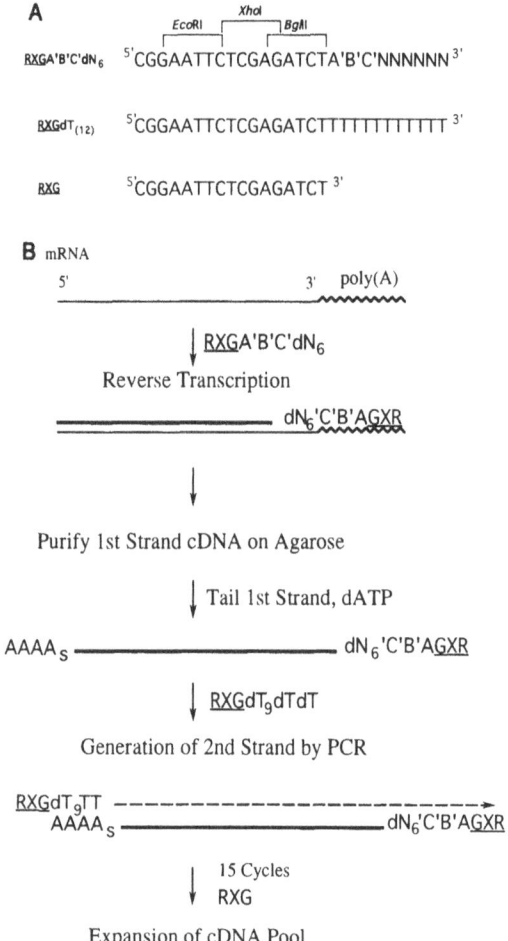

A

RXGA'B'C'dN$_6$ ^{5}'CGGAATTCTCGAGATCTA'B'C'NNNNNN $^{3'}$
(EcoRI, XhoI, BglII restriction sites indicated above)

RXGdT$_{(12)}$ ^{5}'CGGAATTCTCGAGATCTTTTTTTTTTTTTT $^{3'}$

RXG ^{5}'CGGAATTCTCGAGATCT $^{3'}$

B mRNA

5'   3' poly(A)

↓ RXGA'B'C'dN$_6$

Reverse Transcription

═══════════════════ dN$_6$'C'B'AGXR

↓

Purify 1st Strand cDNA on Agarose

↓ Tail 1st Strand, dATP

AAAA$_S$ ─────────── dN$_6$'C'B'AGXR

↓ RXGdT$_9$dTdT

Generation of 2nd Strand by PCR

RXGdT$_9$TT – – – – – – – – – – – – – – – – →
AAAA$_S$ ─────────── dN$_6$'C'B'AGXR

↓ 15 Cycles
↓ RXG

Expansion of cDNA Pool

Figure 2. Preparation of primary cDNA pools. A. The cDNA synthesis was designed to generate random mixtures from RNA and did not require the ligation of oligonucleotide linkers at any step. First strand cDNA was prepared with an oligonucleotide that contained sequences that directed random priming (N6) with reverse transcriptase, that identified the cDNA synthesis (A'B'C'), and that directed subsequent expansion of the cDNA mixture (RXG). The synthesized strand was tailed with terminal transferase and dATP to facilitate second strand synthesis with Taq polymerase. Double stranded cDNA was amplified with a limited number of rounds of PCR to expand the primary cDNA source for hybridization. B. The sequences of the oligonucleotides used in the cDNA synthesis, expansion and the subsequent retrieval are shown. The restriction sites that facilitate cloning of retrieved cDNAs are indicated. A'B'C' refers to a trinucleotide combination that identified the specific cDNA synthesis, thus indicating the RNA source of each retrieved cDNA. N refers to any nucleotide.

Isolation and transfer of YAC to membrane. Individual yeast clone cultures (50 ml) were grown for 30-40 h at 30°C, with vigorous aeration. Cells were then pelleted by centrifugation in a 50 ml tube at 2000 rpm, 12 min at 20°C. The cells were resuspended in 10 ml sterile water and again collected by centrifugation. The washed pellet was resuspended in 0.5 ml of SCE (1.0 M Sorbitol, 0.1 M sodium citrate, pH5.8, 10 mM EDTA) with 2500 units of Lyticase (Sigma Chem. Co.) and 0.2% of β-mercaptoethanol. An equal volume of molten LMP agarose (Bethesda Research Labs), prepared in 50 mM EDTA, was added and mixed by pipetting. The solution was poured quickly onto a level glass plate at 4°C. After 10 min, the solid agarose was sliced into blocks (4-5 mm^2) and transferred to a smaller erlenmeyer flask containing 20 ml SCEM (SCE with 0.2% β-mercaptoethanol). Flasks were incubated overnight at 37°C with slight agitation. The blocks were then rinsed three times with a small amount of lysis buffer (50 mM EDTA, 10 mM Tris-Cl, pH8.0) and then soaked in fresh lysis buffer containing 1% dodecyl lithium sulphate (Sigma) overnight at 37°C with mild agitation. The blocks were finally washed with five changes of 50 mM EDTA, pH8 and stored at 4-8°C. Integrity of the YAC DNA was verified by running 1/4 block on a pulsed field gel using separation conditions appropriate for size of the YAC clone (14).

A preparative gel (generally with 2 - 4 lanes fused together) was run to isolate the artificial chromosome (0.1 to 0.2 μg was used per hybridization). The ethidium bromide stained chromosome was cut out of the gel taking care to exclude excess agarose and to prevent the slice from drying. No attempts were made to separate comigrating yeast chromosomes from YAC chromosomes, both chromosomes were used. The slice was then stained in buffer containing loading dye, placed on a mini-gel tray and embedded into fresh molten agarose. Gel was solidified and set up for Southern transfer as described for pulsed field gels (15). DNA was nicked by soaking the gel in 0.25 M HCl for 6 - 7 min. The gel was then soaked in denaturing solution containing 0.4 M NaOH with 0.6 M NaCl for 30 min at 20°C with agitation and finally soaked in neutralization solution for the same duration in 0.5 M Tris-Cl, pH7.5 with 1.5 M NaCl. The gel was then set up for transfer onto nylon membrane (Hybond-N, Amersham) with 10XSSC. Following 16-30 h transfer, gel slice was precisely marked with a needle and the membrane was washed in 2XSSC. DNA was crosslinked to membrane with UV radiation and the membrane was trimmed for the hybridization.

Hybridization of cDNA pools. The concentration of each expanded cDNA preparation was determined by fluorometry and intercalation of Hoechst dye. Equal masses (usually 1 μg) of cDNA from up to ten tissues were mixed and used in single hybridizations.

All prehybridizations and hybridizations were carried out in 0.5 M sodium phosphate buffer at pH7.2 with 7% SDS, 1mM EDTA and 1% BSA (16). The membranes were wetted in water and placed into a hybridization bag or Nunc vial and prehybridized for 30 min at 60°C. Several membranes were hybridized in a single bag or vial.

For each set of 6 to 10 YACs, a total of 10 μg of cDNA was mixed with 200 μg sonicated human placental DNA. The mixture was precipitated overnight at -20°C with the addition of NaOAc, pH7.2 to 0.3 M and 2.5 volumes of ethanol. The nucleic acid was then pelleted by centrifugation at 15,000 g for 12 min at 4°C and resuspended in 100 μl of 0.2M NaOH and 1mM EDTA. To ensure complete denaturation, the samples were heated to 60°C for 10 min, and then neutralized by the addition of 1/10 volume of 2 M NH$_4$OAC, pH5.4. 1/4 volume of 20XSSC was then added and the mixture was pre-annealed at 55°C for 60 min. Following the pre-annealing step, 700-800 μl of pre-heated fresh hybridization solution was added. This solution was used for hybridization such that the final concentration of cDNA was 12 μg/ml with approximately 1.5 cm^2 of membrane/ml. The Nunc vials were placed in a hybridization oven or the hybridization bags (well sealed) were submerged in a water filled 50 ml tube at 60°C and incubated for 3 days.

Elution and second hybridization of cDNAs. Membranes were washed 3 times with 2XSSC, 0.1%SDS at 20°C for 5, 10 and 20 min durations and three times with 0.2XSSC, 0.1%SDS at 60°C for 20 min durations. Membranes were then rinsed twice with 2XSSC and transferred to a 1.5 ml microcentrifuge tube. 250 μl of sterile and distilled water (enough to submerge the filters) was added. The tube was then heated to 95-100°C for 10 min. 200 μl of the solution was removed and replaced with an additional 200 μl of water. The filters were reheated, and both the first and second elutions were pooled in a microcentrifuge tube; the filters were left with some water and stored at 4°C. 20 μg of sonicated human placental DNA, NaOAc (pH7.2) to 0.3M and 2.5 volumes of ethanol were then added to the collected cDNAs for precipitation overnight at -20°C. Following centrifugation, the pellet was resuspended in 20 μl of sterile water. Samples were then denatured for 10 min at 60°C with the addition of up to 0.2 M NaOH and 1 mM EDTA. The pH was neutralized with the addition of 1/10 volumes of 2M NH$_4$OAc at pH5.4 and 1/4 volumes of 20XSSC. Pre-annealing and the second hybridization was then carried out as for the first hybridization.

Final elution and amplification of hybridizing cDNAs. Membranes were washed and rinsed with 2XSSC as described for the first hybridization. Membranes were then transferred to 1.5 ml centifuge tubes and cDNAs were eluted with water. The collected cDNAs for each YAC was applied to a Sephadex (G50 Fine, Pharmacia) spin column. The column was spun at low speed and the material that washed through was collected and stored frozen at -20°C.

Portions of each sample were tested in analytical PCR reactions prior to larger scale reactions. The conditions for PCR included denaturation at 94°C for 45 sec, annealing at 58°C for 45 sec and extension at 72°C for 32 cycles in presence of the RXG oligonucleotide (see Fig. 2B). PCR products were pooled, phenol extracted and precipitated according to standard procedures (17).

Precipitated PCR products were digested with EcoRI and separated on 1.4% agarose gels. The material with sizes 0.4 to 1.2 kb was isolated with glass milk (Geneclean, BioCan Scientific) and ligated into pBluescript (Stratagene) using standard procedures (17). The plasmid vector was digested with EcoRI and treated with Calf alkaline phosphatase (Boehringer Mannheim) as recommended by supplier. Care was taken to ensure that the dephosphorylation step was efficient such that 95% of clones contained insert.

Characterization of Retrieved cDNAs

To obtain libraries of cDNAs from each YAC clone, *E. coli* DH5 cells (Bethesda Research Labs) were transformed with dilutions of the ligation mixes. Ampicillin resistant clones were gridded and arrayed for storage. Filter lifts were also prepared for preliminary screening to test for repetitive sequences and for subsequent cross-screening with individual clones to search for overlapping clones.

Clones that contained repetitive sequences were identified by hybridizing with labelled total cDNA. Plasmid DNAs of the remaining clones were then prepared and used to produce probes for extensive characterization by hybridization to yeast and mammalian DNAs and to RNA. Radio-labeled probes were prepared according to the random priming method (18) with [α-^{32}P]-dCTP at 3000 Ci/mmol and Klenow Polymerase. Clones were sequenced by dideoxy chain termination sequencing with T7 and T3 oligonucleotides, [α-^{35}S]-dATP at 1000 Ci/mmol and Sequenase (U.S. Biochemicals).

RESULTS

Characterization of the Selected cDNA Libraries

The first step in the characterization of the library prepared from each YAC clone involved an assessment of the numbers of clones that contained repetitive sequences. For example, of 200 retrieved clones from E515 (a YAC of 540 kbases), 15 positive signals were observed when hybridized to radiolabeled total cDNA. Testing of these positive clones indicated that they were frequently of rRNA origin, but that some were derived from Alu, Kpn or Kpn-like repeat sequences. Generally, the number of clones containing repetitive sequences was found to vary considerably (4%-45%, usually 8 - 10%) between libraries and depended largely upon the nature of genomic sequences present in the original YAC clone. The use of cDNA, prepared from total RNA by the method described, was the most reliable method to detect all of these clones as human genomic DNA or repeat specific probes for Alu or Kpn sequences alone were not comprehensive (9). Based on observations with a number of YACs, a key factor contributing to the number of clones containing repetitive sequences appeared to be the number of genes actually present on the YAC. The libraries of gene rich YACs (data not shown) appeared to contain fewer repetitive clones (unpublished data).

Individual clones that did not hybridize with labeled total cDNA were hybridized to genomic DNA of the starting YAC. Of twelve individual libraries tested, 65-100% of the clones were found to map back to the appropriate YAC. The individual clones were also hybridized to each other to test uniqueness, to YACs containing overlapping human genomic sequences for physical mapping, to mammalian DNAs to verify that the cDNA originated from human chromosome 7 and to RNA to determine mRNA sizes.

Physical Mapping

Examples of hybridization of five cDNAs generated from the E515 YAC to yeast DNA are shown in Fig. 3A. All of the clones hybridized to yeast DNA of the E515 YAC. Three clones, E515cDA-3, E515cDA-21 and E515cDB-3 also hybridized to yeast DNA of the E481 YAC, thus indicating that these cDNAs mapped to within the overlapping portions of E515 and E481 (1).

The E515cDB-3 cDNA detected a second larger band that corresponded in size to the EcoRI digestion product of the YAC plasmid vector. The larger band was due to contaminating plasmid vector sequences in the cDNA probe as was evident from the E203 YAC. The blot shown for the E515cDB-3 cDNA was exposed much longer than those of the other cDNAs as its hybridization was considerably weaker (see below).

Hybridizations of each of the five cDNA fragments to mammalian DNAs are shown in Fig. 3B. Human and chromosome 7 specific bands were observed for all cDNAs that corresponded in size to those detected in DNA from the YAC in Fig. 2A. E515cDA-3 and E515cDA-21 cDNAs hybridized to two and four EcoRI fragments, respectively, indicating that they contain multiple exons. E515cDA-21 also hybridized strongly to hamster DNA. None of the clones contained repetitive sequences.

The E515cDB-3 gave an unusual pattern that could not be directly interpreted. There were several human bands with only one chromosome 7 specific band that was barely detectable. These results suggested that this cDNA actually originated from an alternate position in the genome but that it was able to hybridize weakly to the DNA of the E515 YAC. This appeared to be an example of a gene that was a member of a large gene family. It apparently was captured by its similarity, and likely did not map to chromosome 7. As shown in Fig. 4, lower right panel, hybridization to RNA did not resolve this issue further as only one mRNA was detected with this segment.

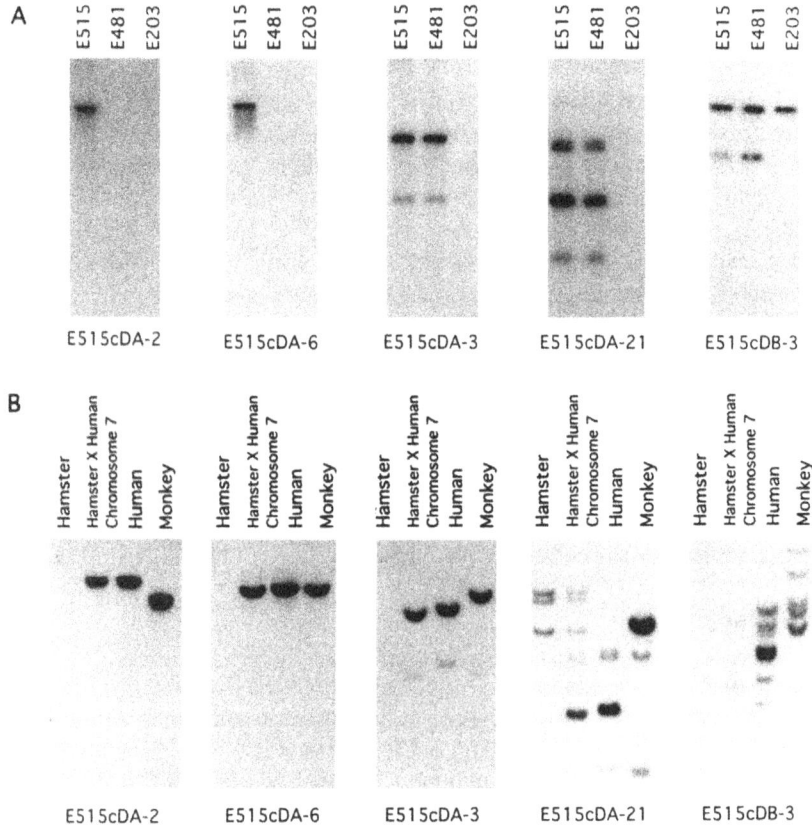

Figure 3. Examples of hybridization of retrieved cDNA clones from the E515 YAC to genomic DNAs. A. DNA (0.5 μg) was prepared from each yeast clone and digested with EcoRI as described previously (14). DNAs of E515, E481 and E203 were separated on 0.7% agarose gels and transferred to nylon membrane (Hybond-N, Amersham). E515 and E481 are partially overlapping and originate from 7q22 (1). E203 does not overlap and is derived from 7q21. Autoradiograms of the membranes hybridized with the clones E515cDA-2, E515cDA-6, E515cDA-3, E515cDA-21 and E515cDB-3 are shown as indicated. The stringency of the washing following hybridization with labeled probes was that achieved with 0.2XSSC with 0.1% SDS at 60°C. The exposure times for the autoradiograms shown were for 2-12 h at 20°C, except for clone E515cDB-3 for which a 40 h exposure is shown. B. The identical set of probes were also hybridized to mammalian DNAs as indicated. DNA (10 μg) was prepared from hamster, hamster-human hybrid 4AF1/106/KO15, from COS cell line and from human peripheral blood and digested with EcoRI as described previously (19). DNAs were separated on 0.7% agarose gels and transferred to nylon membrane (Hybond-N, Amersham). The hamster-human hybrid 4AF1/106/KO15 (20) contains human chromosome 7 as its only human material.

Assembly into Transcription Units

The overlapping and non-overlapping portions of the YACs defined physical intervals in which to group the cDNA fragments and provided a preliminary localization. Transcript analysis by RNA hybridization of representative clones provided additional information. Based on the expression pattern and size of transcripts detected, the minimum number of genes could be determined (9).

Examples of hybridization of RNA are shown in Fig. 4. E515cDA-2 and E515cDA-3 detected mRNAs of discrete size and specific tissue distribution patterns. E515cDA-6 did not detect a mRNA in total RNA preparations. Although E515cDA-21 did not overlap with E515cDA-3 by sequence, it did recognize a mRNA of the identical size (data not shown). Although more refined studies are necessary to confirm that these clones are portions of the same gene, this possibility would be consistent with both clones mapping to the same interval.

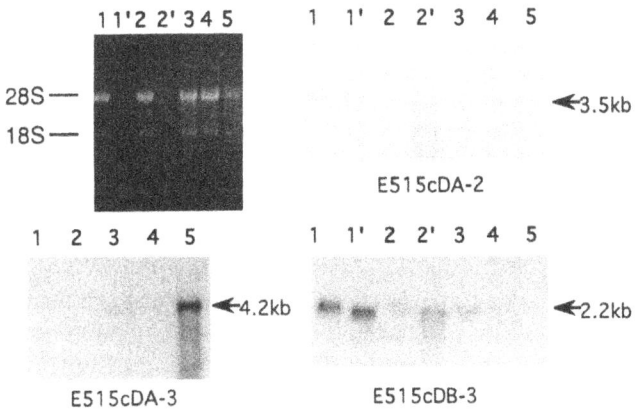

Figure 4. Examples of hybridization of retrieved cDNA clones to RNA. Total RNA (10 µg) from each cell line or tissue: 1) Caco-2 (intestinal cell line), 2) HL60, 3) lymphoblast, 4) fibroblast, 5) liver, or poly A+ RNA (1 µg) from the cell lines 1') Caco-2 and 2') HL60 were prepared by standard procedures (10). RNA was separated on 1% agarose gels containing 0.6 M formaldehyde and transferred onto nylon membrane (Hybond-N, Amersham). The integrity of the RNA is shown on the ethidium bromide stained gel in the left upper panel. Autoradiograms of the membranes hybridized with the clones E515cDA-2, E515cDA-3 and E515cDB-3 are shown in the upper right, lower left and lower right panels, respectively. Washing conditions were as described in Fig. 3. The size of the message detected with each clone is indicated to the right of each panel in kilobases (kb).

Sequence Analysis

Examples of the cDNA fragments recovered from the E515 YAC have illustrated how the retrieved cDNAs have been analyzed and characterized. Each of the cDNA clones provided probes for screening cDNA libraries but could also be sequenced directly to examine their coding potential and to search databases of sequenced genes for identity or similarity. Of the clones mapping to the E515 YAC that were sequenced (six to date), all were found to be independantly derived. Sequencing and sequence analysis of retrieved cDNAs have been described in a previous report (9). Sequencing information was particularly useful for clones that did not detect messages by RNA analysis due to detection limitations (data not shown). E515cDA-4, for example, displayed significant (but not identical) similarity to a portion of the human thyroid peroxidase (TPO) gene. Another clone, E515cDA-30, showed striking similarity to the rat carnitine octanoyltransferase mRNA.

Summary of Strategy to Build the Transcription Map of the 7q21-q22 Region

In order to build an extended transcriptional map of the entire 7q21-q22 region, cDNAs are being selected from the minimal number of YAC clones that provide coverage of the genomic DNA. The chromosome 7 specific YAC library and several established overlapping contigs from the q22 region have been described previously (1,21). In total, over 220 YACs have been mapped to the q22 region by hybridization to random probes, known genes and genetic markers from this region. An additional 80 YACs from this library have been mapped to this region by *in situ* hybridization (22). By extrapolating from the mapping of clones over the q21-q22 region which are currently documented and aligned (S. Scherer and L.-C. Tsui, unpublished data), the cDNAs of 120 YAC clones will have to be analyzed. The average insert size of the YAC library is 475 kilobases and the total region to be covered is 30 Megabases.

A summary of our current strategy to build the transcript map is shown in Fig. 5. We have divided our efforts into three phases. In phase one, libraries of cDNA fragments are prepared from the retrieved material from the hybridization of cDNA pools of ten tissues to immobilized YACs. The next two phases include some overlap but involve characterization of the selected cDNAs to achieve incremental refinement of cDNA position and transcriptional unit alignment.

In the second phase, 10 - 20 unique clones per 100 kbases will be characterized. As the average size of each cDNA is 0.5 kb, 5 - 10 kb of transcribed sequence will be examined in this phase. Based on previous observations of the variable incidence of genes, of the redundancy of clones, of problems with gene families or of mildly repetitive transcribed sequences, it is anticipated that these characterizations will be sufficient to produce a crude map. It will be most appropriate to examine small numbers of cDNA clones and extend these studies based on the outcome of the initial analyses.

The basis for the assembly of the cDNA fragments into a transcription map will rely heavily on physical mapping within the YAC clones. The YAC clones from the chromosome 7 library have been shown not to be plagued by significant chimerism artefacts (21). Detailed mapping is further facilitated by the availability of overlapping YACs providing four fold coverage of most of the q22 region.

The third and last phase extends beyond physical mapping information to include characterization of the genes that may be encoded by the transcriptional units. This latter characterization will be achieved by RNA hybridization, cDNA library screening and by sequence analysis.

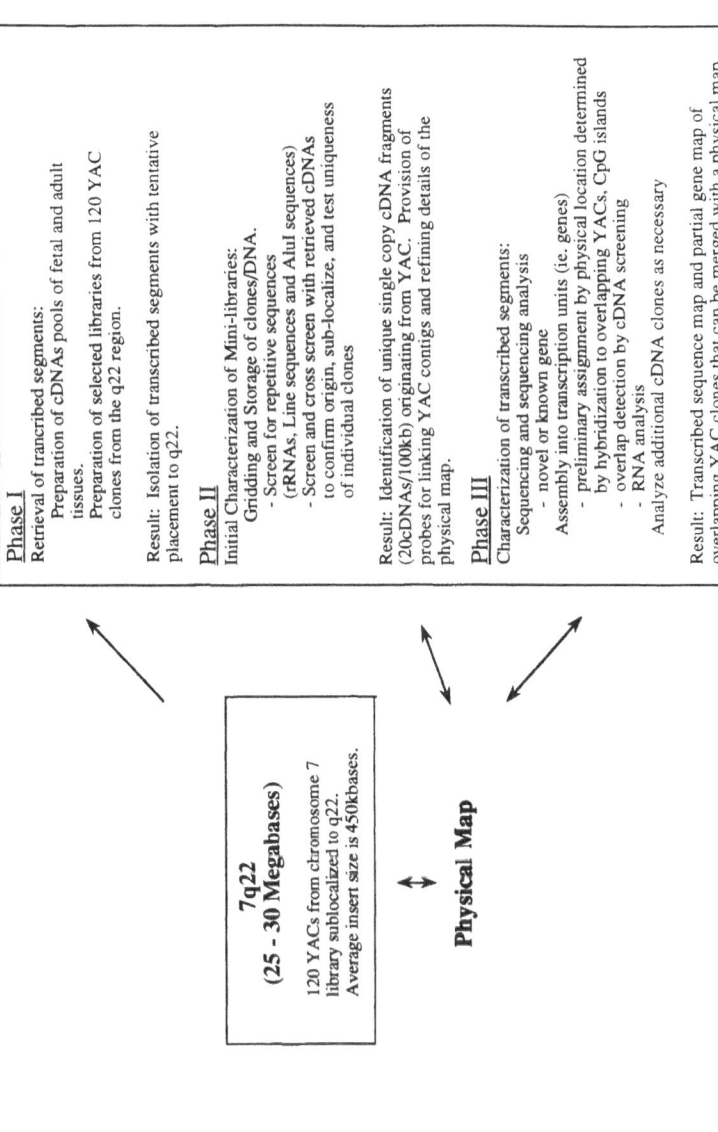

Figure 5. Scheme for generation of transcribed sequence map. The initial phases of the construction of the transcribed sequence map involves the generation of selected mini-cDNA libraries from 120 YAC clones from the q21-q22 region of chromosome 7. The mapping of YAC clones to the q22 region of chromosome 7 containing portions of human chromosome 7 (1,20). Continued characterization has revealed that these hybrids define an interval that includes q21. Primary cDNA sources from ten tissues have been prepared and are being pooled and used for the selection. The libraries are characterized to ensure origin of 10 - 20 cDNA segments per 100 kbases and to determine location on the starting and available overlapping YACs. The physical location provides the first outline of the transcript map. Subsequent characterization including hybridization to RNA and to cDNA libraries as well as sequence analysis will allow for assembly of cDNAs into transcription units.

75

DISCUSSION

A comparable strategy has recently been used to generate a map of a minimum of nine transcription units from 1 Megabase of genomic DNA from 4p16.3 (9). It was clear from the results obtained from this limited region that application to relatively large regions would also be possible. While we are currently in the initial stages of this larger project, we anticipate to be able to obtain at least a crude transcription map of the 7q21-q22 region by direct selection or gene tracking of approximately 120 YACs.

With direct selection procedures, there is no need for the subcloning of YAC clones or for initial screening for CpG rich or conserved phylogenetic sequences. The procedures are rapid and clones are obtained with coding sequences that can be directly subjected to sequence analysis. This meets important criteria for the generation of a transcription map. Further, the cDNAs provide excellent probes for generating extensions of YAC contigs for the ongoing physical mapping and YAC alignment in the q22 region.

A concern with the strategy proposed is related to the comprehensiveness of the derived map as genes may have restricted expression patterns. This has been addressed by the use of pools of cDNAs prepared from several tissues, including those from both fetal and adult stages. Tissues were chosen that are complex in origin and are thought to express large numbers of genes (23,24). They include among others fetal and adult brain, fetal liver, reproductive organs and placental tissues. As evident by the RNA hybridization shown in Fig. 4, cDNAs can be identified that have broad or restricted patterns of expression. The E515cDA-3 cDNA detected a mRNA most strongly expressed in the liver while the E515cDA-2 identified mRNA that appeared expressed in several tissues.

Several features were incorporated into the cDNA synthesis to optimize retrieval. First, we have observed that the best cDNA source for selection procedures includes primary preparations of cDNA. The advantages have also been observed by others and have been discussed (9,25). Further, as these pools are not contaminated by yeast sequences, it has not been necessary to include yeast DNA with the hybridizations. Second, the cDNA synthesis procedure produces fragments that can be retrieved by PCR and, as preparation did not involve the ligation of oligonucleotides, all cDNAs retrieved were necessarily derived from RNA. Also to ensure that loss of representation of a given species does not occur, the reverse transcription was carried out by random priming. Further, the expansion step by PCR was restricted to fifteen cycles. Finally, the information of tissue source is retained by the incorporation of a trinucleotide characteristic for the cDNA preparation. While the success of retrieving cDNA clones that map back to the appropriate YAC varies depending upon the nature of sequences present in the YAC, it is acceptable as it consistently approaches 70 - 95%, following an initial screen to remove repetitive sequences.

We have occasionally detected clones that do not appear to originate from the YAC but do hybridize with less intensity than normal. The E515cDB-3 cDNA, which did detect mRNA and contained open reading frame, was one such example. Examining the hybridization patterns carefully suggested that the cDNA was captured because it resembled sequences present in the tested genomic DNA but that it was likely a gene family member that originated from an alternate location of the genome. These types of clones and those detecting pseudogenes are problematic in our strategy. This problem can be recognized, as has been shown, by the hybridization of the cDNAs to human-hamster hybrid and human DNAs.

Previously, it was noted that using total RNA sources, while adequate for selection, does lead to problems with the retrieval of clones that are derived from unprocessed RNA (9). This problem was likely due to the prolonged hybridizations of cDNA to genomic DNA as duplexes with longer complementarity would be favoured. We have observed, however, that this problem could be reduced with the use of cDNA prepared from purified mRNA. Further, the pooling of cDNA preparations from several tissues would reduce the abundance of cDNAs of tissue specific mRNAs, and in principle it would be most appropriate to limit this dilution.

The practical concern for the problem of abundance of different messages has not proven to be unmanageable. First, this problem is limited to the genes present on each specific genomic clone being analyzed and second, there is some normalization effect gained by the use of two rounds of hybridization and preannealing steps (9,25).

PROSPECTS

In the directed gene searches of positional cloning, gene identification strategies are driven by the endpoint, that is, of obtaining candidate genes. A primary difference influences the assembly of a transcription map of regions of the genome compared to positional cloning. Transcription maps should attempt to include all transcribed sequences originating from a genomic region regardless of when or where that sequence is expressed. This will yield the maximum useful output from the derived map, but also poses the biggest challenge for map construction (26).

By the direct selection methodology that we have described, cDNAs will be captured only if they were expressed in the source tissues. The detailed transcription map that we have proposed will have a bias in this respect. We are aiming to minimize this bias by using mixtures of cDNA prepared from several different tissues. By crude estimation, we feel that this bias is not seriously limiting as it has been observed over short regions (1 - 1.5 Megabase stretches), that there are correlations between the transcribed sequences identified and the numbers of CpG rich islands present in the genomic DNA. The map will become more rigorously tested, however, as it becomes aligned by overlap and as known genes or sequenced cDNAs map to within the regions studied. The proposed map should not have a bias in gene structure *ie*. with respect to presence or absence of exons.

To further address the issue of coverage, we are currently exploring the limits of detection by selection. It has been observed that clones can be recovered that are derived from messages expressed at extremely low levels (27,28). Specifically, we have been able to retrieve cDNA segments corresponding to the CFTR mRNA from genomic YAC clones with cDNA prepared from total lymphoblast RNA (data not shown). CFTR mRNA is expressed in lymphoblasts only at ectopic levels (29) but sufficient depth was provided by the primary cDNA libraries prepared to generate these clones. They were retrieved only in low numbers (0.1%), but these results strongly suggest that combinations of tissues sources and controled normalization and hybridization steps minimize the concern of the coverage issue.

In addition to a large number of interesting new genes that will be discovered, the immediate prospects of a transcription map will include the availability of candidate genes for conditions or diseases that are found to map to within the q22 region of chromosome 7 (30). In the longer term it will also provide relevant information for the next generation of the map which will include gene organization as well as an emerging and detailed picture of chromosome structure and function.

ACKNOWLEDGEMENTS

This work is funded by the Canadian Genome Analysis and Technology Program.

REFERENCES

1. S.W. Scherer, J.M. Rommens, S.Soder *et al.*, Refined localization and yeast artificial chromosome (YAC) contig-mapping of genes and DNA segments in the 7q21-q32 region, *Hum. Mol. Genet.* 2:751 (1993).

2. K. Gardiner, M. Horisberger, J. Kraus *et al.*, Analysis of human chromosome 21: correlation of physical and cytogenetic map: gene and CpG island distributions, *EMBO J.* 9:25 (1990).

3. NIH/CEPH Collaborative Mapping Group. A comprehensive genetic linkage map of the human genome, *Science* 258:67 (1992).

4. A.J. Buckler, D.D. Chang, S.L. Graw *et al.*, Exon amplification: a strategy to isolate mammalian genes based on RNA splicing, *Proc. Natl. Acad. Sci. USA* 88:4005 (1991).

5. G.M. Duyk, S. Kim, R.M. Myers, and D.R. Cox, Exon trapping: a genetic screen to identify candidate transcribed sequences in cloned mammalian genomic DNA, *Proc. Natl. Acad. Sci. USA* 87:8995 (1990).

6. M. Lovett, J. Kere, and L.M. Hinton, Direct selection: A method for the isolation of cDNAs encoded by large genomic regions, *Proc. Natl. Acad. Sci USA* 88:9628 (1991).

7. S. Parimoo, S.R. Patanjali, H. Shukla, D.D. Chaplin, and S.M. Weissman. cDNA selection: Efficient PCR approach for the selection of cDNAs encoded in large chromosomal DNA fragments. *Proc. Natl. Acad. Sci USA* 88:9623 (1991).

8. B. Korn, Z. Sedlacek, A. Manca *et al.*, A strategy for the selection of transcribed sequences in the Xq28 region, *Hum. Mol. Genet.* 1:235 (1992).

9. J.M. Rommens, B. Lin, G. Hutchinson *et al.*, A transcription map of the region containing the Huntington disease gene, *Hum. Mol. Genet.* 2:901 (1993).

10. R.J. MacDonald, G.H. Smith, A.E. Przybyla, and J.M. Chirgwin, Isolation of RNA using guanidinium salts, *Meth. Enzymol.* 152:219 (1987).

11. M.A. Frohman, M.K. Dush, and G.R. Martin, Rapid production of full-length cDNAs from rare transcripts: amplification using a single gene-specific oligonucleotide primer, *Proc. Natl. Acad. Sci. USA* 85:8998 (1988).

12. A. Oshima, J.W. Kyle, R.D. Miller *et al.*, Cloning, sequencing and expression of cDNA for human β-glucuronidase, *Proc. Natl. Acad. Sci. USA* 84:685 (1987).

13. L.T. Bracey and K. Paigen, Changes in translational yield regulate tissue-specific expression of β-glucuonidase, *Proc. Natl. Acad. Sci. USA* 84:9020 (1987).

14. S.W. Scherer and L.-C. Tsui, Cloning and analysis of large DNA molecules, *in* "Advanced Techniques in Chromosome Research" K.W. Adolph, ed. Marcel Dekker, Inc., New York. (1991), pp 33-72.

15. J. M. Rommens, S. Zengerling-Lentes, B.Kerem *et al.*, Physical localization of two DNA markers linked to the cystic fibrosis locus by pulsed-field gel electrophoresis, *Am. J. Hum. Genet.* 45:932 (1989). 16. G.M. Church and W. Gilbert. Genomic sequencing. *Proc. Natl. Acad. Sci. USA* 81:1991 (1984).

17. T. Maniatis, E.F. Fritsch, and J. Sambrook, "Molecular Cloning: A Laboratory Manual", Cold Spring Harbour Laboratory, Cold Spring Harbour, N.Y. (1992).

18. A.P. Feinberg and B.Vogelstein, A technique for radiolabeling DNA restriction endonuclease fragments to high specific activity, *Anal. Biochem.* 132:6 (1983).

19. S.A. Miller, D.D. Dykes, and H.F. Polesky, A simple salting out procedure for extracting for extracting DNA from human nucleated cells, *Nucl. Acids Res.* 16:1215 (1988).

20. J.M. Rommens, S. Zengerling, J. Burns *et al.*, Identification and regional localization of DNA markers on chromosome 7 for the cloning of the cystic fibrosis gene, *Am. J. Hum. Genet.* 43:645 (1988).

21. S. Scherer, B.J.F. Tompkins and L.-C. Tsui, A human chromosome 7-specific genomic DNA library in yeast artificial chromosomes, *Mammal. Genome* 3:179 (1992).

22. J. Kunz, S.W. Scherer, I. Klawitz *et al.*, Regional localization of 725 human chromosome 7 specific yeast artificial chromosome (YACs) clones, Manuscript submitted.

23. M.D. Adams, J.M. Kelley, J.D. Gocayne *et al.*, Complementary DNA sequencing: expressed sequence tags and human genome project, *Science* 252:1651 (1991).

24. M.D. Adams, M. Dubnick, A.R. Kerlavage *et al.*, Sequence identification of 2375 human brain genes, *Nature* 355:632 (1992).

25. J.G. Morgan, G.M. Dolganov, S.E. Robbins, L.M. Hinton, and M. Lovett, The selective isolation of novel cDNAs encoded by the regions surrounding the human interleukin 4 and 5 genes, *Nucl. Acids Res.* 20:5173 (1992).

26. U. Hochgeschwender, Toward a transcriptional map of the human genome, *Trends Genet.* 8:41 (1992).

27. G. Sarkar and S. Sommer, Access to a messenger RNA sequence or its protein product is not limited by tissue or species specificity, *Science* 244:331 (1989).

28. J. Chelly, J.-P. Concordet, J.-C. Kaplan, and A. Kahn, Illegitimate transcription: transcription of any gene in any cell type, *Proc. Natl. Acad. Sci. USA* 86:2617 (1989).

29. N. Fonknechten, J. Chelly, J. Lepercq *et al.*, CFTR illegitimate transcription in lymphoid cells: quantification and applications to the investigation of pathological transcripts, *Hum. Genet.* 88:508 (1992).

30. L.-C. Tsui, Genetic markers on chromosome 7, *J. Med. Genet.* 25:294 (1988).

DIRECT cDNA SELECTION USING HUMAN AND MOUSE cDNAs: APPLICATION TO Xq13.3 CHROMOSOMAL REGION

J. Gecz[1], L. Villard[2], A.M. Lossi[2], and M. Fontes[2]

[1]Institute of Molecular Physiology and Genetics, Department of Genetics
Slovak Academy of Sciences, Vlarska 5, 83334, Bratislava, Slovakia
[2]INSERM U242, 27 Boulevard J. Moulin 13385 Marseille, Cedex 5
France

ABSTRACT

The capability of the direct cDNA selection technique to work in heterologous conditions is described. A human YAC (4551) positive for DXS56 from the Xq13.3 DXS56-PGK1 region was successfully scanned for the presence of conserved expressed sequences exploiting a mouse brain cDNA library. Clones from a known, conserved gene xnp were identified. In addition, one new gene was isolated. The sequence comparison between human and mouse XNP/xnp selected clones revealed at least 75% homology on the nucleotide level. Although the efficiency is limited by the level of sequence homology, the direct cDNA selection is applicable to heterologous conditions.

INTRODUCTION

The recent advent of new techniques for gene identification (1) greatly improved the efficiency and speed of isolation of new genes, and hence the construction of transcription maps of human chromosomes. New transcription units are then being positioned on the evolving physical, genetic and cytogenetic maps. Such an integrated map considerably facilitates the isolation and identification of any disease gene and so leads to a better understanding of the morbid anatomy of the human genome.

Generally, transcription maps are being constructed via two complementary approaches, global and regional (2). In a global plan, randomly picked cDNAs from various tissues are sequenced (3,4,5) and positioned on the physical map (6,7), while in the regional approach, the known particular physical region is scanned for expressed sequences. A package of classical and new techniques is available depending on the particularities of the region. Larger genomic clones are used either directly for screening of cDNA libraries (8,9), identification of evolutionary conserved sequences (10), isolation of CpG islands (11), exon trapping (12) or direct cDNA selection techniques (13,14). Among them, presently only the direct cDNA selection seems to have the potential to be applied to contigs covering large portions of the genome (15).

In our previous transcription mapping effort in Xq13.3, using the direct cDNA selection, we recloned three known genes from the DXS56-PGK1 1Mb region and isolated two new ones (16). In order to extend these studies, other human cDNA libraries and a mouse brain cDNA library were used. In particular, the possibility of using heterologous cDNAs as a target for direct cDNA selection experiments using large human genomic clones (YACs) was tested. Here, we present our first results of human YAC clones used for cDNA selection in combination with mouse brain cDNA library. One known gene and at least one new one have been successfully selected.

METHODS

Direct cDNA Selection on Membrane Discs

Direct cDNA selection was performed according to two published protocols (13,14) which we slightly modified (16). DNA of both YACs were prepared from the preparative PFGE gels using the GeneClean kit (B101). Membrane discs were prepared by spotting 10 ng of YAC DNA together with the carrier DNA (13,14) per 4 mm^2 of the membrane (Hybond N$^+$, Amersham). The membranes were prehybridized in 40 μl of solution containing 5XSSPE, 5XDenhard's solution, 0.5% SDS and a mixture of quenching agents (cotI DNA (BRL), 0.025 mg/ml; sonicated yeast DNA (AB1380), 0.1 mg/ml; and pYAC4 vector DNA, 0.025 mg/ml for 24 h in 65°C under 40 μl of mineral oil. PCR amplification of cDNA inserts from the λgt10 human skin fibroblast and mouse brain cDNA libraries (Clontech) was carried out with oligonucleotide primers a and b (Set1) under conditions described (14). PCR amplified cDNA fragments (a smear of the average size 1 to 1.5 kb) were GeneClean purified. Prior to hybridization with the YAC DNA fixed on the membranes, cDNAs (both human and mouse) were preblocked with the same quench as described above, at 65°C for 1 h. Hybridization of preblocked cDNA insert pools (1 mg/ml) on membrane discs was carried out in 40 μl volume at 65°C for 40 h. After hybridization, the membranes were washed briefly in 5XSSPE at room temperature and then sequentially in 2XSSC, 0.2XSSC and 0.1XSSC containing 0.1% SDS for 20 - 30 min at 65°C. Membranes which were used to select mouse brain cDNAs were finally washed with 0.2XSSC with 0.1% SDS for 20 min at 65°C. After the final wash, the membranes were briefly rinsed with 0.2XSSC at room temperature. The bound cDNAs were eluted from the membranes by heating them at 98°C for 6 min in 40 μl of sterile water. PCR amplification for the second round of selection was performed using nested primers c and d (Set 2; Ref. 17) for 40 cycles. The second cycle of selection was performed as described for the first round, except that 10-times less DNA from PCR amplification after the first cycle, than the amount of the cDNA PCR DNA for the first cycle, was used. PCR amplified products after the second round of selection were directly cloned to T-tailed pCRII plasmid vector (Invitrogen) and 200 recombinants/YAC were picked and further analyzed.

Probe Preparation and Hybridization Conditions

cDNA inserts were amplified directly from the bacterial cells via PCR in the following manner. A bacterial colony (1-2 mm in diameter) from the master plate was transferred to 20 μl of sterile water and mixed well. One microliter from this mixture was then used directly in the standard PCR mix and cycled for 30 PCR cycles. 10 μl (1/10) of the amplification reaction was analysed on a TAE gel. The PCR product was excised, purified by GeneClean (B101) and resuspended in 30 μl of sterile water. Three μls were directly used for labeling.

Probes (original probes as well as PCR amplified cDNAs) were labeled by the random priming method of Feinberg and Vogelstein (18). All Southern blot hybridizations were carried out in standard SSC buffers (13) for 16-18 h in a rotary hybridization oven (Hybaid). After washing the membranes were exposed to Hyperfilm MP (Amersham) films for times according to probe and blot used (1 h to 5 days).

DNA Sequencing

Plasmid DNA was prepared from 10 ml overnight culture using the Qiagen kit (tip20, Diagen) and sequenced by the dideoxynucleotide chain termination method with fluorescent primers either on Pharmacia ALF or Applied Biosystems 373A DNA sequencers. Both strands of an insert were sequenced using vector primers. Sequencing reactions were done as described (19).

RESULTS

The general strategy of the hybrid selection approach we used is outlined in Fig. 1.

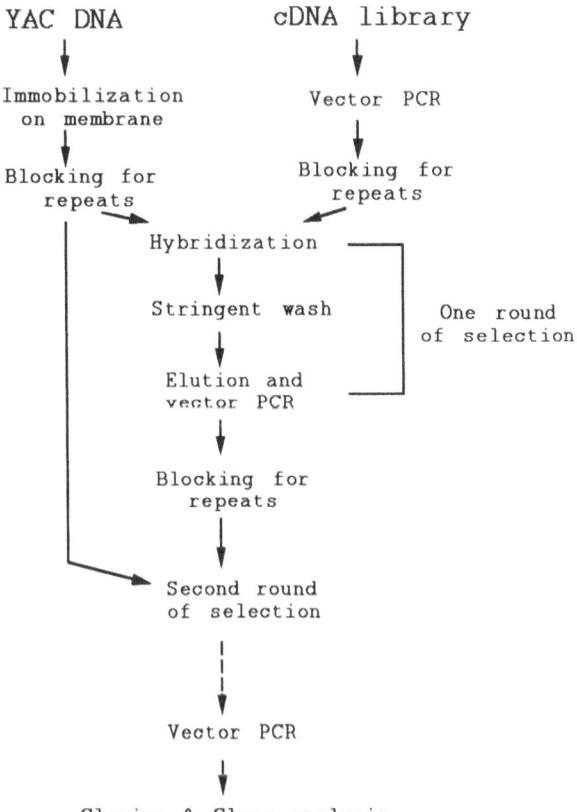

Figure 1. General strategy of the cDNA selection used in this study. 'Vector PCR refers' to PCR amplification using the oligonucleotides flanking the cloning site of a cloning vector (in our case λgt10).

The YACs used in this study were screened from the ICRF (YAC4551;20) and CEPH (YAC326B10;21) YAC libraries using DXS56 and PGK1 probes from the Xq13.3 juxtacentromeric region of the human X chromosome, as already described (16). This region was shown to contain three known genes: Menkes syndrome gene (MNK; 22;23;24), the gene for phosphoglycerate kinase 1 (PGK1; 25), and the XNP (X-linked Nuclear Protein) gene, which we cloned via two different approaches (16; G. Consalez, in preparation). These genes were chosen to serve as internal controls for the hybrid selection experiments. Initially we used two YACs from the region, YAC4551, 550 kb and DXS56-positive, and 326B10, 330 kb and PGK1-positive. These YACs cover 90% of the 1Mb region but they don't overlap (Fig. 2). YAC326B10 harbors the PGK1 and MNK genes, and YAC4551, the XNP gene. The YACs were purified via PFGE electrophoresis to avoid the contamination with the yeast DNA and fixed on membrane discs (see Methods). Initially, a human skin fibroblast oligo(dT) primed cDNA library was used (purchased from Clontech). This was further extended to three more human cDNA libraries: placenta, fetal brain and kidney (results not shown), and a heterologous mouse brain random/oligo(dT) primed cDNA library (Clontech).

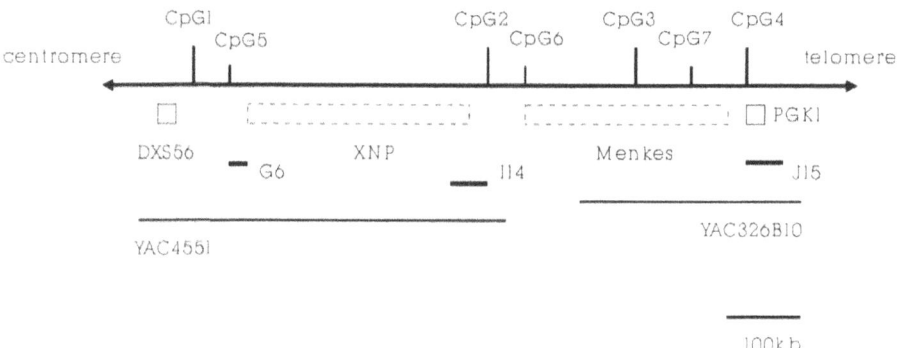

Figure 2. Schematic presentation of DXS56-PGK1 region in Xq13.3. Several putative CpG islands were identified. The known genes (Menkes-MNK, PGK1, XNP) and the new genes (G6, I14 and J15) were placed on the map (16, this study).

cDNA Selection

The hybrid selection scheme was basically the same when using either the human or the mouse cDNA libraries. We performed only two cycles of selection, after that the determining third round didn't improve the enrichment (16). One difference was the washing of the discs, in that the mouse cDNAs were finally washed with 0.2XSSC at 65°C, and the human cDNAs were washed with 0.1XSCC at 65°C (see Methods for details). The PCR amplification products after first and second cycle of selection using human skin fibroblast cDNAs and mouse brain cDNAs were electrophoresed in agarose (see Fig. 3A) and hybridized with a control probe from the XNP gene (Fig. 3B).

The XNP gene shows strong conservation between human and mouse, at least 90% at the protein level (26). A probe from the conserved region (HXNP23.1E6) was used to hybridize the amplified products as shown in Fig. 3B. After the first round, appropriate selection was obtained only from the human fibroblast cDNA library, while the mouse library gave signal only after prolonged exposure (4 h; results not shown). Hybridization of the second round products documented higher enrichment from the mouse library, but still considerably lower then that with the human cDNA. However, the amount (and complexity) of the PCR amplified products was lower for mouse in both rounds (Fig. 3A). In both cases the amplified original cDNA libraries which served as reference controls were negative (Fig. 3B).

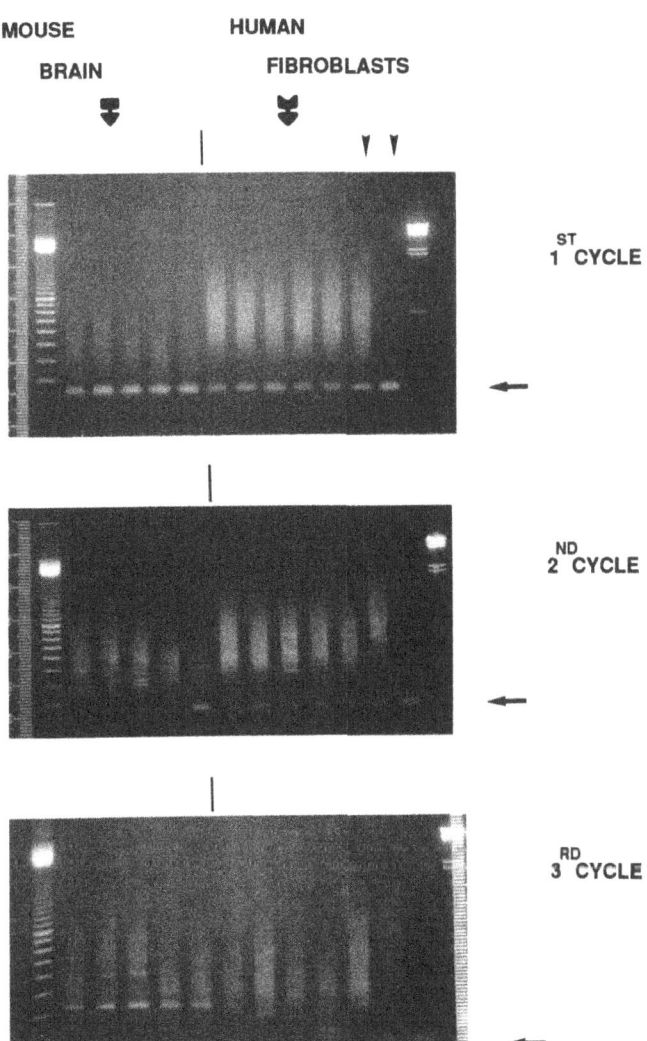

Figure 3A. Agarose gel electrophoresis of once, twice and three times PCR amplified selected cDNAs from human fibroblast and mouse brain cDNA libraries. From left to right after the 100 bp ladder: YAC326B10, YAC4767, YAC4551, lambda phage pool and carrier ϕX174 DNA (negative control for selection experiments) for mouse brain and for human fibroblast cDNA respectively. Arrowheads indicate the lanes with the following controls: amplified human fibroblast cDNA library and negative control of PCR (1st cycle gel); amplified mouse brain cDNA and negative control of PCR (2nd cycle gel); and human fibroblast cDNA with PCR control, again (3rd cycle gel). In the far right lane, lambda phage HindIII marker. Arrows indicate the λgt10 vector PCR products obtained in all rounds of nested PCRs.

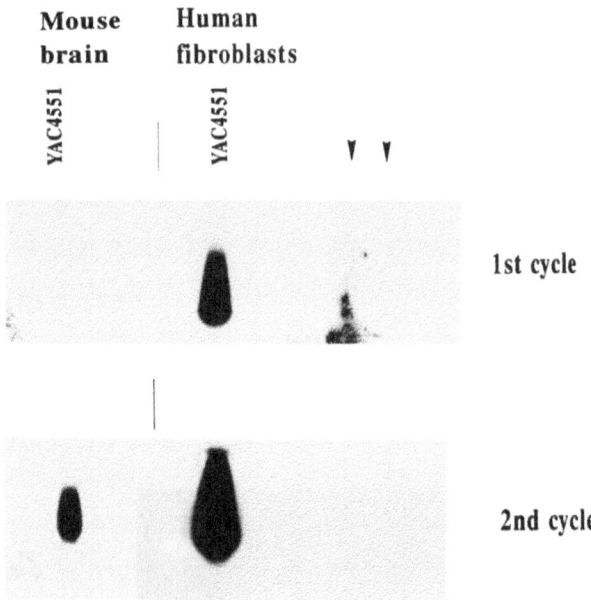

Figure 3B. Hybridization of the 1st and 2nd round gels as in (A) with the HXNP23.1E6 probe. Exposure time was 1 h (-80°C). Hybridization signal after the first round in mouse brain was detectable only after a prolonged exposure (not shown). In both cases either the mouse brain or the human fibroblast cDNA library (vector PCR products) products were negative, as was the control PCR (lanes with the arrowheads).

Clone Analysis

In principal, cDNA selection is very simple and rapid, it is possible to generate in a short time a considerable amount of YAC/cosmid selected cDNA products which have to be cloned and analyzed. For these reasons we chose to clone the selected PCR cDNA products directly to a T-tailed plasmid vector (pCRII; Invitrogen). Two hundred clones were picked for each YAC and 4 replica filters were prepared. These were hybridized with the following probes: human rDNA; human cotI; original cDNA PCR product (as used for the 1st round of selection), and finally with the HXNP23.1E6 probe (results not shown). The results of this negative/positive screening are summarized in Table 1.

To screen out all positive cDNA clones, we tried to hybridize the gridded clones with the PFGE prepared YAC DNA preblocked for repeats. We failed to obtain specific signals and found that all positive clones were repetitive (results not shown). One possible solution would be to hybridize not the bacterial clones, but the PCR products of the clones, and in this way to increase the sensitivity. Our preliminary results show this could work and one can isolate at least some true positives, which map to the YAC of origin (results not shown). However, in order to select new cDNAs from the region, routinely the negative clones are tested via Southern blot hybridization on a YAC panel. Using this strategy, we isolated 2 new genes from the region, I14 from the YAC4551 and J15 from the YAC326B10 (16), in addition to 3 already known ones (MNK, PGK1 and XNP). These genes were isolated from the human fibroblast cDNA library. As mentioned above, the mouse brain cDNA library was also used and results compared to the fibroblast experiment. Of the resulting minilibraries, only the YAC4551 has been analyzed in detail. The human XNP 'pooled' cDNA probe identified 7 clones (see Table 1). Three of these were sequenced, and the sequence compared to that of the human XNP. Homology comparison revealed at least 75% homology at the nucleotide level (75 - 95%; results not shown). Two of these three clones were identical (F7 and G5). The third clone (A1) was from the adjacent region, downstream of the previous. The remaining four clones were not analysed. The majority of the rest of the negative clones (47.5%) were tested on YAC panel,

Table 1. Summary of the analysis of the twice selected and cloned YAC4551 cDNAs from the mouse brain and human fibroblast cDNA libraries. For both experiments the rDNA and cDNA probes were mixed. 'No insert' refers to clones bearing the vector PCR contaminating product as shown in Fig. 3A.

	Human fibroblast cDNA	Mouse brain cDNA
Probe	YAC4551	YAC4551
cot1	12 (6%)	2 (1%)
rDNA cDNA	120 (60%)	65 (32.5%)
XNP	2 (1%)	7 (3.5%)
no insert	32 (16%)	38 (19%)

sequenced and compared with the XNP sequence and databases, via Blastx search (27). Three additional clones from the mouse xnp have been identified. In addition, one clone (Y4551MBG6) showed no homology to XNP. When hybridized to the YAC panel the clone maps distal to DXS56 locus (Fig. 2; results not shown) and may represent a clone corresponding to a new gene from the region. To isolate a human homologue, this clone was hybridized against the first and second round cDNA selection products from the five cDNA libraries we subselected with the 4551 YAC (Fig. 4; results not shown). A strong signal was found in the lane where the human fetal brain cDNA PCR products were applied (lanes HB, Fig. 4). Compared with mouse brain (lanes MB, idem), the signal was much stronger, indicating higher level of enrichment. A faint smear is visible also in the control cDNA HB PCR lane indicating relative high abundance of this gene in the library.

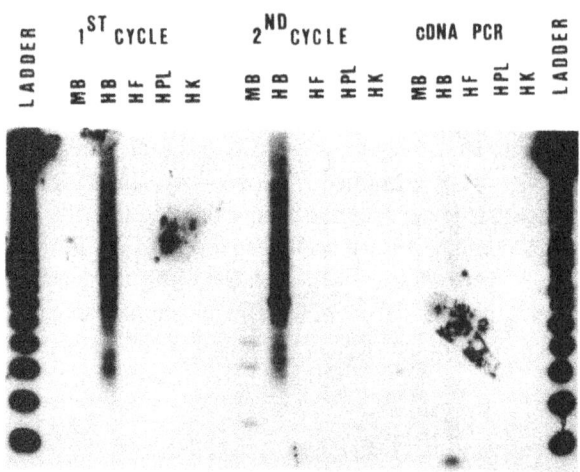

Figure 4. Hybridization of the 1ˢᵗ and 2ⁿᵈ round selected YAC4551 cDNAs from five cDNA libraries: MB, mouse brain; HB, human fetal brain; HF, human fibroblasts; HPl, human placenta; and HK, human adult kidney with the YAC4551MBG6 probe. cDNA PCR refers to vector PCR amplified products from the libraries analyzed. 100 bp ladder was used as a molecular marker.

The rest of the PCR cDNA products analyzed were either negative ($>15\%$; false positives), or not tested to completion ($\sim 10\%$). During the characterization of the I14 gene which was originally isolated from the human fibroblast cDNA library using the 4551YAC (16), we found it showed conservation with mouse (results not shown). However, when a probe of the I14 gene was tested on the mouse brain experiment selected PCR products, no signal was present. This would indicate that either the sensitivity of the approach was insufficient, or the level of sequence homology too low to permit its selection.

DISCUSSION

Recent improvements in gene isolation techniques considerably speeded up the identification of positionally mapped genes (22,23,24,28,29, review 30). Exon trapping (12) and direct cDNA selection techniques (13,14) proved to be very fruitful when applied to large genomic regions (cosmids, YACs). Separately, or in combination with the other technique(s) (30), almost complete transcription maps are being generated (2,9,31). However, both major techniques have some 'temporary' limits: for exon trapping, the length of the scanned region (applicable at present only to low complexity clones, e.g. cosmids), and for cDNA selection, the tissue source of starting cDNA.

In our previous transcriptional studies in the Xq13.3 chromosomal region, we successfully selected 5 genes (2 new) from one tissue source: human skin fibroblast cDNA (16). We recently complemented these studies with several other tissue cDNAs, among them a heterologous cDNA from the mouse brain. This was principally of interest to determine whether the cDNA selection technique could work in heterologous conditions. Our results, as presented above, basically confirm the applicability of the technique to cross-species systems, however with a lower enrichment. As a reporter gene, we used the XNP gene isolated by our group previously (16,26, and G. Consalez, in preparation). Using the 'pooled' human XNP cDNA probe, we identified seven mouse xnp cDNAs in the selected minilibrary. Homology comparison revealed at least 75% homology on the nucleotide level, a probable homology limit necessary for efficient selection under the washing stringency applied. Stringency of wash versus 'stringency of homology' would then be the crucial parameters determining the success of heterologous cDNA selection. Decreased washing stringency resulted in the increased proportion of co-selected clones (false positives, >15%), although the complexity of selected cDNAs was also lower. This would indicate, that only highly conserved cDNAs could be selected with reasonable enrichment. However, our expectations for heterologous selection wouldn't be the exhaustive selection of transcribed sequences from a given region. On the contrary, the isolation of conserved sequences for comparative mapping purposes and/or identification of differentially expressed conserved genes would be of interest. The G6 mouse cDNA clone may represent such a gene.

More detailed analyses will have to be carried out in order to elucidate the feasibility of the heterologous cDNA selection in both directions. However, we can conclude that the cross-species selection of cDNAs is applicable more for special purposes, than for standard transcription mapping.

ACKNOWLEDGEMENTS

We would like to thank to A.P. Monaco and D.Le Paslier for providing us with YACs and to S. Parimoo for the quench materials. This work was supported by grants from MRT (ACC 'Genome humain'), EEC ('Human genome analysis program') and Association Francaise Contre les Myopathies, J. Gecz was supported by a fellowship from CREBIOP.

REFERENCES

1. U. Hochgeschwender, Toward a transcriptional map of the human genome, *Trends Genet.* 8:41 (1992).
2. Z. Sedlacek, B. Korn, D.S. Konecki, R. Siebenhaar, J.F. Coy, P. Kioschis, A. Poustka, Construction of a transcriptional map of a 300kb region around the human G6PD locus by direct cDNA selection, *Hum. Mol. Genet.* 2:1865 (1993).
3. J.M. Sikela and C. Auffray, Finding new genes faster than ever, *Nature Genetis* 3:189 (1993).
4. M.D. Adams, M. Dubnick, A.R. Kerlavage, M. Moreno, J.M. Kelley, T. Utterback, J.W. Nagle, C. Fields, and J.C. Venter, Sequence identification of 2,375 human brain genes, *Nature* 335:632 (1992).
5. J. Takeda, H. Yano, S. Eng, Y. Zeng, and G.I. Bell, A molecular inventory of human pancreatic islets:sequence analysis of 1000 cDNA clones, *Hum. Mol. Genet.* 2:1793 (1993).
6. M.H. Polymeropoulos, H. Xiao, A. Glodek, M. Gorski, M.D. Adams, R.F. Moreno, M.G. Fitzgerald, J.C. Venter, and C.R. Merril, Chromosomal assignment of 46 brain cDNAs, *Genomics* 12:492 (1992).
7. M.H. Polymeropoulos, H. Xiao, J.M. Sikela, M.D. Adams, J.C. Venter, and C.R. Merril, Chromosomal distribution of 320 genes from a brain cDNA library, *Nature Genetics* 4:381 (1993).
8. P. Elvin, G. Slynn, D. Black, A. Graham, R. Butler, J. Riley, R. Anand, and A.F. Markham, Isolation of cDNA clones using yeast artificial chromosome probes. *Nucl. Acid Res.* 18:3913 (1990).
9. A.E. Kahloun, B. Chauvel, V. Mauvieux, I. Dorval, A.M. Jouanolle, I. Gicquel, J.Y. Le Gall, and V. David, Localization of seven new genes around the HLA-A locus, *Hum. Mol. Genet.* 2:55 (1993).
10. A.P. Monaco, R.L. Neve, C. Colletti-Feener, C.J. Bertelson, D.M. Kurnit, and L.M. Kunkel, Isolation of candidate cDNAs for portions of the Duchenne muscular dystrophy gene, *Nature* 326:646 (1986).
11. S. Lindsay and A.P. Bird, Use of restriction enzymes to detect potential gene sequences in mammalian DNA, *Nature* 327:336 (1987).
12. A.J. Buckler, D.D. Chang, S.L. Graw, D.J. Brook, D.A. Naber, P.A. Sharp, and D.E. Houseman, Exon amplification: A strategy to isolate mammalian genes based on RNA splicing, *Proc. Natl. Acad. Sci. USA* 88:4005 (1991).
13. S. Parimoo, S.R. Pantanjali, H. Shykla, D. Chaplin, and S.M. Weissman, cDNA selection: Efficient approach for the selection of cDNAs encoded in large chromosomal DNA fragments, *Proc. Natl. Acad. Sci. USA* 88:9623 (1991).

14. M. Lovett, Kere, and L.M. Hinton, Direct selection: A method for the isolation of cDNAs encoded by large genomic regions, *Proc. Natl. Acad. Sci. USA* 88:9628 (1991).

15. W. Fan, X. Wei, H. Shukla, S. Parimoo, H. Xu, P. Sankhavaram, and S.M. Weissman, Application of cDNA selection techniques to regions of the human MHC, *Genomics* 17:575 (1993).

16. J. Gecz, L. Villard, A.M. Lossi, P. Millasseau, M. Djabali, and M. Fontes, Physical and transcriptional mapping of DXS56-PGK1 1Mb region: identification of three new transcripts, *Hum. Mol. Genet.* 2:1389 (1993).

17. S.R. Pantanjali, S. Parimo, and S.M. Weissman, Construction of a uniform-abundance normalized) cDNA library, *Proc. Natl. Acad. Sci. USA* 88:1943 (1991).

18. A.P. Feinberg and B. Vogelstein, A technique for radiolabeling DNA restriction endonuclease fragments to high specific activity, *Anal. Biochem.* 123:6 (1983).

19. J. Weissenbach, G. Gaypay, C. Dib, A. Vignal, J. Morissette, Millasseau, G. Vaysseix, and M. Lathrop, A second-generation linkage map of the human genome, *Nature* 359:794 (1992).

20. Z. Larin, A. Monaco, and H. Lehrach, Yeast artificial chromosome libraries containing large inserts from mouse and human DNA, *Proc. Natl. Acad. Sci. USA* 8:4123 (1991).

21. H.M. Albertsen, H. Abderrahim, H.M. Cann, J. Dausset, D. Le Paslier, D. Cohen, Construction and characterization of a yeast artificial chromosome library containing seven haploid genome equivalents, *Proc. Natl. Acad. Sci. USA* 87:4256 (1990).

22. C. Vulpe, B. Levinson, S. Whitney, S. Packman, and J. Gitschier, Isolation of a candidate gene for Menkes disease and evidence that it encodes a copper-transporting ATPase, *Nature Genetics* 3:7 (1993).

23. J. Chell, Z. Tumer, T. Tonnese, A. Petterson, Y. Ishikawa-Brush, N. Tommerup, N. Horn, and A.P. Monaco,. Isolation of a candidate gene for Menkes disease that encodes a potential heavy matal binding protein, *Nature Genetics* 3:14 (1993).

24. J.F.B. Mercer, J. Livingston, B. Hall, J.A. Paynter, C. Begy, S. Chandrasekharappa, P. Lockhart, A. Grimes, M. Bhave, D. Siemieniak, and T.W. Glover, Isolation of a partial candidate gene for Menkes disease by positional cloning, *Nature Genetics* 3:20 (1993).

25. A.M. Michelson, G.A.P. Bruns, C.C. Morton, and S.H. Orkin, The human phosphoglycerate kinase multigene family, *J. Biol. Chem.* 260:6982 (1985).

26. J. Gecz, H. Pollard, G.G. Consalez, L. Villard, C. Stayton, P. Millasseau, M. Khrestchatisky, and M. Fontes, Cloning and expression of a murine homologue of a human X-linked nuclear protein gene close to PGK1 (Xq13.3), *Hum. Mol. Genet.* 3:39 (1994).

27. S.F. Altschul, W. Gish, W. Miller, E.W. Myers, and D.J. Lipman, Basic local alignment search tool, *J. Mol. Biol.* 215:403 (1990).

28. The Huntington Collaborative Research Group, A novel gene containing a trinucleotide repeat that is expanded and unstable on Huntington's chromosomes, *Cell* 72:971 (1993).

29. D. Vetrie, I. Vorechovsky, P. Sideras, J. Holland, A. Davies, F. Flinter, L. Hammarstrom, C. Kinnon, R. Levinsky, M. Bbrow, C.I.E. Smith, and D. Bentley, The gene involved in X-linked agammaglobulinaemia is a member of the src family of protein-tyrosine kinases, *Nature* 361:226 (1993).

30. J.E. Parrish and D.L. Nelson, Methods for finding genes a major rate-limiting step in positional cloning, *Genetic. Anal. Tech. Appl.* 10:29 (1993).

31. J.M. Rommens, B. Lin, G.B. Hutchison, S.E. Andrew, Y.P. Goldberg, M.L. Glaves, V. Lai, J. McArthur, J. Nasir, J. Theilmann, H. McDonald, M. Kalchman, L. Clarke, K. Schappert, and M.R. Hayden, A transcriptional map of the region containing the Huntington disease gene, *Hum. Mol. Genet.* 2:901 (1993).

A SANDWICH-HYBRIDIZATION METHOD FOR SPECIFIC AND EFFICIENT SELECTION OF cDNA CLONES FROM GENOMIC REGIONS

Denise Yan[1,3] and Anand Swaroop [1,2,3]

Departments of [1]Ophthalmology and [2]Human Genetics, and [3]Human Genome Center, University of Michigan, 1000 Wall Street, Ann Arbor, MI 48105

ABSTRACT

We are developing a novel strategy for efficient and specific isolation of cDNA clones encoded by a particular genomic region. This sandwich-selection method consists of: 1. Hybridization of single-stranded (ss) circular cDNA molecules to tagged *in vitro* synthesized RNA from the genomic region; 2. Retention of "genomic RNA" - cDNA hybrids on an avidin-matrix through a biotinylated RNA complementary to the tag; and 3. Elution of specific cDNAs from the avidin-matrix with ribonuclease A. The selected ss cDNAs are directly used for electroporation of competent E. coli cells without requiring any amplification or cloning step. Here, we report the construction of appropriate vectors and results from model experiments. This protocol has been applied for specifically selecting NRL (neural retina leucine zipper) cDNA from a mixture of cDNA clones using a sub-library from the NRL genomic region. The control experiments have allowed us to optimize various steps in the procedure and provided valuable data regarding the yield and specificity of selection process. The strategy is now being used for the isolation of cDNAs from large regions of genomic DNA cloned in cosmids or yeast artificial chromosome (YAC) vectors and from microdissected and flow-sorted chromosomal DNA. We expect that a high degree of specificity and efficiency will permit wider application of this cDNA selection strategy.

INTRODUCTION

The isolation of transcribed sequences from a defined chromosomal region is an essential step for identifying new "disease" genes and for generating the expression map of the human genome. The strategies for identifying expressed sequences, such as interspecies cross-hybridization to find evolutionarily conserved DNA fragments (1), the use of CpG islands as markers of transcriptional units (2), and sequence-based approaches, are not readily applicable if the region of interest spans several hundreds (or thousands) of kb of DNA. Direct screening of cDNA libraries with cosmid or YAC clones (3) is tedious and suffers from being low in sensitivity and reproducibility. The cloning of human transcripts from human-rodent somatic cell hybrids (4-6) is limited by the low (or no) expression of many tissue-specific genes in hybrid cell lines. Exon-trapping strategies (7, 8) and hybridization-based cDNA selection methods (9-13) have recently been used successfully in various laboratories (see several papers in this volume). However, reproducibility, in addition to efficiency and specificity, is still a concern with these methods.

We have optimized a conceptually simple hybridization-based method, called sandwich-selection, for isolating cDNA clones from genomic regions. The method does not utilize polymerase chain reaction (PCR) methodology, thereby avoiding possible bias in selected cDNAs. Using model experiments, we demonstrate highly specific selection of a particular cDNA from complex mixtures using a sub-library from a corresponding 5 kb genomic clone. The specificity of the selection is dictated by the enzymatic elution of cDNAs from the retention-matrix under mild conditions. In addition, a major advantage of the procedure is that it should be independent of the size or complexity of genomic DNA being used for selection.

METHODS

The Method

Figure 1 shows the schematic representation of the sandwich-selection method. In brief, the single-stranded (ss) cDNAs are hybridized to the "SV40-tagged" RNA derived from a library of genomic clones in pSV9Zf9 vector (called "genomic RNA"). The ss cDNA - "genomic RNA" hybrids are captured on an avidin-matrix through a biotin-labeled "capture RNA" (from pSV7Zf3 vector), which has a 225 bp region complementary to the SV40 tag in "genomic RNA". The avidin-matrix is then extensively washed, and the specific ss cDNA molecules, which hybridize to the "genomic RNA", are eluted by incubation with RNase A. The eluted cDNAs are purified by phenol/chloroform extraction and ethanol-precipitated. An aliquot of the cDNA is directly used for electroporation of competent E. coli cells to obtain the library of selected cDNA clones.

Single-Stranded (ss) DNA From cDNA Libraries

The ss DNA from cDNA libraries in Bluescript KSM13(-) vector was generated by using helper phage R408 according to the protocol from Stratagene (La Jolla, CA). The plasmid libraries were derived from those in Charon BS phage vector, as described (14, 15). The cDNA libraries from a number of human tissues and cell lines in Charon BS vector (16) are now available from American Type Culture Collection (ATCC, Rockville, MD).

The number of colonies obtained upon electroporation of competent E. coli XL1Blue cells (Stratagene) using ECM 600 (BTX Inc., San Diego, CA) ranged between 10^5 to 10^6 per μg of ss DNA.

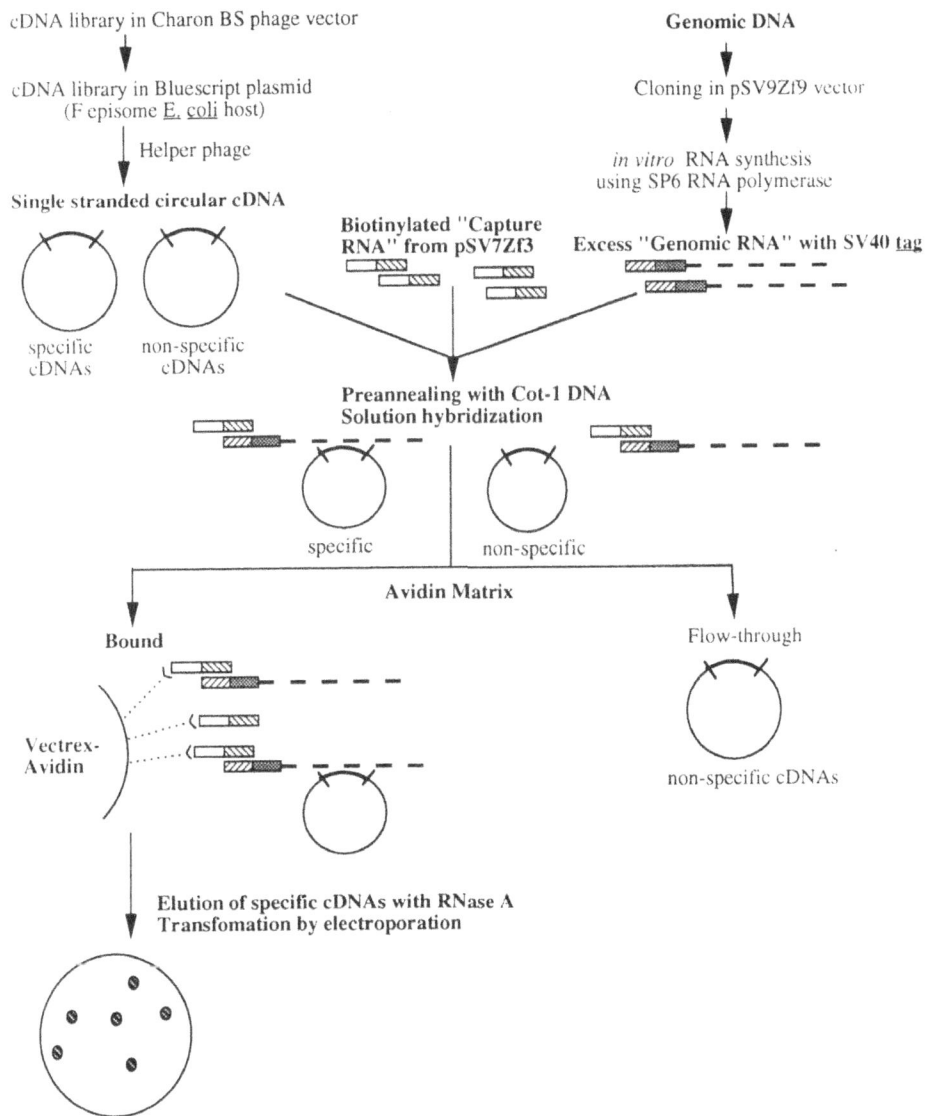

cDNA library in Charon BS phage vector

cDNA library in Bluescript plasmid
(F episome E. coli host)

Helper phage

Single stranded circular cDNA

specific
cDNAs

non-specific
cDNAs

Genomic DNA

Cloning in pSV9Zf9 vector

in vitro RNA synthesis
using SP6 RNA polymerase

Biotinylated "Capture
RNA" from pSV7Zf3

Excess "Genomic RNA" with SV40 tag

Preannealing with Cot-1 DNA
Solution hybridization

specific

non-specific

Avidin Matrix

Bound

Flow-through

Vectrex-
Avidin

non-specific cDNAs

Elution of specific cDNAs with RNase A
Transfomation by electroporation

Library enriched for specific cDNAs

Figure 1. Schematic representation of the sandwich-selection strategy.

Construction of pSV9Zf9 Vector and Synthesis of "Genomic RNA"

pSV9Zf9 vector was constructed by inserting a 375 bp SV40 DNA fragment (PstI-DraI, bp 1988 to 2363) into pGEM-9Zf(-) vector (Promega, Madison, WI) at the HindIII site in the multiple cloning region after generating blunt ends (Fig. 2). The genomic DNA, used for selecting cDNA clones, is cloned in pSV9Zf9 vector at the EcoRI site after digestion with a 4-cutter enzyme (e.g., RsaI, MboI, or BstUI). The plasmid DNA from the resulting library of clones is digested with SacI or SalI, and used for synthesizing RNA *in vitro* (called "genomic RNA") with SP6 RNA polymerase (Promega protocol). All of the "genomic RNA" molecules should have a SV40 tag sequence at the 5' end.

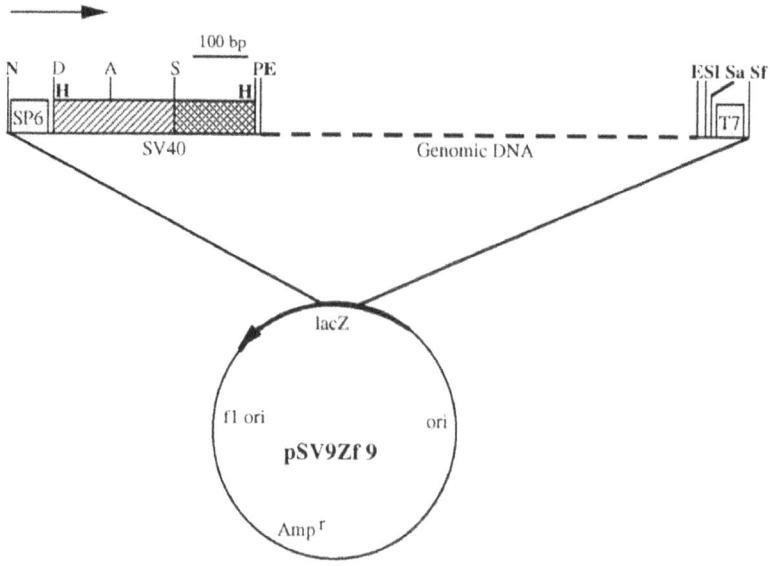

Figure 2. pSV9Zf9 vector for cloning genomic DNA and subsequent synthesis of tagged "genomic RNA". The restriction enzyme abbreviations are: N, NotI; D, DraI; H, HindIII; A, ApaI; S, Sau3A; P, PstI; E, EcoRI; Sl, SalI; Sa, SacI; and Sf, SfiI. The sites in the vector are shown in bold. HindIII site is no longer available in the construct. The arrow indicates the orientation of *in vitro* synthesized "genomic RNA" with SP6 RNA polymerase.

Construction of pSV7Zf3 and Synthesis of Biotinylated "Capture RNA"

pSV7Zf3 (shown in Fig. 3) was constructed by cloning a 395 bp SV40 fragment (Sau3A-BamHI, bp 2138 to 2533) into pGEM-7Zf(+) vector (Promega). A region of 225 bp in this fragment is identical to the SV40 sequence present in pSV9Zf9 vector. The DNA from pSV7Zf3 construct is used for synthesizing RNA *in vitro* with T7 RNA polymerase in the presence of Bio-11-UTP (Enzo Diagnostics, New York). The resulting biotinylated RNA (referred as "capture RNA") contains a 225 bp sequence that is complementary to the tag in "genomic RNA".

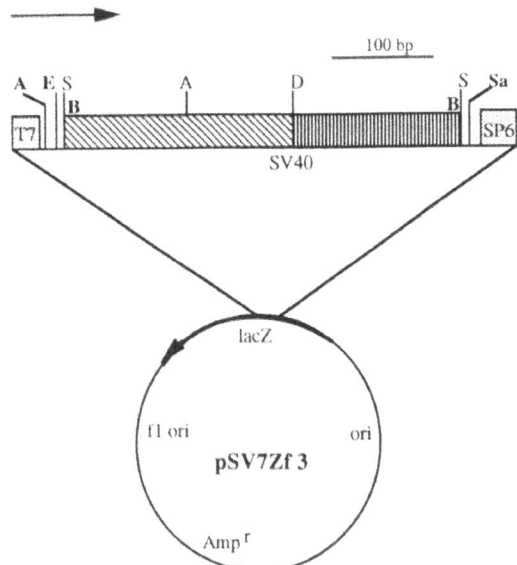

Figure 3. pSV7Zf3 construct for the synthesis of biotinylated "capture RNA". The restriction enzyme abbreviations are: A, ApaI; E, EcoRI; B, BamHI; S, Sau3A; D, DraI; and Sa, SacI. The sites in the vector are shown in bold. Only one BamHI site is regenerated after cloning the SV40 fragment. The arrow indicates the orientation of *in vitro* synthesized biotinylated "capture RNA" with T7 RNA polymerase.

Reagents for Model Sandwich-Selection Experiments

NRL (Neural retina leucine zipper) cDNA clone (17) in Bluescript KSM13(-) vector was used for generating ss DNA with helper phage. A Bluescript plasmid containing the human red pigment cDNA (hs7, 18) was modified by cloning a 1.6 kb NotI fragment that confers kanamycin resistance (kindly provided by H. Arenstorf). This construct, pBhs7kana, provided ss DNA that upon transformation gives both kanamycin and ampicillin resistant colonies. NRL genomic clone of 5 kb (Swaroop *et al.*, unpublished data) was used for generating a sub-library to obtain "genomic RNA". For this, the genomic insert was digested with RsaI and MboI independently, blunt-ended and cloned in pSV9Zf9 vector at the EcoRI site. RsaI has 6 sites in the NRL genomic clone, whereas MboI cuts it at 14 sites. The gene consists of three exons. The largest single exon fragment in the sub-library should be about 475 bp.

Hybridization, Washing and Elution Conditions for cDNA Selection

The NRL ss DNA (1 μg or less) was combined with ss DNA from other clones (e.g., pBhs7kana or retinal cDNA library) in varying amounts to obtain different ratios of target to non-specific clones. The ss DNA mixture was hybridized to SP6 polymerase-generated "genomic RNA" (generally between 10-40 μg, 10 to 100 fold molar excess compared to ss DNA) from genomic clones in pSV9Zf9 vector and to the biotinylated "capture RNA" from pSV7Zf3 construct (generally between 20-50 μg, 2-10 fold molar excess compared to "genomic RNA"). The annealing was performed in a sealed capillary at 60°C for about 16 h in the presence of 0.5 M sodium phosphate pH 7.2, 10 mM EDTA and 0.5% SDS in a final volume of 50 μl.

Before binding, the avidin matrix (Vectrex-Avidin from Vector Laboratories,

Burlingame, CA) was equilibrated at room temperature with binding buffer (100 mM Tris. Cl pH 7.5, 150 mM NaCl, 0.5% SDS) containing sonicated denatured herring-sperm DNA (50-100 μg/ml) and washed twice in the same buffer without herring-sperm DNA. The hybridization mixture was incubated with equilibrated Vectrex-Avidin (100-250 mg) in binding buffer for a total of about 60 min with intermittent mixing. The avidin-matrix was then washed twice in binding buffer, twice each with buffer A (10 mM Tris-Cl pH 7.5, 0.5% SDS, 2 mM EDTA and 100 mM NaCl) and buffer B (same as A except 50 mM NaCl), and equilibrated in 10 mM Tris.Cl pH 7.5, 80 mM NaCl, 2 mM EDTA (pre-elution buffer). The bound ss cDNAs were eluted from the matrix at 37°C in 600 μl of elution buffer (pre-elution buffer containing RNase A (125 μg/ml)) for two successive periods of 30 min each. The eluants were extracted with phenol/chloroform, ethanol precipitated in the presence of glycogen, and used for electro-transformation of XL1Blue cells.

RESULTS

To test the feasibility of the approach and to optimize various steps, we decided to perform model experiments using a well-defined system. Our laboratory has recently identified a human retinal leucine zipper gene, NRL, and obtained the sequence and structure of NRL cDNA and corresponding gene (17, Swaroop et al., unpublished data). In model experiments, the purpose was to select NRL cDNA from a mixture of clones (as in a cDNA library) using the NRL "genomic RNA". As described in Methods section, a sub-library of genomic clones was generated in pSV9Zf9 vector from a 5 kb NRL genomic clone and used for preparing "genomic RNA". In the first set of experiments, a mixture of ss DNA for the red pigment cDNA (containing both kanamycin and ampicillin resistance markers) and NRL cDNA (ampicillin - resistance marker only) was hybridized to the NRL "genomic RNA" and the biotinylated "capture RNA". After hybridization, the mixture was incubated with Vectrex-Avidin. The matrix was then washed extensively. The bound cDNAs were eluted with RNase A and used for electroporation. The transformed XL1Blue cells were then spread on kanamycin and ampicillin - containing LB plates, in parallel. The results are summarized in Table 1. Before hybridization, most of the cDNA clones were for red pigment cDNA (the ratio of Kana^r + Amp^r to Amp^r only clones in the mixture was 40:1). However, after selection with NRL "genomic RNA", almost all of the clones were for NRL cDNA (the ratio of Kana^r + Amp^r to Amp^r only clones in elution 1 was 1:20), indicating an enrichment of at least 800 fold.

Table 1. Enrichment of NRL cDNA after selection

Experimental steps	only amp^r colonies	kana^r + amp^r colonies
Hybridization (before selection)	3,500	137,500
E1 (RNase elution, #1)	1,000	50
E2 (RNase elution, #2)	100	0

In another experiment, the NRL ss cDNA was spiked in ss DNA from a retinal cDNA library. The "genomic RNA" derived from a sorted X-chromosome-library (generated by transferring LAOXNLO1 library from ATCC to pSV9Zf9 vector) was then used for hybridization to ss DNAs in the presence of biotinylated "capture RNA". The

NRL gene maps to human chromosome 14 (19), and should not be selected by the X-chromosome "genomic RNA" used in this experiment. Quantitative assessment of the NRL cDNA was performed by hybridizing the colony lifts from the initial (before hybridization) and selected cDNA clones with NRL probe. The results of this selection are summarized in Table 2. As expected, the % of NRL positive clones is reduced from 26% to 2-4% after selection.

Table 2. Reduction of NRL cDNA in clones selected with unrelated "genomic RNA" (negative selection)

Experimental steps	NRL positive / total clones	% Positive
Hybridization (before selection)	95 / 364	26
E1 (RNase elution, # 1)	2 / 43	4
E2 (RNase elution, # 2)	2 / 86	2.3

Essentially similar experiments (NRL ss DNA spiked in ss DNA from a retinal cDNA library) were then performed using NRL "genomic RNA", instead of X-"RNA", for selection. The ratio of NRL versus other unrelated clones was altered in two independent experiments. The selection efficiency was analyzed by hybridizing the filters lifted from the colonies obtained with starting and selected DNA using NRL probe. The results are summarized in Table 3. In one case (Table 3A), 12% of the clones in the starting material were for NRL, whereas almost 100% of the selected clones were NRL positive. In another experiment (see 3B), NRL positive clones were about 1.7% of the total cDNAs before hybridization. Selection with NRL "genomic RNA" recovered almost all of the NRL clones in the RNase-eluted material.

Table 3. Selection of NRL cDNA from a spiked retinal cDNA library using NRL "genomic RNA" (positive selection)

A.

Experimental steps	NRL positive / total clones	% Positive
Hybridization (before selection)	6,000 / 50,000	12
E1 (RNase elution, # 1)	3,600 / 3,650	99 (recovery: 60%)

B:

Experimental steps	NRL positive / total clones	% Positive
Hybridization (before selection)	400 / 23,600	1.7
E1 (RNase elution, # 1)	500 / 1,500	33 (recovery: 100%)

DISCUSSION

The identification of transcribed sequences within large genomic regions is an essential step towards the isolation of "disease" genes and for construction of the human gene map. Different approaches currently in use for detecting expressed sequences are still far from being optimal. In this report, we describe a novel method for selecting cDNAs from genomic regions. Like several other hybridization-based selection methods, it utilizes biotin-avidin interaction for capturing the cDNA clones. However, the use of a sandwich for retention of cDNAs on avidin-matrix and of enzymatic elution with RNase A provide higher efficiency and specificity to this method. The cDNA clones that are non-specifically retained on the avidin-matrix are not eluted under the low stringency conditions (high salt concentration and low temperature) used for obtaining specific hybridizing cDNA clones. After optimizing various steps of the protocol individually, we have now demonstrated the feasibility of the method in model experiments.

The method was applied to select NRL cDNA from a complex mixture with a sub-library of genomic clones from the NRL region. Although only 5 kb in size, this genomic region contains stretches of repetitive elements and an AluI repeat (based on sequence data). The results (shown here) demonstrate the high yield and specificity of selection process when target and non-specific ss cDNAs were used in various ratios. This method is now being applied to select cDNAs encoded by YAC clones using human retina, fetal brain and fetal eye cDNA libraries. Initial results (not shown here) are very promising and once completed will be published elsewhere.

This approach has several advantages over other selection methods. Once the genomic library is prepared in pSV9Zf9 vector, the cDNA selection can be realized in less than one week. The use of PCR-based methodology is not necessary at any step of the strategy. The sequence alterations and selective amplification of short inserts during selection can, therefore, be avoided. The limitations of exon-trapping system can be overcome since the method is insensitive to the size and number of the introns, or activation of cryptic splice sites. However, some of the potential limitations associated with other hybridization-based cDNA selection methods are not alleviated in this strategy. Availability of high complexity cDNA libraries from various tissues and cell lines and from different stages of development will be required for complete saturation of the genomic region. However, the use of whole fetus, fetal brain and teratocarcinoma cell line cDNA libraries (16) can enhance the probability of finding most genes. The problem of repetitive sequences in some cDNA clones can be partially overcome by preannealing the "genomic RNA" with human Cot-1 DNA (GIBCO-BRL, Gaithersburg, MD). A related concern is the co-selection of pseudo-genes or members of a multigene family.

In addition to specificity, another major advantage of this approach is that it can be applied to select cDNAs from genomic source of any complexity. It would, therefore, provide a convenient method to identify cDNAs encoded by large YAC clones or microdissected chromosomal region. We also believe that it will also be possible to use flow-sorted chromosomal DNA for sandwich-selection to isolate all the cDNAs encoded by a particular chromosome.

ACKNOWLEDGEMENT

We thank Prof. Sherman M. Weissman for his generosity and several fruitful discussions during the initial stages of the conception of this strategy. Thanks are also due to Ms. Dorothy Giebel for typing the manuscript. This research is supported in part by grants from NIH (EY07961) and RP Foundation, Baltimore, and by pilot grant from Michigan Human Genome Center (NIH P30 HG00209).

REFERENCES

1. A.P. Monaco, R.L. Neve, C. Colletti-Feener, C.J. Bertelson, D.M. Kurnit, and L.M. Kunkel, Isolation of candidate cDNAs for portions of the Duchenne Muscular Dystrophy gene, *Nature* 323:646 (1986).

2. A. Bird, CpG-rich islands and the function of DNA methylation, *Nature* 321:209 (1986).

3. P. Elvin, G. Slynn, D. Black, A. Graham, R. Butler, J. Riley, R. Anand, and A.F. Markham, Isolation of cDNA clones using yeast artificial chromosome probes, *Nucl. Acids Res.* 18:3913 (1990).

4. P. Liu, R. Legerski, and M.J. Siciliano, Isolation of human transcribed sequences from human-rodent somatic cell hybrids, *Science* 246:813 (1989).

5. L. Corbo, J.A. Maley, D.L. Nelson, and C. Caskey, Direct cloning of human transcripts with hnRNA from hybrid cell lines, *Science* 249:652 (1990).

6. K. W. Jones, M. Chevrette, M.H. Shapero, and R.E.K. Fournier, Generation of region- and species-specific expressed gene probes from somatic cell hybrids, *Nat. Genet.* 1:278 (1992).

7. G.M. Duyk, S. Kim, R.M. Myers, D.R. Cox, Exon trapping: a genetic screen to identify candidate transcribed sequences in cloned mammalian genomic DNA, *Proc. Natl. Acad. Sci. USA* 87:8995 (1990).

8. A.J. Buckler, D.D. Chang, S.L. Graw, D.J. Brook, D.A. Haber, P.A. Sharp, and D.E. Housman, Exon amplification: a strategy to isolate mammalian genes based on RNA splicing, *Proc. Natl. Acad. Sci. USA* 88:400 (1991).

9. M. Lovett, J. Kere, and L.M. Hinton, Direct Selection: a method for the isolation of cDNAs encoded by large genomic regions, *Proc. Natl. Acad. Sci. USA* 88:9628 (1991).

10. S. Parimoo, S.R. Patanjali, H. Shukla., D.D. Chaplin, and S.M. Weissman, cDNA Selection: efficient PCR approach for the selection of cDNAs encoded in large chromosomal DNA fragments, *Proc. Natl. Acad. Sci. USA* 88:9623 (1991).

11. D.A. Tagle, M. Swaroop, M. Lovett, and F.S. Collins, Magnetic bead capture of expressed sequences encoded within large genomic segments, *Nature* 361:751 (1993).

12. J.G. Morgan, G.M. Dolganov, S.E. Robbins, L.M. Hinton, and M. Lovett, The selective isolation of novel cDNAs encoded by regions surrounding the human interleukin 4 and 5 Genes, *Nucl. Acids Res* 20:5173 (1992).

13. B. Korn, Z. Sedlacek, A. Manca, P. Kioschis, D. Konecki, H. Lehrach, and A. Poustka, A strategy for the selection of transcribed sequences in the Xq28 region, *Hum. Mol. Genet.* 1:235 (1992).

14. A. Swaroop and S.M. Weissman, Charon BS (+) and (-), versatile phage vectors for constructing directional cDNA libraries and their efficient transfer to plasmids, *Nucl. Acids Res.* 16:8739 (1988).

15. A. Swaroop, Construction of directional cDNA libraries from human retinal tissue/cells and their enrichment for specific genes using an efficient subtraction procedure, *in*: "Photoreceptor Cells," P. Hargrave, ed., Methods Neurosci.: 15:285 (1993).

16. A. Swaroop and J. Xu, cDNA libraries from human tissues and cell lines, *Cytogenet. Cell. Genet.* 64.292 (1993).

17. A. Swaroop, J. Xu, H. Pawar, A. Jackson, C. Skolnick and N. Agarwal, A conserved retina-specific gene encodes a basic motif/leucine zipper domain, *Proc. Natl. Acad. Sci. USA* 89:266 (1992).

18. J. Nathans, D. Thomas, and D.S. Hogness, Molecular genetics of human color vision: the genes encoding blue, green, and red pigments, *Science* 232:193 (1986).

19. T.L. Yang-Feng and A. Swaroop, Neural retina-specific leucine zipper gene NRL (D14S46E) maps to human chromosome 14q11.1-q11.2, *Genomics* 14:491 (1992).

NOVEL STRATEGY FOR ISOLATING UNKNOWN CODING SEQUENCES FROM GENOMIC DNA BY GENERATING GENOMIC-cDNA CHIMERAS

Pudur Jagadeeswaran, Michael W. Odom and Edward J. Boland

Department of Cellular and Structural Biology, The University of Texas Health Science Center at San Antonio, 7703 Floyd Curl Drive, San Antonio, Texas 78284

ABSTRACT

A novel strategy for rapid identification of unknown coding sequences from large genomic regions has been developed. It is based on selective *in vitro* recombination of genomic DNA and cDNA followed by Polymerase Chain Reaction. The technique involves generation of cDNA primers, by restriction digestion of cDNA libraries, that hybridize to their cognate genomic DNA sequences. These hybrids are chain elongated using the free 3' end of the cDNA fragment as a primer, then PCR amplified using primers previously attached to genomic DNA and cDNA. Unknown coding sequences are clearly discernable as genomic cDNA chimeras.

INTRODUCTION

The essential information contained in the human genome is encoded in short segments of DNA (genes/exons), which are interspersed within much more abundant non-coding regions. The task of identifying the unknown coding sequences within the morass of the genome using conventional screening methods is difficult for two major reasons. First, only 3% of genomic DNA constitutes coding sequences that are homologous to cDNAs. Second, a significant proportion of cDNAs contain repetitive DNA elements, complicating their use as probes in the screening of genomic libraries. Recently, two important approaches have been described to isolate coding sequences from large genomic regions. They are exon trapping (1-3), and selection by hybridization (4-8). Although these techniques have certain advantages, they still have some drawbacks when used to identify several coding regions from large areas of the chromosome. In this report, we describe a novel strategy for the specific retrieval of coding sequences by the Polymerase Chain Reaction (PCR), using selective *in vitro* recombination, to form a genomic-cDNA chimeric molecule. The fundamental principle is to use the terminal free 3'OH-groups of restriction-enzyme digested cDNA fragments as primers for chain elongation.   The elongation generates genomic-cDNA chimera molecules. These are then amplified by PCR to identify the coding sequences from the genomic DNA.

The details of the principle are summarized in Fig. 1. The large genomic fragment of choice is first converted into a short fragment genomic library by shotgun subcloning into a vector. The inserts are flanked by vector specific primers P1 and P2. At the same time, a cDNA library is generated in a different vector. The inserts in this vector are flanked by primers X1 and X2. The cDNA is then digested by a restriction enzyme that

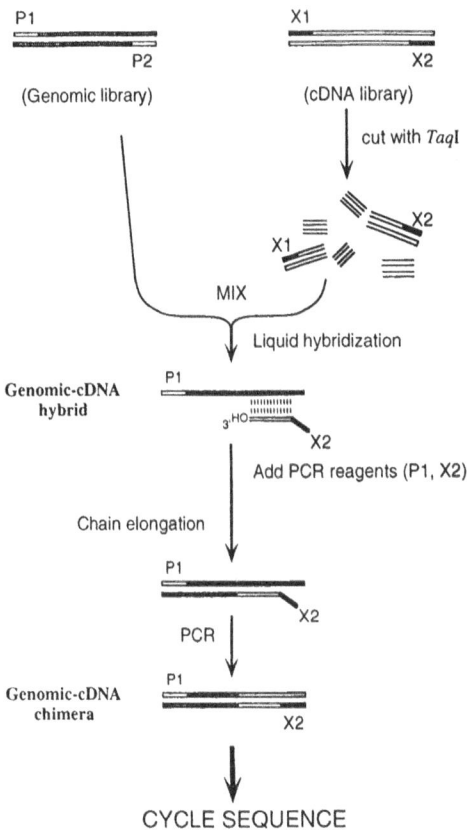

Figure 1. General scheme for the selection of genomic-cDNA chimeras by a novel PCR strategy. P1, P2, X1 and X2 represent vector-specific primers.

cuts infrequently in the repetitive DNA, but does not cut between the primers and the insert and cleaves most unique cDNA sequences at least once. When the genomic DNA and the restriction enzyme-digested cDNAs are mixed under conditions to allow hybridization, genomic-cDNA hybrids are formed. The hybrid molecule is then amplified by primers, one flanking the cDNA and the other flanking the genomic DNA. Thus, coding sequences can be retrieved as genomic-cDNA chimeras.

MATERIALS AND METHODS

Plasmids and Libraries

A short fragment genomic clone having a 625 bp insert (nucleotides from 30780 to 31404 in reference (9)) containing the terminal portion of the seventh intron and the coding portion of the eighth exon of the human factor IX (FIX) gene was generated by PCR (9). This fragment was ligated into the SmaI site of pUC19 to generate a FIX.8 genomic clone. The FIX cDNA clone, containing a truncated cDNA fragment of the FIX gene inserted into the Pst1 site of pBR322, was previously isolated in our laboratory and the sequence has been published (10). A random genomic library (average insert size 400 bp) was generated by digesting human genomic DNA with Sau3A (Promega) and ligating the resulting fragments into the BamHI restriction site of pUC19. High efficiency *E. coli* DH5-α cells (Bethesda Research Laboratories) were transformed with the ligated DNA. The colonies were plated on LB plates containing ampicillin. The human liver cDNA library was the same used in our laboratory for isolation of the factor IX cDNA except it underwent two amplifications. The colonies were plated on LB plates containing tetracycline (LB-tet). For these experiments, all DNA was isolated by alkaline lysis (11) followed by purification with NACS-52's (Gibco BRL) according to the recommended protocol. Restriction enzyme digestions were according to manufacturer's recommendations (New England Biolabs).

In-vitro Recombination by Polymerase Chain Reaction

Genomic and cDNA hybridizations were performed in 7.5 μl of 1.0 M NaCl containing 1 μg/μl of DNA. The sample was heated to 95°C for 10 min and allowed to reanneal at 65°C overnight (approximately at 120 C_ot value when 83 μg/ml is considered to give C_ot of 1). It was then diluted 100-fold and 1 μl was used in a standard PCR reaction described below which was preceded by a 5 min chain elongation step at 72°C. Standard Polymerase Chain Reactions were performed in 25 μl containing 10 mM Tris.HCl pH9.0 at 25°C, 50 mM KCl, 0.1% Triton X-100, 1.5 mM MgCl$_2$, 200 μM (each) dATP, dGTP, dCTP, and dTTP, 1 μM of each primer (New England Biolab pBR322 primer #1240 5'-ATTGTTGCCGGGAAGCTAGAGTAA-3' designated as X2 in Fig. 1 and the M13 primer #1224 5'-CGCCAGGGTTTTCCCAGTCACGAC-3' designated as P1 in Fig. 1), 1-10 ng of DNA and 0.5 unit of Taq DNA Polymerase (from Promega) in thin-walled microtubes using a thermal cycler (Perkin Elmer Cetus GeneAmp PCR System 9600). The DNA was amplified for 35 cycles (each cycle including denaturation at 95°C for 30 sec, reannealing at 54°C for 20 sec and chain elongation at 72°C for 30 sec). Subsequent to PCR, approximately 10-15 μl were loaded on the 5% polyacrylamide gels, and electrophoresis was performed at 10 v/cm of the gel using 40 mM Tris.acetate buffer pH8.0, 2 mM EDTA. MspI digest of pBR322 were used as size markers. Cycle sequencing was performed using kits from Perkin Elmer Cetus.

RESULTS

In order to test the feasibility of the strategy proposed above we performed three experiments. In the first experiment, we used a sample of negligible complexity to test the ability of the approach to form and amplify a known chimera. The FIX cDNA was cleaved by TaqI and added to an aliquot of the FIX.8 DNA. *In vitro* recombination PCR was performed as described above except the hybridization and chain elongation steps were omitted. A band of 325 bp in length was amplified (Fig. 3A). The eluted fragment was then digested by the enzymes TaqI, RsaI, HindIII and EcoRI, and fractionated by gel electrophoresis (Fig. 3B). TaqI digestion gave 166, 110 and 49 bp fragments; RsaI resulted in 143, 129 and 59 bp, and EcoRI yielded 280 and 45 bp. HindIII did not digest the amplified band. The nucleotide sequence and restriction map for the predicted 325 bp FIX genomic-cDNA chimera are shown in Fig. 2 and are derived from published sequences (9,10). The data obtained above match the predicted sizes of the restriction fragments given in Table 1. The size and restriction map data conclusively demonstrate that the amplified fragment is the predicted FIX genomic-cDNA chimeric molecule.

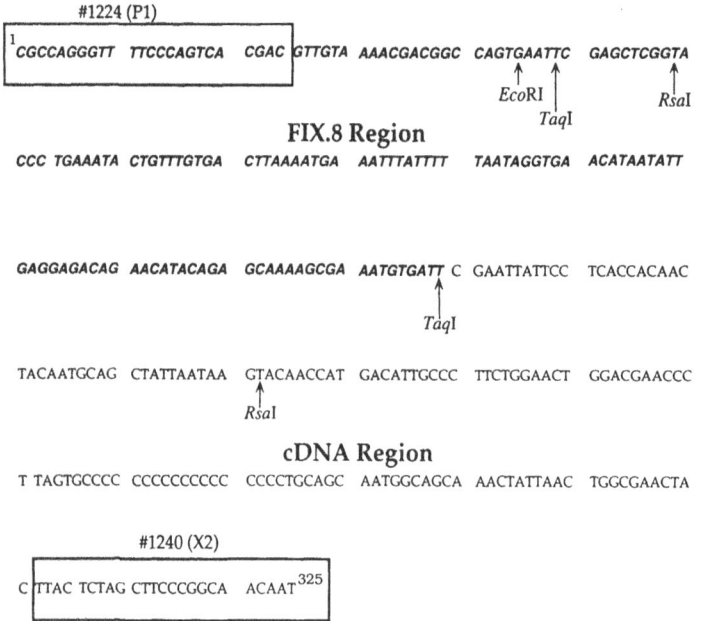

Figure 2. Predicted sequence and partial restriction map of FIX.8 genomic-cDNA chimera. This control data was generated by restriction analysis of published data on the FIX cDNA and FIX.8 clones. The portions of the chimera derived from the FIX.8 genomic clone are in bold print and the portion derived from the FIX cDNA are in normal print. The M13 primer #1224 and pBR322 primer #1240 are enclosed in boxes. EcoRI, RsaI and TaqI restriction sites are indicated by arrows.

Table 1. Predicted Restriction Data for The FIX.8-FIX cDNA Chimera

Enzyme	No. of Sites	Fragment Sizes bp
EcoRI,	1	45,280
HindIII	0	-
RsaI	2	59,123,143
TaqI	2	49,110,166

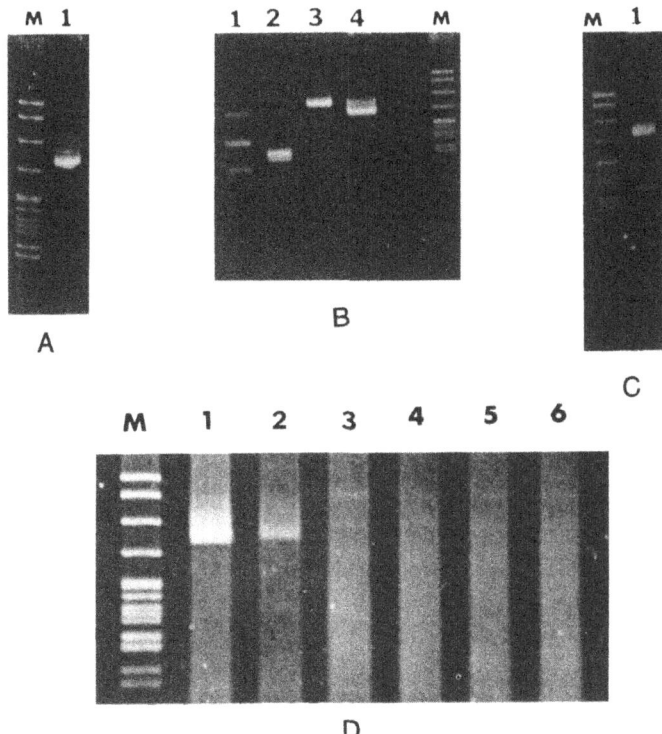

Figure 3. Photograph of ethidium bromide stained, 5% polyacrylamide gel of genomic-cDNA chimera and its restriction digestions. Lane M in all panels contains MspI cut pBR322 size marker. Starting from top of each marker lane the bands are 622, 527, 404, 309, etc. bp. Panel A: lane 1 is FIX.8-cDNA chimera (325 bp) generated by mixing FIX cDNA with FIX.8 DNA. Panel B, single digestions of the genomic-cDNA chimera were performed by TaqI, RsaI, HindIII, and EcoRI (lanes 1-4 respectively). Lane 1 contains, from the top, a partial digestion product, and 166, 110 and 49 bp fragments; lane 2 contains 143, 129 and 59 bp; lane 3 contains an undigested 325 bp chimera; lane 4 contains an undigested 325 bp chimera and 280 and 45 bp fragments. Panel C: lane 1 is the FIX.8-FIX cDNA chimera obtained from purified FIX.8 and the total cDNAs containing FIX cDNA. Fainter bands are unidentified products. Panel D: PCR products from DNA, generated from pools of colonies. Lanes 1-3 have DNA from genomic pool with FIX.8 DNA, lanes 4-6 have DNA from genomic pool only, lanes 1 and 4 have cDNA from 10K pool, lanes 2 and 5 have cDNA from 50K pool, lanes 3 and 6 have cDNA from 250K pool. All the cDNA pools contain FIX cDNA.

In our second experiment, we tested the capacity of the strategy to retrieve the FIX genomic-cDNA chimera from a sample with increased complexity in one component, the cDNA library (10^6 clones). At this level of complexity, the reassociation should be 80% complete at a C_0t value of approximately 300 (13). We used a Taq1 digested human liver cDNA library cloned in pBR322. This is an amplified library and may contain cDNAs corresponding to partial FIX cDNAs and other homologous cDNAs such as factor VII and factor X. As an internal control, the FIX cDNA was added into the total cDNA at a ratio of $1:10^4$ (w:w). The initial hybridization was at a 1:10 (w:w) ratio of FIX.8 genomic DNA to total cDNA. Upon *in vitro* recombination PCR the results showed a major product of 325 bp (Fig. 3C). There were also several faint bands (not characterized). The 325 bp band was eluted, cycle sequenced and demonstrated to be the control FIX chimera (data not shown).

Theoretically, one could use the DNA from a total genomic library and a cDNA library. At this complexity, genomic-cDNA hybrids should reassociate to 80% completion at 10^4 C_ot values (13). However, chimeric molecules would form in large numbers which would only be differentiated by cloning. This would essentially generate a partially normalized library of chimeric molecules with little or no utility. More practically, a variety of genomic sources of lesser complexity can be employed to restrict the number of chimeras formed from a cDNA library. These sources include large genomic DNA cloned in cosmids or Yeast Artificial Chromosomes (YACs), and microdissected chromosome libraries, among others. Under these reduced complexities, the hybridization would still be essentially 80% complete at approximately 300 C_ot. For individual cosmid or YAC clones' short fragment libraries, the number of hybrids produced would be limited because of a lesser number of genes in these clones. These hybrids might be differentiated by gel electrophoresis. However, in the situation of microdissected chromosome libraries, where as many as 300 genes may be present, there may be more PCR products than can be separated by gel electrophoresis. Therefore, to be applicable for microdissection libraries, we reasoned that if we split the cDNA library into pools of approximately 10^4 colonies and the genomic library into pools of 500 colonies, it would reduce the complexity and thereby reduce the number of products per reaction tube. This would allow for clear separation and direct sequencing of the chimeras.

In order to test this strategy, a third experiment was performed. From the genomic library, 500 antibiotic resistant colonies were selected; this was designated the Genomic Pool (GP). The human liver cDNA library (10) was plated on LB-tet plates and three pools were generated: one containing 10^4 independent colonies, a second with 5×10^4 independent colonies and a third with 2.5×10^5 independent colonies. These pools were designated 10K, 50K and 250K respectively. One FIX.8 colony (above) was added to GP to generate a positive control FIX containing genomic pool. One colony of the human FIX cDNA clone was added to each of the cDNA pools to generate positive control FIX cDNA pools. These would allow for the retrieval of the known control chimera at a level of complexity equivalent to a cosmid sublibrary. DNAs from the above positive and negative control pools were prepared using the alkaline lysis method (11) and processed through the scheme outlined in Fig. 1 at a 10:1 ratio of genomic to cut cDNA. Figure 3D shows the results. The control 325 bp fragment was retrieved from the 10K cDNA library. At a complexity of 50K colonies, the control chimera was easily identified. At a complexity of 250K colonies, the chimera was not clearly visible although an extremely faint signal can be seen. The chimera was eluted and was subjected to cycle sequencing. The sequence generated matched the predicted chimeric sequence (data not shown).

DISCUSSION

The results obtained in this study demonstrate the feasibility of isolating unknown coding sequences with genomic-cDNA chimeras. When the initial hybridization and chain elongation were employed, it was possible to retrieve the internal FIX cDNA control as a chimera from a total cDNA library. It is likely that the other faint bands are different chimeras formed because of the presence of several partial cDNAs, however we have not tested this. The above approach was also able to isolate the FIX sequence from a mixture of 500 genomic and up to 5×10^4 cDNA colonies. The fact that the 325 bp band was retrieved over the background confirms the feasibility of this pooling strategy. In our second experiment, we expected additional bands of comparable intensity due to cDNAs homologous to FIX cDNA (for example, factor VII and X cDNAs); however, we observed none.

Approximately 3-10% of cDNAs (14) contain repetitive DNA; therefore, cleavage with a restriction enzyme within the repetitive region in the cDNA will generate 3'OH groups which may prime various repetitive DNA sequences within the genomic clones and thus cause background amplification. In order to minimize this background, a restriction enzyme that cleaves repetitive DNA infrequently must be chosen. Additionally, the restriction site for the selected enzyme must also be present in all the cDNAs so that 3'OH groups within the coding regions can be generated to initiate the genomic-cDNA formation. TaqI was found to satisfy these conditions. According to GenBank release 74.0, one out of three members of highly repetitive (AluI) sequences (15) is cut by TaqI. There is one TaqI site for every 800 bp of the reported cDNA sequences, therefore, all cDNAs are likely to have at least one site, assuming an average size of 1 kb. Because 1/3 of the repetitive sequences found in cDNAs may be cleaved by TaqI, we anticipated a significant background with the use of this enzyme. Surprisingly, we did not generate a high background using pooled genomic DNA and cDNAs. This may be because some of the 3'OH ends generated by the digestion of repetitive sequences of the cDNAs may bind to the uncut repetitive region found in 2/3 of the cDNA, thus reducing the background in pools (which otherwise may generate high noise) by acting as primers on the genomic repetitive DNA.

Future investigators may want to try other restriction enzymes such as RsaI and MaeII to generate the 3'OH group in the cDNA. The possible advantage of these enzymes over TaqI is that the sites are not only less frequently present in repetitive DNAs but, in addition, they occur more frequently in the cDNAs. For example, an RsaI site occurs in one out of ten members of AluI family DNA and in 1/500 bp of cDNA. Conversely, restriction digestion of genomic DNA generates 3'OH groups which can then act as primers for uncleaved cDNA sequences.

The novel PCR strategy that we developed to isolate coding sequences from the given genomic areas has a number of inherent advantages: [1] Because a limited number of products should be generated in each reaction, cycle sequencing (16,17) or chemical sequencing (18,19) can be used to sequence the chimeras, following PCR and subsequent purification of the products by gel electrophoresis. Therefore, it will not be essential to subclone chimeras in order to simply generate sequences. [2] Similar to other procedures, there is some degree of normalization (20,21) in this method. This clearly enhances the selection of low abundant cDNAs. [3] The proposed approach is highly efficient, allowing the screening of roughly 1 Mb of genomic DNA (corresponding to 5×10^4 genomic clones with an average insert size of 200 bp) against a cDNA library (approximately 1×10^6 cDNA clones) with 100 sets of 96 reactions (standard on the 9600 thermalcycler) without the need for cloning. Because this approach is essentially a simple PCR-based procedure, it is far less laborious and more cost effective in the isolation of low abundant cDNAs. [4] Interference from pseudogenes, contaminating non-human sequences, and probably most homologous genes, is greatly reduced because the restriction sites are probably not conserved; therefore the 3' terminal bases will not hybridize to prime in the chain elongation. [5] Both the 5' and 3' ends of a given cDNA molecule can be obtained. [6] If different tissue-specific cDNA libraries are made in different vectors, amplifying the pools with a mixture of vector specific primers would increase the chance of identifying coding segments which are not expressed in all tissues. [7] In procedures where cDNA is cleaved with TaqI, if the genomic library is prepared by digesting the original genomic DNA with TaqI instead of chimeras, only cDNA portions can be generated. [8] The genomic-cDNA chimera can be digested further with the restriction enzyme that was used to cleave the original cDNA and which was subsequently used as a primer on an uncleaved cDNA library to obtain longer cDNAs. [9] If a cDNA library from a different tissue is used instead of a genomic library, it will generate chimeras of housekeeping genes. These sequences can then be used as probes to eliminate the housekeeping sequences when screening a cDNA library to obtain tissue-specific cDNAs. [10] Adapters can be ligated to the genomic fragments instead of preparing the libraries; however, the ligated sample must be restricted and used as primers against the cDNA library. [11] If genomic fragments are used as primers against a cDNA library, and the Vent DNA polymerase is used, full length cDNAs can potentially be amplified and isolated without screening.

We have used restriction enzyme digestion to generate the 3'OH group within the cDNA. We have other suggestions on possible ways to generate such groups. For example, the cDNAs can be partially digested by DNaseI; however, the clones would all have to be in the same orientation. If a population of clones exists in both orientations, it will generate more background amplification because the original clones in the library will be amplified with a single primer. Therefore, a library generated by the forced cloning method may be required when one uses DNaseI. Repetitive DNA may cause more interference; however, it can be suppressed by the use of the $C_o t1$ DNA fraction.

A second and similar approach is possible by simply mixing DNAs from the two libraries and adding a limited amount of dideoxy nucleotides in the same PCR reaction. This would generate multiple-nested, dideoxy-chain-terminated products. Subsequently, they would be allowed to reassociate to form hybrids and be extended by Vent DNA Polymerase (22). This enzyme removes the terminal dideoxy nucleotide and generates the chimera after the chain elongation. This has been verified in our laboratory by using the two control plasmids, FIX.8 and FIX cDNA. The results of this experiment yielded a chimera identical to the one obtained by the TaqI digestions (data not shown). The above techniques have an advantage of selecting homologous genes from different genomes, because of the generation of multiple primers.

A third possible approach would be to generate the 3'OH primers by reverse transcribing poly A RNA either in the presence of dideoxy nucleotides or by partial extensions. However, this approach would be unlikely to obtain 5' ends of the cDNA.

In summary, the method described in this report is a powerful tool for retrieving the coding sequences from unknown areas in genomic DNA and should be useful in isolating genes from YAC or cosmid clones and from chromosomal specific libraries.

REFERENCES

1. D. Auch and M. Reth, Exon trap cloning: using PCR to rapidly detect and clone exons from genomic DNA fragments, *Nucl. Acids Res.* 18:6743 (1990).
2. A.F. Buckler, D.D. Chang, S. K. Graw, J.D. Brook, D.A. Haber, P.A. Sharp, and D.E. Housman, Exon amplification: A strategy to isolate mammalian genes based on RNA splicing, *Proc. Natl. Acad. Sci. USA* 88:4005 (1991).
3. M. Hamaguchi, H. Sakamoto, H. Tsuruta, H. Sasaki, T. Muto, T. Sugimura, and M. Terada, Establishment of a highly sensitive and specific exon-trapping system, *Proc. Natl. Acad. Sci. USA* 89:9779 (1992).
4. S. Parimoo, S.R. Patanjali, H. Shukla, D.D. Chaplin, and S.M Weissman, cDNA selection: Efficient PCR approach for the selection of cDNAs encoded in large chromosomal DNA fragments, *Proc. Natl. Acad. Sci. USA* 88:9623 (1991).
5. M. Lovett, J. Kere, and L.M. Hinton, Direct selection: A method for the isolation of cDNAs encoded by large genomic regions, *Proc. Natl. Acad. Sci. USA* 88:9628 (1991).
6. J.G. Morgan, G.M. Dolganov, S.E. Robbins, L.M. Hinton, and M. Lovett, The selective isolation of novel cDNAs encoded by the regions surrounding the human interleukin 4 and 5 genes, *Nucl. Acids Res.* 20:5173 (1992).
7. B. Korn, Z. Sedlacek, A. Manca, P. Kioschis, D. Konecki, H. Lehrach, and A. Poustka, A strategy for the selection of transcribed sequences in the Xq28 region, *Hum. Mol. Genet.* 1:235 (1992).
8. D.A. Tagle, M. Swaroop, M. Lovett, and F.S. Collins, Magnetic bead capture of expressed sequences encoded within large genomic segments, *Nature* 361:751 (1993).
9. S. Yoshitake, B.G. Schach, D.C. Foster, E.W. Davie, and K. Kurachi, Nucleotide sequence of the gene for human factor IX (antihemophilic factor B), *Biochemistry* 24:3736 (1985).
10. P. Jagadeeswaran, D.E. Lavelle, R. Kaul, T. Mohandas, and S.T. Warren, Isolation and characterization of human factor IX cDNA: identification of *TaqI* polymorphism and regional assignment, *Somat. Cell. Mol. Genet.* 10:465 (1984).

11. H.C. Birnboim, A rapid alkaline extraction method for the isolation of plasmid DNA, *Meth. Enzymol.* 100:243 (1983).

12. R.K. Saiki, B. Hildebrand, S. Roberts, and P. Jagadeeswaran, Isolation and characterization of human blood-coagulation factor X cDNA, *Science* 239:487 (1988).

13. B.D Young and M.L.M. Anderson, Quantitative analysis of solution hybridization, *in*: "Nucleic Acid Hybridization-A Practical Approach" B.D Hames and S.J Higgins, ed., IRL, Oxford (1985).

14. J.M. Crampton, K.E. Davies, and T.F. Knapp, The occurrence of families of repetitive sequences in a library of cloned cDNA from human lymphocytes, *Nucl. Acids Res.* 9:3821 (1981).

15. J. Jurka and E. Zuckerkandl, Free left arms as precursor molecules in the evolution of Alu sequences, *J. Mol. Evol.* 33:49 (1991) and GenBank Release: 74.0 (1993).

16. F. Sanger, S. Nicklen, and A. R. Coulson, DNA sequencing with chain-terminating inhibitors, *Proc. Natl. Acad. Sci. USA* 74:5463 (1977).

17. M.A. Innis, K.B. Myambo, D.H. Gelfand, and M.A.D. Brow, DNA sequencing with *Thermus aquaticus* DNA polymerase and direct sequencing of polymerase chain reaction- amplified DNA, *Proc. Natl. Acad. Sci. USA* 85:9436 (1988).

18. A.M. Maxam and W. Gilbert, A new method for sequencing DNA, *Proc. Natl. Acad. Sci. USA* 74:560 (1977).

19. P. Jagadeeswaran and R.K. Kaul, Use of reverse-phase chromatography in the Maxam-Gilbert method of DNA sequencing: a step toward automation, *Gene Anal. Tech.* 3:79 (1986).

20. M.S. Ko, An 'equalized cDNA library' by the reassociation of short double-stranded cDNAs, *Nucl. Acids Res.* 18:5705 (1990).

21. S.R. Patanjali, S. Parimoo, and S.M. Weissman, Construction of a uniform-abundance (normalized) cDNA library, *Proc. Natl. Acad. Sci. USA* 88:5066 (1991).

22. P. Mattila, J. Korpela, T. Tenkanen, and K. Pitkanen, Fidelity of DNA synthesis by the *Thermococcus litoralis* DNA Polymerase-an exteremely heat stable enzyme with proofreading activity, *Nucl. Acids Res.* 19:4967 (1991).

IDENTIFYING AND DIRECTLY PURIFYING TRANSCRIBED ELEMENTS BY COINCIDENT SEQUENCE CLONING

A. J. Brookes

MRC Human Genetics Unit, Western General Hospital, Crewe Road, Edinburgh, United Kingdom, EH4 2XU

ABSTRACT

Coincident Sequence Cloning (CSC) is a strategy for genome analysis which involves the selective isolation of sequences on the basis of their shared existence in a pair of otherwise distinct DNA mixtures. Two related methodologies, Hybrid Fishing CSC (HF-CSC) and End Ligation CSC (EL-CSC), have been developed which are able to handle highly complex DNAs and can enrich for coincident elements by 10^3-10^5 fold and 10^6-10^7 fold respectively. Using these methods to 'integrate' genomic DNA with cDNA provides an effective way of identifying transcribed sequences within defined genomic domains. The immense enrichment power of the EL-CSC procedure permits target genes to be directly purified to homogeneity.

INTRODUCTION

Background

Coincident Sequence Cloning (CSC) is a general term for methodologies that attempt to selectively clone DNA sequences by virtue of their shared presence in a pair of otherwise distinct DNA mixtures (1-6). This basic concept is illustrated in Fig. 1a. Effective CSC procedures could be applied to address numerous challenging problems. By 'integrating' genomic DNAs and cDNAs in various combinations and by focusing studies both within and between species one could rapidly isolate many types of important sequences including novel markers, genes and conserved motifs. The application of CSC to the integration of genomic DNA with cDNA provides an empirical approach to the recovery of transcribed sequences from a given physical domain (7,8). This idea is illustrated in Fig. 1b. This report describes two intimately related and technically simple CSC methodologies, termed Hybrid Fishing CSC (HF-CSC) and End Ligation CSC (EL-CSC), and demonstrates their utility as methods for gene identification.

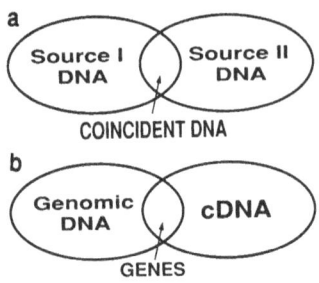

Figure 1. The principle of coincident sequence cloning. a) the general principle, and b) the application to gene identification.

Human genome analysis has progressed to a stage where gene 'hunting' has become a high priority, both in order to find specific disease genes as well as to construct long range transcription maps. The limited efficiency of conventional technologies has necessitated the development of more powerful alternatives. Two new and mutually complementary strategies, exon 'trapping' (9-13) and cDNA hybrid selection (14-19), promise to be of great utility. These systems attempt to identify and recover the genic elements from genomic DNA. While the first endeavours to find the genes containing introns, the second isolates genes on the basis of their representation in a chosen cDNA resource. Both HF-CSC and EL-CSC are forms of cDNA hybrid selection, the principle of which involves using a genomic resource to physically isolate (fish out), by base pairing, encoded genes from a mixed cDNA resource. It is now clear, as illustrated by the considerable range of published effective cDNA selection procedures (8, 14-19), that this approach to gene finding is extremely robust. Hereafter, we call cDNA selection, cDNA 'fishing.'

The HF-CSC procedure is designed to physically isolate coincident restriction enzyme digestion products that contain any significant degree of sequence homology. EL-CSC additionally requires such elements to have both length similarity and absolute end sequence identity. This practical difference is reflected in the enrichment factors achieved by the two methods, namely 10^3-10^5 fold and 10^6-10^7 fold respectively. As detailed below, the greater specificity of EL-CSC results from the inclusion of a high stringency end ligation step. This has two consequences. Firstly, it permits the recovery of genes in an essentially <u>pure</u> form and contrasts with HF-CSC (and all other cDNA "fishing" methods) where genes are significantly <u>enriched</u> but still have to be identified amongst a background of non-coincident products. Secondly, in order to be successful, the method requires that there be an identical restriction fragment within genomic and cDNA versions of a target gene. This problem can be overcome, and the recovery of intra-exonic fragments maximised, by performing parallel studies utilising several restriction enzymes that have 4 bp recognition sequences. In this way experiments are deliberately focused upon DNA fragments of 0.1-0.5 kb. These are short enough to be contained within single exons (3' untranslated regions can be many kb in length and are usually greater than 0.6 kb (20)) and are ideally sized for the polymerase chain reaction (PCR) steps of the procedure.

Considerations and Expectations

The detailed design of the HF-CSC and EL-CSC procedures is given in Methods. The underlying principle involves the selective isolation of inter-resource duplexes (heteroduplex) formed between the input genomic and cDNAs. One of the most important factors in a CSC based study will be the nature and quality of the input DNAs. This material must be modified as described below. The primary genomic resource can be comprised of phage, cosmid, plasmid or YAC recombinants and even uncloned DNA. The removal of cloning vehicle sequences is unnecessary but one must obviously avoid using genomic and cDNA resources cloned in sequence related vectors. Once modified, the concentration per unit length of the genomic component needs to be high in comparison to that of the cDNA in order to effectively drive the formation of heteroduplex. For this reason, we would normally limit studies to the order of 1-2 Mb. For chromosome scale studies, one could attempt to capture single exons for many different genes by analysing pools of random clones (possibly microdissected from a single region), alternatively the DNA could be fractionated either by size or sequence content. The choice of cDNA resource is critical because its content will predetermine the range of genes available for isolation. Our best results have been obtained using uncloned oligo-dT primed cDNA made from poly-A selected mRNA. This needs to be bulked up to several micrograms by some form of universal PCR (21-23). For global searches we recommend that mRNA from many different sources should be combined; however, when a specific disease gene is being sought, an affected tissue might provide a better starting point for mRNA isolation.

Possibly the most difficult aspect of a CSC based gene hunt is the phase of product analysis. In common with other cDNA "fishing" procedures, HF-CSC will inevitably give rise to a product library within which true genes are represented at anywhere from less than 1% to maybe as much as 50%. Positive control genes within the genomic resource are easy to locate and thereby assess for levels of enrichment; however, this does not assist in the identification of novel products. Although the pro's and con's of alternative schemes for product analysis are discussed below, it is worth highlighting at this point the spectre of low copy dispersed repeats (24,25). These repeats comprise many different elements which are cumulatively abundant in the genome and often represented in mRNA/cDNA (probably in 5' and 3' untranslated regions). Unlike their high copy number counterparts, their low representation prevents effective blocking by repeat sequence preannealing prior to heteroduplex formation. Consequently, alleles of these 'coincident by default' sequences, originating from various genomic loci, may be isolated in hybrid "fishing" studies and can be confused with real products. Hybridisation analysis using a product probe containing a low copy repeat will probably reveal a 'unique' localisation within the genomic resource, a positive signal in the cDNA and a multiple band whole genome pattern easily mistaken for that of a gene family. Therefore, before declaring a product to be real, it is essential that it and its genomic counterpart be sequenced to enable pairwise and database comparisons. Alternatively, an independent high stringency (sufficient to give one signal) genome mapping study should be performed.

Problems such as those described above can be dramatically reduced by employing the EL-CSC procedure. Low copy repeats, as well as background products generally, are selected against by the greater stringency and increased enrichment afforded by this method. The implications for simplified product analysis and more rapid gene identification are obvious. To date, the results of initial applications of this procedure support our belief that EL-CSC could emerge as a major force in the armory of procedures for transcribed sequence identification.

METHODS

DNA Integrations

Schematic representations of the HF-CSC and EL-CSC procedures are shown in Figs. 2a and 2b, respectively. These diagrams summarise the four steps involved in producing enriched or purified gene sequences (in the form of a PCR reaction product) from input DNA resources. The two procedures are very simple to perform and are methodologically extremely similar. For this reason they shall be described together.

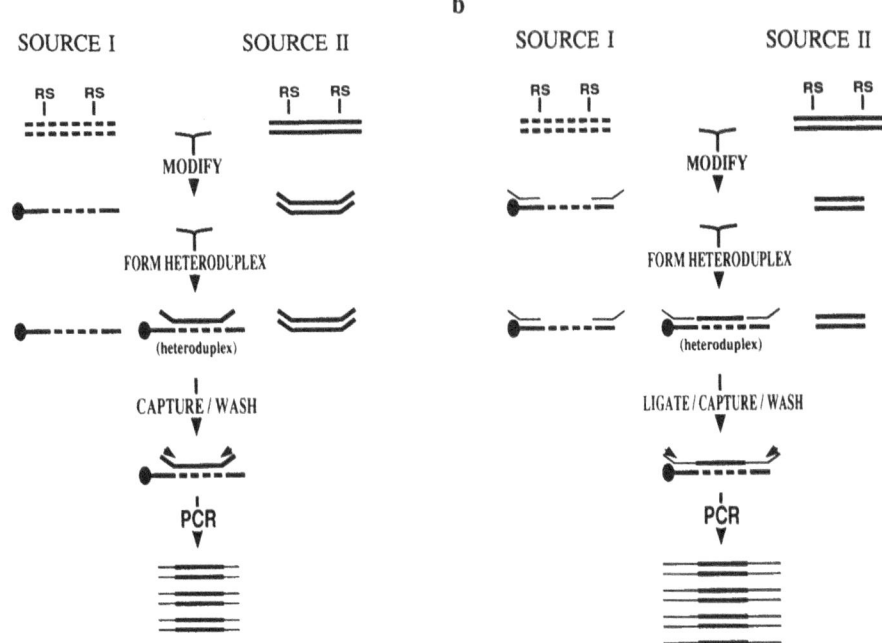

Figure 2. Coincident sequence cloning methodologies. a) Hybrid Selection CSC, and b) End Ligation CSC. Following inter-resource duplex formation, non-coincident DNAs are shown in the outer panels and coincident DNAs in the center panel. Abbreviations and symbols are as follows: long dotted lines, input source I DNA; long solid lines, input source II DNA; RS, restriction site; solid ellipse, biotin; horizontal and sloping short thick lines, distinct catch-linker sequences; angled thin lines, capture oligos; and arrow heads, PCR primers.

The first step of both procedures involves input DNA modification. This entails the linking of biotin moieties to the molecules of the source I DNA (genomic DNA). This is conveniently achieved by ligating synthetic oligonucleotides (linkers) to restriction fragment ends and then performing PCR via these sequences using 5' biotinylated primers. Source II DNA (cDNA) is cleaved with the same enzyme as was used on source I. In HF-CSC only, the cDNA is ligated to a second set of linkers or, if using a cloned resource, it is PCR amplified using vector sequence primers. The second step is heteroduplex formation for which both modified resources are denatured, mixed and left to reanneal to a C_ot of 50 - 100. Preblocking of high copy repeat sequences is required and may be achieved by preincubating the genomic and/or cDNAs with human C_ot1 DNA before mixing the two resources. In the third and fourth steps, coincident sequences are specifically recovered. Source I material is captured by physical purification using its biotin moieties along with streptavidin coated magnetic beads (Dynal M-280 beads). Following appropriate stringency washes to remove source II DNAs that are weakly bound to source I molecules, those cDNA sequences present as strong heteroduplexes are isolated by denaturation and elution.

In HF-CSC studies, the common end sequences of the eluted source II molecules are used to direct a PCR which specifically amplifies the small amount of recovered coincident DNA. Problems may occur because some source II elements contain motifs related, but not identical, to sequences of source I DNAs (e.g. low copy repetitive elements). These sequences are potentially able to form imperfect heteroduplexes that can be sufficiently stable to survive the stringent washes. Furthermore, cDNAs may interact with non-specific binding sites on the M-280 beads and, by this route, become eluted as if correctly recovered. Being cDNA derived, all such falsely recovered molecules become

PCR amplified in direct competition with genuine products. In practice, this procedure typically produces a complex cDNA product enriched for coincident gene sequences by 10^3-10^5 fold.

In EL-CSC the source II DNA molecules are devoid of common end sequences for PCR amplification. Instead, linkers (capture oligos) which are to perform this role, are specifically ligated to coincident source II molecules while present as perfectly end-matched heteroduplexes. In an additional step over HF-CSC, before product elution is performed, the capture oligos are appropriately positioned by base-pairing with the common end sequences of the source I molecules. A high stringency ligation reaction is then used to join these oligos to their cDNA target. Source II molecules that survive the stringent washes, but are either non-specifically bound or in poorly base paired heteroduplexes, will not acquire the capture oligo end sequences. Hence, this simple additional manipulation furnishes EL-CSC with significantly higher specificity than HF-CSC. A PCR step, directed towards the capture oligo sequences, finally recovers coincident gene fragments. Typical enrichment factors achieved by this procedure are in the range 10^6-10^7.

PCR products from CSC studies can be remodified as necessary and reprocessed through additional enrichment cycles. When this is done, the further level of enrichment obtained is trivial in comparison to that of the original cycle. Typical increases are often only a few fold and occasionally there may be no or even a negative change. However, what invariably does change (decreases) is the average size of the recovered material. Smaller fragments have an intrinsic advantage in PCR and amplify at the expense of larger molecules. Furthermore, internal non-specific PCR priming events tend to cause some shortening of longer molecules. These effects can bring the average product size down dramatically (making subsequent analysis difficult) and may delete genuine target sequences. Arguably, these disadvantages outweigh the minimal gains of multiple cycling.

Product Cloning and Analysis

To simplify the cloning of CSC products it is usual to employ PCR primers that contain internal restriction sites that are relatively rare in mammalian DNA. It is important to clone using these engineered sites, rather than the sites of initial linker addition, since during the various PCR steps there is always a degree of non-specific internal priming, as described above. This phenomenon can eliminate a linker site, and adjacent sequences, from a considerable proportion of the larger amplifying molecules. In both HF-CSC and EL-CSC studies, it is usual to clone the total amplified material to produce a 'product library'. This will generally be of considerable complexity (gel analysis can allow an estimation of this parameter) including genes of interest at an unknown frequency. An ordered reference set ('gridded product library') of 100-500 such clones is sometimes prepared. Further analysis then involves the rationalisation of these primary libraries into a minimal set of non-redundant recombinants. Ways in which this might be accomplished are considered in the Discussion along with strategies for the identification of genuine products.

A more direct approach to product cloning is available in the case of EL-CSC studies. Since the PCR product of this method often comprises a series of discreet bands upon agarose gel electrophoresis, it is a trivial task to directly isolate these fragments for ligation into a plasmid vector. This surmounts the difficulties of library assortment since individual recombinants from different clonings immediately represent distinct gene candidates. In some situations it might even be possible to analyse gel purified EL-CSC products directly, thus completely bypassing the cloning obstacle.

RESULTS

The power and utility of HF-CSC and EL-CSC are best illustrated by a brief description of a series of typical applications. The following examples serve to highlight some important considerations.

HF-CSC:Enrichment Potential

The 'enrichment factor' of a cDNA selection study is a measure of the efficiency of recovery of a target gene sequence relative to the average efficiency for other non-coincident cDNAs. It does not consider representation levels in the genomic DNA. The enrichment factor is calculated by dividing the fractional mass of a target cDNA in the total CSC product by its fractional mass in the input cDNA. For low level products this approximates to the percentage of target cDNA in the total CSC product divided by the percentage in the input cDNA.

Observed enrichment factors for different gene fragments will often vary considerably, even within a single experiment. Possible reasons for this might include PCR biases, differences in initial representation and variations in heteroduplex stability. Observed enrichments for HF-CSC are of the order of 10^3-10^5 fold.

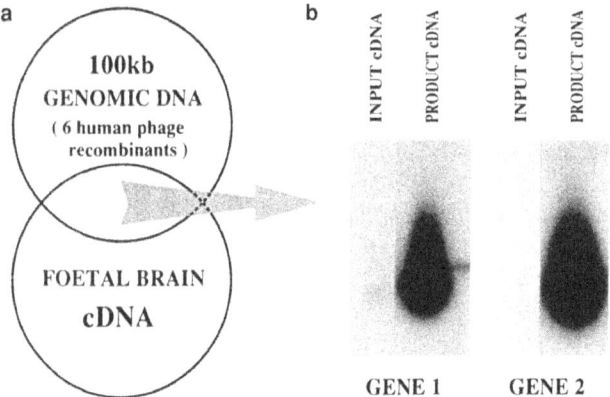

Figure 3. Example enrichments for an HF-CSC based gene hunt. a) Representation of the experiment, and b) Autoradiographic data obtained when comparable amounts of input and product cDNA mixtures were hybridised with probes for two genes encoded by the input genomic DNA.

Figure 3 shows what might be expected from a typical HF-CSC study in which a pool of 6 human phage recombinants (~ 100 kb combined insert sequence, vector arms not removed), are integrated with cDNA from a tissue such as the human foetal brain. The autoradiographs of Fig. 3 show the enrichment attained for fragments of two target genes. Parallel integrations, employing different genomic DNA, did not enrich these sequences.

HF-CSC:Isolation of Novel Genes

It is difficult to estimate the number of genes one might recover from any given genomic resource because chromosomal regions vary enormously in their density of encoded genes (26). As a global average, there might be one gene every 30 kb which in

50% of cases could be represented in a given cDNA resource. In cDNA selection experiments, genes will often be recovered as several fragments that may or may not be overlapping. By examining a dispersed set of genomic clones it is likely that each recovered fragment will originate from a different gene. In this way one can maximise the number of genes identified for a given quantity of genomic DNA. Applying this principle to 85 dispersed short fragment clones gives the result shown in Fig. 4. Here the genomic resource contained about 120 kb of non-overlapping DNA. Integration was with primary human foetal brain cDNA using the HF-CSC methodology. Figure 4b indicates various PCR product regions from which an incomplete analysis has so far identified 5 novel genes.

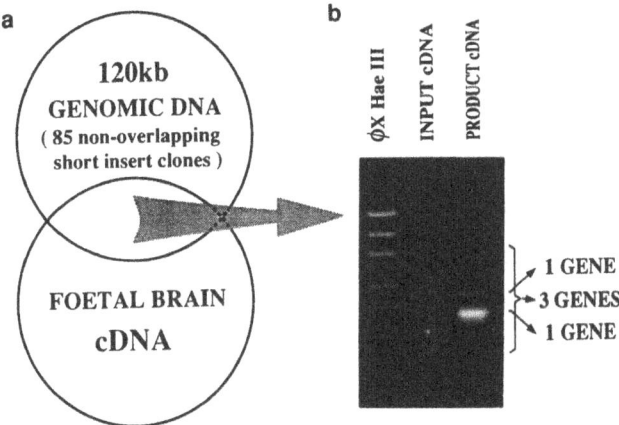

Figure 4. Example identification of novel genes by HF-CSC. a) Representation of the experiment, and b) PCR product of the study showing band and smear origins for 5 novel coincident genes.

EL-CSC:Enrichment Potential

The great enrichment potential of EL-CSC is best illustrated by a model experiment which evaluates specific and non-specific recovery rates for both EL-CSC and HF-CSC. This experiment attempts to recover a λ phage fragment spiked into complex DNA mixtures. Source I DNAs comprise two 40 kb human cosmids and an equal mass of total human DNA. 'Positive' source I samples also include the target λ fragment at a molarity per kb equivalent to that of the cosmid sequences. This resource therefore represents the full complement of human repeat sequences. Source II DNAs comprise total human genomic DNA plus the target λ fragment (added such that it constituted 10^3 of the total resource). The source II material was restriction enzyme digested to permit EL-CSC recovery. For HF-CSC integrations the target λ DNA of source II, but not the genomic material, was ligated to PCR linkers.

'Positive' (λ DNA in both resources) and 'negative' (λ DNA in source II only) integrations were performed using the HF-CSC and EL-CSC methodologies in parallel. Figure 5 shows the final PCR products. Although the target fragment is recovered by both positive integrations, the selectivity of this recovery (the difference in signal between the positive and negative integrations) is vastly superior for EL-CSC. Further hybridisation based analysis of the positive EL-CSC integration shows that the ratio of recovered target to recovered background is at least 10^3. This is in perfect accord with observations from actual investigative applications which indicate that the EL-CSC methodology achieves an enrichment factor of 10^6-10^7.

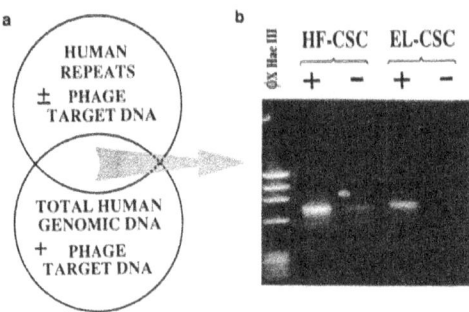

Figure 5. Comparison of HF-CSC and EL-CSC enrichment. a) Representation of the experiment, and b) PCR products for 'positive' (+) and 'negative' (-) integrations as described in the text.

HF-CSC ; Isolation of Novel Genes

Due to the substantial enrichment power of EL-CSC, studies employing this methodology are found to give rise to product libraries in which the vast majority of clones represent genuine coincident recombinants. This is the case, both in cDNA selection studies and in genomic-genomic integrations. The PCR product of a typical EL-CSC based gene hunting experiment is shown in Fig. 6. This investigation employed 230 kb of genomic sequence within plasmid and cosmid recombinants. Integration was performed with human foetal brain cDNA. This example demonstrates nicely the typical appearance of EL-CSC PCR products when analysed by agarose gel electrophoresis. Distinct bands are seen with strikingly little, if any, background smear. Upon cloning and analysis, the bands of the given example were found to be coincident gene fragments which had been essentially <u>purified</u> by the EL-CSC methodology.

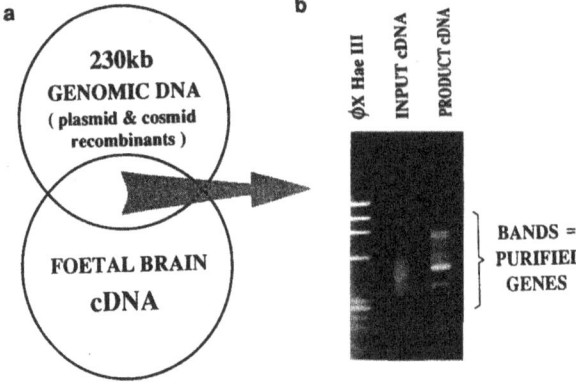

Figure 6. Example identification of novel genes by EL-CSC. a) Representation of the experiment, and b) PCR product of the study showing distinct bands that represent purified novel coincident genes.

118

DISCUSSION

Product Analysis

Practical applications have shown that the HF-CSC and EL-CSC methodologies are effective and robust. Of the various steps they entail, it is the process of product analysis that can be the most troublesome. This is particularly so in the case of HF-CSC because of its lower enrichment potential. There are several alternative strategies one might employ for product library rationalisation and subsequent genuine product identification. These shall be considered.

A complete analysis of a product library involves the sorting of all clones into groups of identical and overlapping members. An effective but laborious way to accomplish this is to employ randomly chosen isolates as probes for serial exclusion hybridisations of the whole library. When dealing with complex libraries, an exhaustive analysis by this approach would not be a reasonable option. A simpler and more effective route to the same end is PCR based fingerprint analysis of 200 library members (27). This permits unique clone sets to be rapidly identified and their representation scored at and above the level of 0.5%. Having defined a representative set of clones one then has to examine their individual characters. This can involve a range of standard technologies for gene analysis (zoo blots, Northern blots and cDNA library screening) and mapping (Southern blotting). However, each candidate 'real' product must be convincingly shown a) to be a unique mRNA sequence, and b) to exist in both starting DNA resources. As discussed in the Introduction, sequencing and independent mapping studies are critical in these respects. Finally, to conclusively demonstrate the genic nature of a recovered product (and similarly for any conventional cDNAs subsequently isolated) one needs to demonstrate either intron splicing or the presence of a polyA tail.

A far simpler strategy for product analysis does not involve the sorting of a product library. Efforts can be directly focused upon candidate genic clones. These are identify by simple hybridisation of primary product library platings with probes derived from the input genomic DNA. To obtain a clear signal to noise in this type of screening each target sequence needs to represent at least 1% of the complex probe. Genuine recovered fragments could be as short as 0.1 kb and so it is best to conduct serial screenings using probes of 10 kb total length. Any hybridising recombinants should be immediately sequenced to enable low copy repeat identification, database searching and the detection of duplications and overlaps between individual isolates.

Artefacts

It is inevitable that some false products will be derived in any type of cDNA selection study. Some of these are isolated by virtue of imperfect but significant homologies between the input DNAs. These would include pseudogenes and, as discussed above, repeats elements. When employing YAC genomic material, ribosomal sequences will also be recovered because of their high degree of evolutionary conservation. This applies even if the representation of yeast DNA is minimised by excising the YAC from a preparative pulsed field gel. Preblocking of modified input DNAs with $C_o t1$ human DNA, and yeast DNA where appropriate, is essential in order to reduce this type of background.

Other false products can arise through accidental contaminations during sample handling. These can have two origins. Firstly, a lack of care in setting up PCRs or handling amplified material can cause DNA sequences from completed experiments to reappear in ongoing studies. Secondly, the contamination of any DNA whatsoever, into solutions used to process both input resources, will directly lead to there being false coincident molecules available for recovery. This second type of artefact could be particularly difficult to distinguish from genuine clones and might lead to the wasting of considerable time and effort. Routine sequencing of CSC products greatly simplifies the identification of artefactual products since most laboratory contaminations and previously recovered clones will be immediately recognisable.

CONCLUSIONS

HF-CSC and EL-CSC represent two empirical approaches to the identification of transcribed sequences within genomic DNA. The EL-CSC methodology provides both greater specificity and higher enrichment than the HF-CSC alternative but depends upon the existence of a common restriction fragment within genomic and cDNA versions of a target gene. Appropriate experimental design can overcome this limitation in most circumstances. Neither strategy has yet been pushed to its limit with respect to input genomic DNA complexity. There are several novel applications yet to be exploited, for example, directed searches for genes belonging to families and possible studies based upon conservation between species. A CSC based approach to genome analysis, both for gene identification and in a more global sense, has an enormous potential which can now begin to be realised. To this end further technical details of the described methods will be made available upon written request.

ACKNOWLEDGEMENTS

I wish to express my gratitude to those who provided various resources described in this report, in particular Drs K. Davies, K. Morrison, D. Blake, A. Tunnacliffe, B. Ponder, C. Gosden, W. Muir, A. Weith and V. VanHeyningen. I gratefully acknowledge the important contributions made by E. M. Slorach, S. J. Qureshi, R. Devon and B. J. Stevenson. Finally I would especially like to thank Dr D.J. Porteous, in whose laboratory this work was conducted, for his valuable guidance and enthusiastic support.

REFERENCES

1. A.J. Brookes and D.J. Porteous, Coincident sequence cloning: A new approach to genome analysis, *Trends BioTechnol.* 10:40 (1991).
2. A.J. Brookes and D.J. Porteous, Coincident sequence cloning, *Nucl. Acids Res.* 19:2609 (1991).
3. A. Charalampos and P.J. deJong, Coincidence cloning of Alu PCR products, *Proc. Natl. Acad. Sci.* 88:6765 (1991).
4. D.M.D. Bailey, N.P. Carter, D. deVos, M.A. Leversha, M.T. Perryman, and M.A. Ferguson-Smith, Coincidence painting: A rapid method for cloning region specific DNA sequences, *Nucl. Acids Res.* 21:5117 (1993).
5. A.J. Brookes and D.J. Porteous, Coincident cloning and gene targeting applied to human chromosome 11, *Am. J. Hum. Genet.* 47:abs.965 (1990).
6. A.J. Brookes, E.M. Slorach, S.J. Qureshi, R. Devon, W. Muir, C. Gosden, J. Moore, A. Tunnacliffe, B.A.J. Ponder, and D.J. Porteous, Gene identification and genome analysis by coincident sequence cloning, *Am. J. Hum. Genet.* 53:abs.1272 (1993).
7. J.E. Parish and D.L. Nelson, Methods for finding genes a major rate limiting step in positional cloning, *GATA.* 10:29 (1993).
8. A.J. Brookes, E.M. Slorach, S.J. Qureshi, R. Devon, M. Thompson, A. Weith, W. Muir, C. Gosden, and D.J. Porteous, Isolation of bBrain expressed sequences from around a balanced t(1;11)(q43,q21) translocation linked to schizophrenia, *Psych. Genet.* 3:122 (1993).

9. G.M. Duyk, S. Kim, R.M. Myers, and D.R. Cox, Exon trapping: A genetic screen to identify candidate transcribed sequences in cloned mammalian genomic DNA, *Proc. Natl. Acad. Sci.* 87:8995 (1990).

10. A.J. Buckler, D.D. Chang, S.L. Graw, J.D. Brook, D.A. Haber, P.A. Sharp, and D.E. Housman, Exon amplification: A strategy to iIsolate mammalian genes based on RNA splicing, *Proc. Natl. Acad. Sci.* 88:4005 (1991).

11. D. Auch and M. Reth, Exon Trap Cloning: Using PCR to rapidly detect and clone exons from genomic DNA fragments, *Nucl. Acids Res.* 18:6743 (1990).

12. G.B. Hutchinson and M.R. Hayden, The prediction of exons through an analysis of spliceable open reading frames, *Nucl. Acids Res.* 20:3453 (1992).

13. M. Hamaguchi, H. Sakamoto, H. Tsuruta, H. Sasaki, T. Muto, T. Sugimura, and M. Terada, Establishment of a highly sensitive and specific exon-trapping system, *Proc. Natl. Acad. Sci.* 89:9779 (1992).

14. S. Parimoo, S.R. Patanjali, H. Shukla, D.D. Chaplin, and S.M. Weissman, cDNA selection: efficient PCR approach for the selection of cDNAs encoded in large chromosomal DNA fragments, *Proc. Natl. Acad. Sci.* 88:9623 (1991).

15. M. Lovett, J. Kere, and L.M. Hinton, Direct selection: A method for the isolation of cDNAs encoded by large genomic regions, *Proc. Natl. Acad. Sci.* 88:9628 (1991).

16. K. Abe, Rapid isolation of desired sequences from loner linker PCR amplified cDNA mixtures: Application to identification and recovery of expressed sequences in cloned genomic DNA, *Mammal. Genome* 2:252 (1992).

17. B. Korn, Z. Sedlacek, A. Manca, P. Kioschis, D. Konecki, H. Lehrach, and A. Poustka, A strategy for the selection of transcribed sequences in the Xq28 region, *Hum. Molec. Genet.* 1:235 (1992).

18. A. Forster and T.H. Rabbitts, A method for identifying genes within yeast artificial chromosomes: Application to isolation of MLL fusion cDNAs from acute leukaemia translocations, *Oncogene* 8:3157 (1993).

19. D. Vetrie, I. Vorechovsky, P. Sideras, J. Holland, A. Davies, F. Flinter, L. Hammarstrom, C. Kinnon, R. Levinsky, M. Bobrow, C.I. Edvard Smith, and D.R. Bentley, The gene involved in X-linked agammaglobulinaemia is a member of the src family of protein-tyrosine kinases, *Nature* 361:226 (1993).

20. R.C. Levitt, Polymorphism in the 3' untranslated region of eukaryotic genes, *Genomics* 11:484 (1991).

21. M.S.H. Ko, S.B.H. Ko, N. Takahashi, K. Nishiguchi, and K. Abe, Unbiased amplification of a highly complex mixture of DNA fragments by "lone linker" tagged PCR, *Nucl. Acids Res.* 18:4293 (1990).

22. D.H. Johnson, Molecular cloning of DNA from specific chromosomal regions by microdissection and sequence-independent amplification of DNA, *Genomics* 6:243 (1990).

23. L. Zhang, X. Cui, K. Schmitt, R. Hubert, W. Navidi, and N. Arnheim, Whole genome amplification from a single cell: Implications for genetic analysis, *Proc. Natl. Acad. Sci.* 89:5847 (1992).

24. D.J. Kaplan, J. Jurka, J.F. Solus, and C.H. Duncan, Medium reiteration repetitive sequences in the human genome, *Nucl. Acids Res.* 19:4731 (1991).

25. J. Jurka, D.J.Kaplan, C.H. Duncan, J. Walichiewicz, A. Milosavljevic, G. Murali, and J.F. Solus, Identification and characterisation of new human medium reiteration repeats, *Nucl. Acids Res.* 21:1273 (1993).

26. J.M. Craig and W.A. Bickmore, Chromosome bands - Flavours to savour, BioEssays,:15:349 (1993).

27. B.J. Stevenson, D.J. Porteous, and A.J. Brookes, A simple strategy for rapidly fingerprinting large numbers of small clones, *Techniq.* 3:167 (1991).

FINDING CANDIDATE GENES BY PREPARATIVE IN SITU HYBRIDIZATION

J. C. Hozier[1], L. M. Davis[1], P. D. Siebert[2], K. Dietrich[3] and M. C. Paterson[3]

[1]Applied Genetics Laboratories, Melbourne, Florida
[2]Clontech Laboratories, Palo Alto, California
[3]Cross Cancer Institute and University of Alberta, Edmonton, Alberta, Canada

ABSTRACT

We have developed preparative *in situ* hybridization (Prep-I.S.H.) of complex DNA populations to mitotic chromosomes as a means of generating chromosome region-specific cDNA subpopulations. Prep-I.S.H. is a combination of two cytogenetic techniques: hybridization of DNA molecules to mitotic chromosomes, and chromosome microdissection. Here we provide technical details of the procedure and describe its application to the human chromosome 11q22-23 region containing the ataxia telangiectasia genes. Prep-I.S.H. has a number of applications in studies of gene expression and genome organization, including efficient cytogenetic sorting of tissue-specific cDNAs and in dramatically reducing the number of candidate genes to aid in gene discovery. Prep-I.S.H. provides a technical approach to a "positional candidate" strategy for gene discovery that differs from positional cloning by focusing on analysis of the genes in a chromosomal region of interest rather than on detailed analysis of physical and genetic maps.

INTRODUCTION

Gene discovery has, over the past decade, become an important function of molecular genetics. This is especially true in the context of human genetic diseases, since the discovery of their underlying gene defects has the potential for defining all of the physiological targets for diagnostic, preventive and therapeutic medicine in the future.

The chain of events leading to the discovery of a new disease gene usually begins with the assignment of the gene to a single chromosome or to a specific sub-chromosome region (usually a cytogenetically distinguishable chromosome band) by studying the segregation of the disease phenotype with selected genetic markers in affected families. From this point (Fig. 1) it has proven possible to narrow the region of the search for a gene from the 10 million base pairs (Mb) of the average chromosome band, down to a region of 100 kilobases or less containing only the gene of interest. This process, now

Identification of Transcribed Sequences, Edited by
U. Hochgeschwender and K. Gardiner, Plenum Press, New York, 1994

known as positional cloning, can proceed "unadulterated by any influences of biochemistry, cell biology or physiology" (1) and eventually can find a gene based on position in the genome alone. The successes of positional cloning have been reviewed recently (1,2) and the approach works well when combined with techniques at the end of the positional cloning search to identify transcribed sequences in the resultant genomic clone(s). The techniques that are useful for identifying transcribed sequences include direct selection of cDNAs on immobilized YACs, exon trapping from cosmids, computer searching from sequence data for a gene-like DNA sequence, and/or zoo blotting. These techniques then supply candidate sequences for studies of alterations that segregate with the disease phenotype. Proof that the isolated candidate gene is in fact the gene of interest usually relies on analyzing mutations, which must be present in affected individuals and absent in normals. Though straightforward in concept, the successful application of positional cloning to actual gene discovery projects has been helped by the presence of disease-related gross genetic rearrangements to guide the search for the specific genomic region of interest, and by the availability of large pedigrees that allow very precise genetic mapping (1,2). There also are recent examples of special rearrangements (ex: augmentations of trinucleotide repeats in Huntington's disease and fragile-X syndrome) that directly mark genes in the positional approach as likely genes of interest. These special cases may become less prevalent with time.

We have developed a strategy, which we call "preparative *in situ* hybridization" (Prep-I.S.H.), for isolating the cDNA clones in a given cDNA library that map to a particular chromosome region. With this strategy, isolating the gene responsible for a particular disease requires only a chromosome location, a tissue expressing the gene of interest, and a reasoned ability to identify the gene of interest from a small collection of genes sharing the same tissue and chromosome location. The Prep-I.S.H. procedure offers distinct advantages over other gene-finding strategies.

Preparative In Situ Hybridization

The procedure that is outlined here incorporates useful elements of current gene identification strategies (e.g.: direct selection of cDNAs by genomic cDNA, except that selection is done by mitotic chromosomes, as explained below), without requiring expression by target DNA, to produce cDNA clones mapping to cytogenetically defined regions. In essence, the procedure allows for the selection of a subset of clones from a cDNA library by the chromosome region to which they hybridize. The outcome of the procedure is a selection of clones with a cytogenetically defined location.

We call this new technology preparative *in situ* hybridization to distinguish its purpose from the analytical technique of fluorescent *in situ* hybridization (FISH). Prep-I.S.H. is the merging of two existing technologies into a uniquely powerful tool for selecting a subset of clones from a cDNA library by virtue of their hybridization to specific regions of the genome. The first of these existing technologies is essentially identical to the fluorescent *in situ* hybridization (FISH) procedures that are commonly used for analytical gene mapping purposes. The second procedure, microdissection of chromosomes and chromosome bands, is somewhat more specialized, although practiced in a few labs, including ours. The Prep-I.S.H. technology relies on the ability of the insert fraction of a cDNA library or other complex population of cDNAs to hybridize to denatured mitotic chromosomes fixed on slides, on the ability of the chromosome segments to be physically dissected using micro techniques, and on the ability of the polymerase chain reaction to amplify small quantities of DNA in-vitro. Essentially, the method (Fig. 2) provides chromosome region-specific cDNA libraries by: 1) preparing chromosome spreads suitable for hybridization and micro-dissection; 2) PCR amplifying the insert fraction of a cDNA library using vector primers, or a complex pool of uncloned cDNAs that are tailed with PCR primer binding sites; 3) hybridizing the insert fraction of these libraries to the chromosomes under conditions suitable for subsequent identification by banding and microdissection; 4) removing unhybridized and spuriously hybridized cDNA fragments following hybridization by washing; 5) microdissecting the particular chromosome region of interest, which will include the hybridized cDNA fragments; 6) amplifying the microdissected cDNA fragments using the PCR and primer pairs flanking the cloning site; and 7) recloning the amplification products to propagate the chromosome region-specific library.

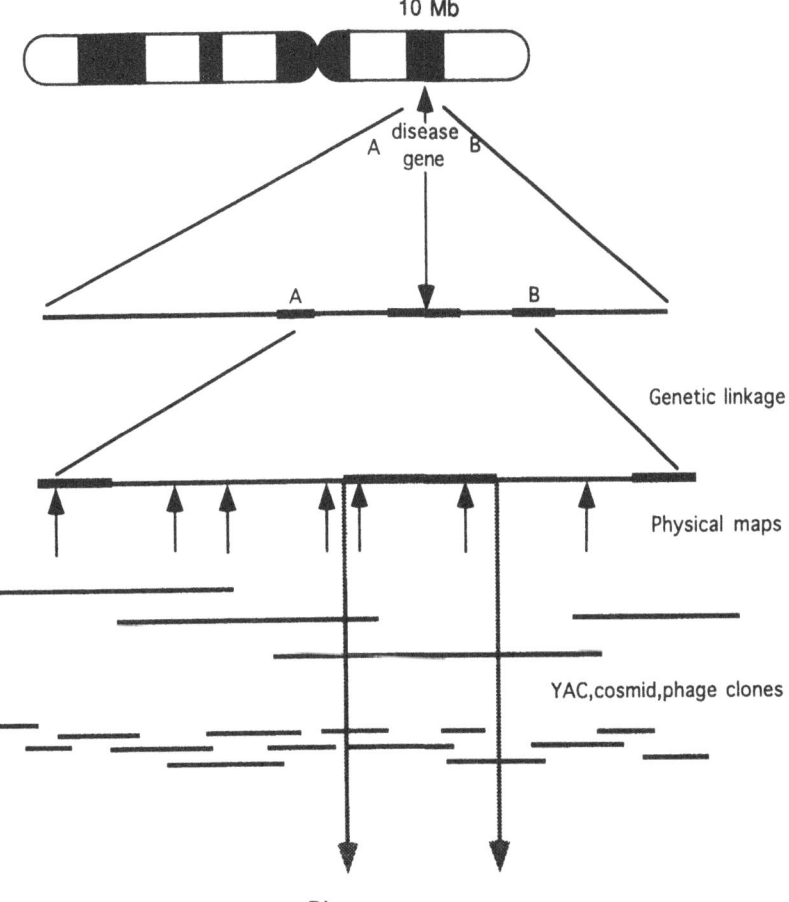

Figure 1. The process of locating a disease gene by position in the genome. Below the chromosome, on which a disease gene has been genetically mapped relative to DNA markers (A,B) to within about 10 Mb, we show the typical mapping (e.g.: genetic linkage and physical maps) and cloning (e.g.: yeast artificial chromosomes (YACs) and cosmids or phage) strategy used to reduce the search to a small segment of DNA containing the gene of interest. Adapted from (3).

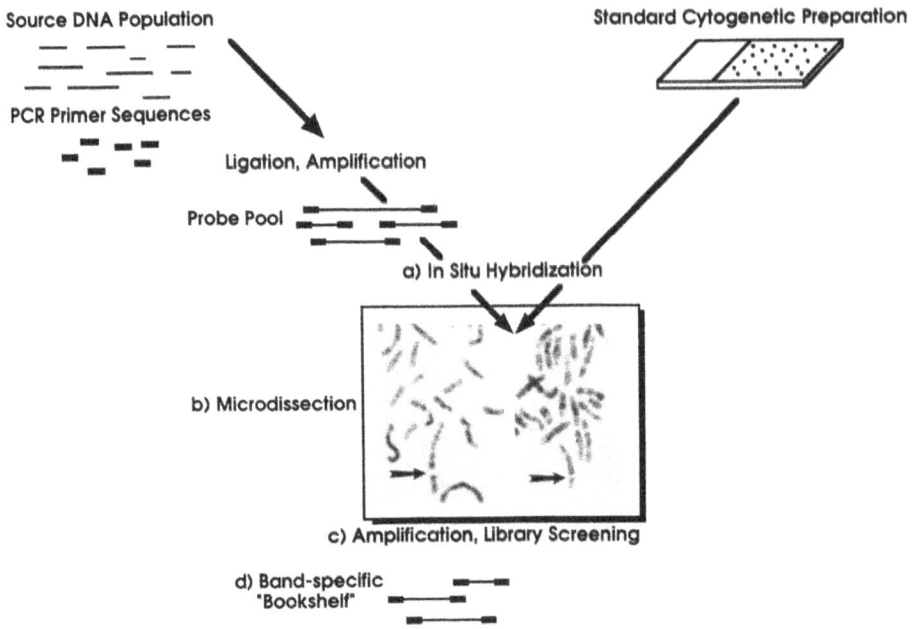

Figure 2. A typical Prep-I.S.H. strategy: cDNA hybridized to mitotic chromosomes. a) Mitotic chromosome spreads are prepared for hybridization on glass slides using standard cytogenetic procedures. The "probe" is the insert fraction of a cDNA library or a primary cDNA with adapters containing PCR binding sites at the termini of each molecule. Prep-I.S.H. hybridization follows standard protocols for F.I.S.H. hybridization of complex probes to chromosomes (as described in Methodology), except that the cDNA probe is not labeled with any detectable "reporter" molecule. b) After hybridization, the chromosomes are stained to produce the characteristic banding pattern that makes individual chromosomes distinguishable. The chromosome region of interest is microdissected from the rest of the genome. Here we show human (left) and mouse (right) chromosomes 2, each with a specific band removed (2q31 and 2F1, respectively). c) The microdissected material, containing the hybridized DNA, is amplified in-vitro using the PCR. d) The amplified material is cloned into a suitable vector to form a "bookshelf" sub-library for further analysis and propagation. Adapted from (11).

This Prep-I.S.H. technology offers important advantages over conventional strategies for gene mapping and gene finding. The first advantage is the opportunity to directly target certain regions of the genome, since the strategy utilizes microdissection of any cytogenetically definable region of the genome. A major advantage of this new Prep-I.S.H. technology is that the end-product is a set of chromosome band-specific cDNA clones, eliminating the need for the tedious search for coding regions contained within large segments of genomic DNA, as described above. These cDNA clones could also be used as genetic markers, if advantage were taken of the reported high incidence of polymorphisms in the 3' untranslated regions of cDNAs (4). Additionally, the clones can be used as markers for the construction or filling in of physical maps, by creation of sequence-tagged sites (STSs) or expressed sequence tags (ESTs) for each clone (5-7). Of course, the final and perhaps most important advantage is that the procedure offers the opportunity to isolate, specifically, coding regions of the genome, which can then be further characterized and/or sequenced directly without sequencing all the noncoding intergenic and intronic DNA. In addition, the coding regions that are isolated by Prep-I.S.H. will be classified according to their tissue of expression, as well as their physical location. This is a tremendous advantage over the procedures currently in use for isolating coding sequences from particular regions of the genome.

Present cytogenetic techniques can reliably distinguish about 300 bands per haploid genome at low cytogenetic resolution, and 800 or more bands at high resolution in human and other mammalian species (8-10). It has been estimated that the entire human genome contains about 100,000 expressed genes, and each low resolution chromosome band will therefore contain on average 300 expressed genes. Obviously this estimate does not take into consideration the differential distribution of genes between bands, nor does it take into consideration the difference in gene expression among different tissues and developmental stages. When these factors are taken into consideration, along with an estimate that the average cDNA library contains 10-30,000 genes, the average chromosome band is expected to contain 30-100 tissue-specific clones, depending on the tissue of origin and the developmental stage of the tissue. In sum, we describe a strategy for rapidly producing chromosome band-specific cDNA libraries, reducing the complexity of the average cDNA library from 10,000 clones to 100 or fewer genes.

In the following sections we describe the technical aspects of Prep-I.S.H. to a degree of detail permitting its application in other labs (Methods); present the outcomes of experiments to show the feasibility of the Prep-I.S.H. approach (Results) and after pointing out some useful technical improvements, we compare Prep-I.S.H. with positional cloning as a means of gene discovery (Discussion). We use a specific example, ataxia telangiectasia (A-T), a gene that has thus far benefitted from neither of the advantages typically used in positional cloning including linkage analysis of large families, nor specific chromosome rearrangements, to make a case for the "positional candidate" strategy for gene discovery recently proposed by Ballabio (2). Humans afflicted with A-T, a complex neurovascular, immune and endocrine disorder with an autosomal recessive mode of inheritance, are cancer-prone and radiotherapy-sensitive (14,15). Pronounced radiosensitivity extends to the cellular level in vitro. On exposure to ionizing radiation or free radical-generating chemicals (e.g. bleomycin, neocarzinostatin), cultured A-T cells invariably display impaired colony-forming ability and excessive chromosomal changes (15-19).

The disease is genetically heterogeneous as evidenced by the identification of four clinically indistinguishable complementation groups [A(55%), C (28%), D(14%), E(3%)] and two clinical variants (V1 and V2) from the closely related Nijmegen breakage syndrome (19,20). Genetic linkage and whole chromosome transfer studies have mapped A-T A, A-T C, and A-T D loci to chromosomal region 11q22-23, with groups A and C localized to a 5-cM interval between STMY and NCAM/DRD2 and group D some 20-30 cM telomeric near THY1 (17, 20-27) (see Fig. 3). Accordingly, the products of these genes might either participate in a common metabolic pathway or perhaps form parts of a multiheteromeric protein. The proximity of A-T A and A-T C loci may indicate a single gene with structural alterations in distinct intragenomic functional domains accounting for the two complementations groups (18). Recently, Kapp and coworkers (28) have identified in a cosmid a candidate cDNA that partially overcomes the A-T radiosensitivity phenotype when transfected into an A-T group D cell line.

METHODS

There is now a wide variety of background information on the "precursor" techniques of *in situ* hybridization and microdissection and the basic methodology for Prep-I.S.H. has been published recently elsewhere (11). There is also an extensive literature on PCR amplification of small amounts (<1 pg) of target material, taking into consideration the critically important first few cycles of amplification (12, 13). What follows therefore is a set of technical guidelines for adapting these techniques to

Prep-I.S.H. and some guidelines for those unfamiliar with cytogenetic techniques required to perform Prep-I.S.H. All these procedures have been applied successfully in our laboratories for Prep-I.S.H.

Amplification of cDNA Insert Fractions

DNA from the libraries used in this procedure can be prepared by any of several procedures. A very easy method is to freeze and thaw a small aliquot of the library, and use the DNA released with the cell lysate as template for the PCR reaction. This procedure is especially useful when the vector is a lambda phage such as gt10 or gt11. The procedure that we use for the amplification of the insert fraction of the cDNA library is as follows. The cDNA libraries were prepared in the vectors λgt10 and λgt11 (Clontech, Inc. Palo Alto, CA), and are received as phage preps. Phage are lysed by diluting 50 μL into 250 μL water and then freezing and thawing them twice. Samples are heated to 94°C for 30 min to denature any contaminating proteins, to inactivate any contaminating enzymes, and to disrupt any secondary or tertiary DNA structure. Lambda gt10 or gt11 forward and reverse insert screening primers (Clontech, Inc.) are added to the lysed phage. A typical 50 μL reaction contains 25 μL of lysed phage, and 1.25 units of Taq polymerase (Perkin Elmer) in 10 mM Tris pH8.3, 1.5 mM $MgCl_2$, 50 mM KCl, 200 μM each dATP, dCTP, TTP, and dGTP, and each primer at .1-5 μM. After denaturation at 94°C for 5 min, the Taq polymerase is added, and the template is amplified through 35 cycles of denaturation for 1 min at 94°C, 1 min of reannealing at 55°C, and 1.5 min of extension at 73°C. A Uni-Amp cDNA primer-ligation cassette system for PCR amplification of uncloned primary cDNAs has been described previously (11). Conditions for PCR amplification with the Uni-Amp primers are similar (11), except that the annealing temperature is 60°C, and the extension is performed at 72°C.

Amplification is assessed by comparing the PCR products to the unamplified library, with appropriate controls, by gel electrophoresis through 1% agarose, followed by ethidium bromide staining. Generally, when inserts have been successfully amplified, the pattern of amplification products will resemble the pattern of unamplified insert from the original library. Successful reactions typically produce a broad smear of DNA in the size range of 300 bp to 2.0 kb, with the highest concentration of DNA in a size range depending on the average size of the inserts in the library. Negative controls are set up the same way; except that the template is the vector alone.

Chromosome Preparations

Chromosomes from practically any source can be prepared for hybridization by a variety of well known cytological methods. Typically, metaphase chromosomes will be preferred for microdissection, but the techniques used for metaphase chromosomes can also be adapted to chromosomes in earlier stages of mitosis, particularly late prophase, prometaphase, and early metaphase. Chromosomes in these earlier stages of mitosis offer the possibility of higher resolution microdissection than typical mitotic spreads. The extended morphology of chromosomes during these earlier stages of mitosis results in more highly resolved banding patterns that can be dissected more precisely. Thus, chromosomes in these earlier stages offer the possibility of creating cDNA libraries that map to more finely defined chromosome regions, such as chromosome sub-bands.

Cytological preparations typically are made from peripheral blood, because it can be obtained without harm to most mammals, and because standard techniques have been developed for use in clinical hematology for making very high quality blood derived chromosome preparations. In some other cases it could be preferable to use cells cultured in vitro that have been arrested in mitosis, for preparing chromosomes for microdissection. A number of compounds including Colcemid, and colchicine, are useful and available for arresting both plant and animal cells in mitosis. Typically, cells growing in culture are allowed to proceed through the cell cycle to mitosis. After the cells are blocked in mitosis, they are treated with a hypotonic solution such as 75 mM KCl, which causes the cells to swell but not burst. Following hypotonic treatment, the cells typically are fixed by exposure to 3:1 methanol:glacial acetic acid. Fixation is repeated three times to insure proper spreading of the chromosomes. Droplets of the fixed cells, applied to clean microscope slides or coverslips, are air dried. Chromosomes prepared by this procedure are suitable for hybridization, banding, and microdissection.

Hybridization of Amplified cDNA to Metaphase Chromosomes

The procedure we use for hybridization is very similar to procedures typically used for analytical fluorescent *in situ* hybridization of complex probes, except that the probe is not labeled, and the probe detection steps are therefore eliminated. The procedure calls for removing the RNA and denaturing the chromosomal DNA as follows. Slides are treated with 70 μL of 100 μg/ml RNAase A in 2XSSC under a coverslip for one hr at 37°C. Cells are then dehydrated by successive rinses in 70%, then 95% ethanol. Slides are submerged in 70% formamide/2XSSC, and incubated at 75°C for 5 min to denature chromosomal DNA, and then quickly quenched and again dehydrated by immersion in a series of ice cold ethanol baths of increasing concentration, as before. Chromosomes are then treated with proteinase K at 60 ng/ml in 20 mM Tris-HCl/2mM CaCl$_2$/pH 7.5 for 12 min at 37°C. The DNA is dehydrated as before in ethanol, then allowed to air dry before addition of the probe in hybridization solution.

The probe mix consists of 50% formamide, 12.5% dextran sulfate, 1.6XSSC, 10 μg/ml salmon sperm DNA, 1x Denhardt's, 0.1% SDS, 40 mM phosphate buffer pH 6.8, and 200 - 400 ng of cDNA probe. The probe volume is typically 50 μL per slide, and a coverslip sealed with rubber cement typically covers the probe mix. The hybridization proceeds at 37°C for 18 h in a moist chamber, and control hybridizations are set up that include no cDNA probe.

After overnight hybridization, the cover slips are removed and spreads are washed twice at 42°C for ten min with 50% formamide, 2XSSC. The spreads are then washed twice with 2XSSC, and then dehydrated in 70% ethanol, followed by a 95% ethanol wash. Finally, the spreads are air-dried, and ready for staining and microdissection.

Chromosome Banding

Chromosomes may be banded with a number of stains, including Giemsa and Wright's stains, which both produce essentially the same characteristic banding patterns. These reproducible patterns serve as a means to distinguish individual chromosomal regions for microdissection. A typical staining procedure is as follows: Wright's stain is mixed 1:3 with phosphate buffer (60 mM, pH6.8), and applied to the chromosomes for 4 to 10 min at room temperature. The slides are washed briefly with water and air-dried.

Microdissection

Following fixation, hybridization and staining, the cytogenetic preparations are ready for microdissection. Typically the chromosomes are on slides or coverslips, which are mounted on a standard microscope stage in a conventional manner. The microdissection procedure uses micromanipulators and other microinstruments, including mechanical, hydraulic, and computerized stepper-motor driven micromanipulators. Heat-drawn glass microdissection needles are mounted in the micromanipulator and applied to chromosomes under high magnification. The banding pattern is used as a guide when removing the chromosome or segment from the slide.

Our current procedure for microdissection uses an inverted microscope fitted with a 3-axis (x, y, and z) stepper-motor driven "coarse" control (>1 μm per step) manipulator that holds a hydraulically driven 3-axis manipulator capable of sub-micron movements. A combination of coarse and fine 3-axis motions positions a heat-drawn glass microdissection needle. A chromosome band or region is cut, picked up, and transported to microcentrifuge tubes containing water or PCR buffer. Microdissection needles are usually broken off in the microcentrifuge tube to ensure transfer.

Post-microdissection Amplification

Following the hybridization and microdissection steps, the microdissected cDNAs are amplified away from the chromosomal material using PCR procedures essentially similar to the original amplification step, except the "heat soak" and"booster" PCR conditions (12,13) are used to accommodate the small number of molecules. (See 11 for further details of PCR amplification.)

Cloning and Characterizing the Amplified cDNAs

Following amplification, the cDNAs are subcloned into the SalI site of the Bluescript II Phagemid vector (Stratagene, La Jolla, CA),using standard cloning techniques. Typically 50-500 ng of the amplified insert cloned into 20 ng vector will result in hundreds to thousands of clones. Individual clones are grown as 1-2 ml "mini-preps", to produce double stranded cDNA clones. The inserts are sized by PCR amplification with T3 and T7 primers, followed by analysis on agarose gels. The DNA sequence at the termini of these clones is determined using primers unique to the cloning vector, flanking the cloning site. Forward and reverse sequencing primers will allow approximately 500 bp of sequence at each terminus of the double stranded clones to be determined, using di-deoxy Sequenase kit (USB Biochemical), provided that 35S, and Hydrolink and Long Ranger gels are used. Alternatively, single stranded DNA can be prepared using the modified single strand rescue protocol. Up to 1000 bp of sequence can be determined from this single strand DNA template, using the same di-deoxy sequencing protocols. Single stranded phage is prepared by inoculating a single colony into 5 ml 2X YT broth containing 75 μg/ml ampicillin and VCSM13 helper phage at 107-108 pfu/ml. The culture is grown at 37°C with aeration 1-2 hr. Kanamycin is added to 70 μg/ml to select for infected cells, and the culture is aerated another 16-24 hr. 1.5 ml of the culture is pelleted in a microcentrifuge, and 1 ml of the supernatant is processed for preparation of single stranded DNA using standard protocols. This template is used directly in a sequencing reaction.

Prep-I.S.H. of Chromosome 11, Bands q22-23

As a specific example of Prep-I.S.H.we have microdissected 11q22-23. 10 ng/μl of an oligo-dT and random oligomer-primed Uni-Amp tailed fetal brain cDNA pool was hybridized to human chromosomes in standard hybridization buffer for 48 hours using 500 ng/μl of Cot 1 DNA as a competitor for repeat sequences. After cytogenetic banding, a single chromosome 11 segment comprising the distal portion of band q22 and the proximal portion of band q23 (see Fig. 3) was dissected, transferred to a dry thin-wall 200 μl PCR tube and amplified for 50 cycles. The amplification products were separated electrophoretically on a 1.4% agarose gel. Six major bands and a background smear were visible. Individual bands or size "cuts" were sliced from the gel and the DNA was electroeluted. The individual bands and size fractions were analyzed separately for the presence of known genes in the 11q 22-23 regions. In addition, because the primary hybridization library included cDNAs that had been primed with a random oligomer, we did not expect to recover a population of completely full length cDNAs. We therefore screened a secondary full length human fetal brain cDNA library in λZap II (Stratagene) with these size fractions of the 11q22-23 Prep-I.S.H. material. Because each size fraction was actually a pool of probes, rather than an individual Prep-I.S.H.product, individual positive clones were picked and used to re-screen the full length fetal brain library. The pattern of signals resulting from each individual full length clone was therefore a subset of the signals produced by the size fraction. Using this method of systematically rescreening the full length cDNA library with individual clones, we were able to characterize individual clones with respect to abundance in the full length library, as well as estimate the number of individual unique PCR products in each size fraction. (Additional details of this cloning strategy are given in the Results section.)

RESULTS

The following is a summary of our results to date on the feasibility of the Prep-I.S.H.technique. We refer the reader to a previous publication (11) for most of the technical details; in other cases a more detailed description of the experiments is given here.

The results of our previous experiments provide evidence that many individual steps of the Prep-I.S.H. technology are feasible, including: (1) the insert fraction of cDNA libraries can be non-selectively amplified; (2) pools of cDNA molecules can be hybridized to chromosome spreads, and they can be detected by F.I.S.H.; (3) chromosomes that have been hybridized with pools of cDNA can be banded; (4) DNA that is hybridized to chromosomes is not removed by the banding procedure; (5) a population of cDNA molecules can be PCR amplified from a hybridized and microdissected chromosome or chromosome band; (6) the same population can be biotin labeled and re-hybridized to a metaphase spread, and show F.I.S.H. specificity only for the chromosome region from which it was dissected; (7) the same population can be cloned in a representative manner and individual clones can be characterized and used successfully as probes on Northern blots; and (8) unique cDNAs can be amplified specifically from the appropriate microdissected chromosome to which the gene has been mapped. In addition, we have demonstrated that quantities of as little as .05 fg of template cDNA can be amplified with the Uni-Amp primers, and as little as .05 pg can be amplified with the λgt10 and gt11 primer pairs, under conditions that produce no contaminant or artifactual DNA in "no template" controls. We have also demonstrated that the primers do not amplify human or mouse genomic DNA, nor do they amplify the carrier or competitor DNA used in the hybridization itself.

Thus the Prep-I.S.H. technology can provide chromosome band-specific, tissue-specific cDNA sub-libraries. Many variations on the basic Prep-I.S.H. theme should be possible. For instance, many different sources of DNA for hybridization should be acceptable as starting material for hybridization to chromosomes. Likewise at the end of the process, there is a growing array of cloning techniques available for the PCR products of Prep-I.S.H.

Application of Prep-I.S.H. to the 11q22-23 Region Containing the AT Gene(s)

Figure 3 is a composite genetic linkage and cytogenetic map of the distal end of human chromosome 11. It concentrates on recent data placing genes and anonymous polymorphic sequences in this region of the genome in order to narrow the likely position of the A-T gene on the cytogenetic map. The emphasis on cytogenetic sublocalization and gene content is related to the Prep-I.S.H. technology which isolates expressed sequences based solely on cytogenetic position.

The bracket on Fig. 3 defines the region dissected and PCR amplified after hybridization of a Uni-Amp tailed fetal brain library. These Prep-I.S.H. products were separated on an agarose gel as shown in Fig. 4. The gel shows six major bands with an underlying smear of DNA ranging in size from about 200 bp to 1000 bp We interpret this result as reflecting the presence of a small number of high abundance cDNAs and a larger number of lower abundance cDNAs from the 11q22-23 region. We do not have any direct evidence that this abundance distribution reflects that in the original cDNA pool, but we note that previous analysis by others of the expression pattern of genes in cDNA libraries typically show a limited number (100-200) of high abundance genes in a much more complex background of low abundance message (37). Thus, the distribution of cDNA species in Fig. 4 may represent a chromosome region-specific (11q22-23) version of that common distribution pattern.

Figure 4 also shows three regions (labelled 1, 6 and 5), surrounding the major bands in the Prep-I.S.H. DNA, that were cut from the agarose gel. DNA present in the gel "cuts" was eluted from the gel and ^{32}P-labelled for use as probes against the λ-based human fetal brain full length cDNA library and a small panel of cloned genes known to map to the 11q22-23 region (Fig. 3). The rationale behind the "cutting" approach to analysis of Prep-I.S.H. DNA is to separate tissue-specific, chromosome region-specific Prep-I.S.H. products by size, thus adding an additional dimension of characterization to the pool of PCR products. This size fractionation also makes it possible to rescue less abundant DNA on the size distribution by separate cutting and amplification. For instance, the region between cuts-1 and -6 can be amplified independently of the two flanking regions, without being overwhelmed by their major bands.

We have screened the λ-ZAP fetal brain library with ^{32}P-labelled cut-1, cut-5 and cut-6 DNA and isolated 9 independent cDNA clones   (A-T 1-5,7,9,11,12).   The independence of the clones was determined directly during the library screening process by observing the pattern of re-hybridization of each clone on replicas of the library. Thus the pattern of fetal brain cDNA library hybridization signals for clone A-T 1 was shown to be different from A-T 2, etc. In addition, comparison of the original probing of the library with each cut to the sum of the signals for the individual isolated clones gives an estimate of the number of remaining individual clones in each cut available for isolation. Most of the clones (6 of 9) were rare in the target library (1-2 per 10^5), two clones were present at 20 copies per 10^5, and one clone was present in 70 copies per 10^5.

Competition with an excess of unlabelled DNA from the abundant clones (derived from the cut-6 probing) before rehybridization with labelled cut-6 DNA should reveal additional clones of low abundance in the target library. Even without this improved strategy of competition with previously isolated clones, we estimate that there are about 3 or 4 additional clones detected by the cut-6 probing of the fetal brain library, giving a total of 10 or 11 clones in cut-6. Assuming that cut-6 is typical in terms of the number of expressed sequences present in the entire Prep-I.S.H. size distribution, we estimate that we should be able to isolate 30-40 expressed sequences from the single Prep-I.S.H. microdissection of 11q22-23.

^{32}P-labelled cut-6 DNA was also used to probe Southern blots containing a set of eight gene sequences known to map in the 11q22-23 region: TYR, CLG, STMY, APOA1, APOC3, CD3D, THY1 and ETS (see Fig. 3). Cut-6 probe hybridized to two of these genes on Southern blots (APOC3 and CD3D; data not shown) indicating the presence of

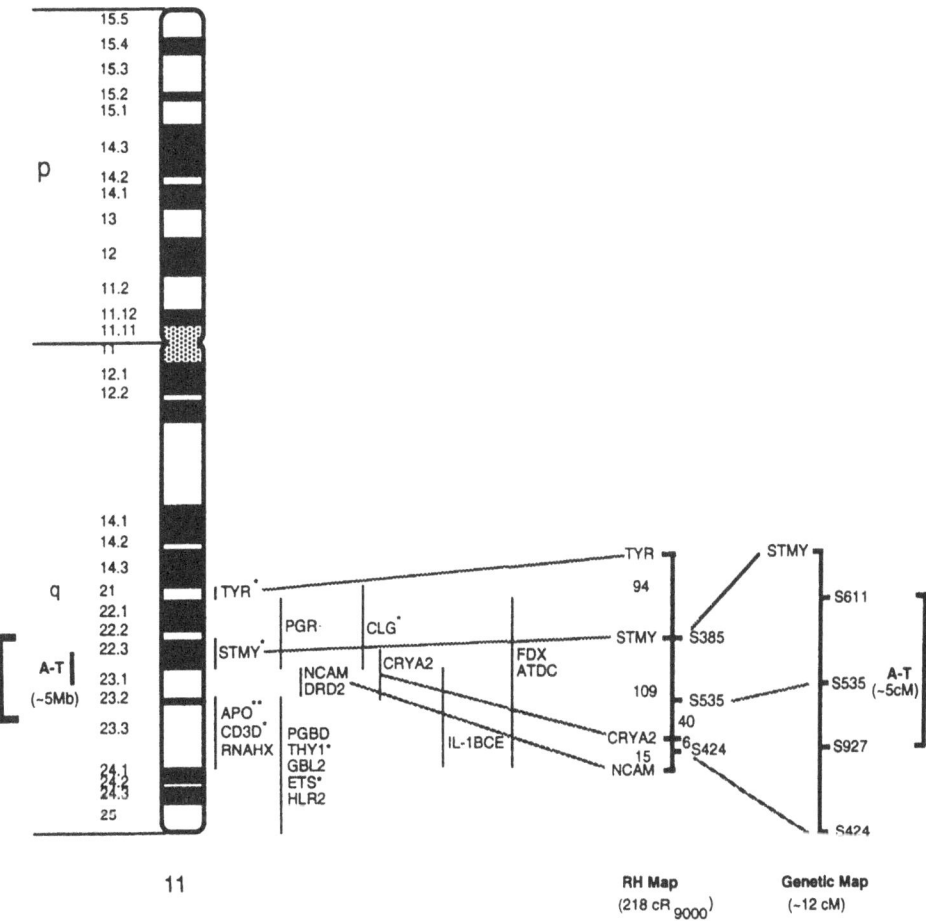

Figure 3. A composite cytogenetic, radiation hybrid (RH), and genetic linkage map of the 11q22-23 region of human chromosome 11 (17, 29-36). The map concentrates on genes in the region,along with some key anonymous DNA linkage markers(17, 29-36). The most likely position for the A-T gene(s) is shown with the wide vertical bar and the region dissected for Prep-I.S.H.analysis described here is shown by the vertical bracket, both at the left of the chromosome 11 diagram. Asterisks to the right of gene designations mark genes probed with Prep-I.S.H. material to demonstrate the presence of coding sequences from 11q22-23 as described in the Results section. The two asterisks associated with APO denote two loci; APOA1 and APOC3.

133

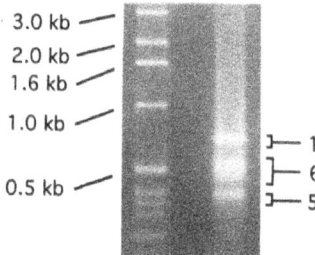

Figure 4. Size distribution of Prep-I.S.H. products from a single microdissection of 11q22-23. Size makers are shown on the left . Brackets to the right show "cuts" used to probe secondary full length cDNA libraries as described in the Results section.

Prep-I.S.H. DNA corresponding to known genes present in the microdissection region. Given that cut-6 probably contains 10-11 genes, the finding that cut-6 as a probe hybridizes to two genes known to be in the target genomic region is an indication of very high specificity of the 11q22-23 Prep-I.S.H. DNA for genes in the target region. Given that these two genes are in the center of the microdissected region, that the other six genes may not reside in the microdissected region, that cut-6 may represent only about 25% of the available expressed sequences in the Prep-I.S.H. DNA, and that not all the mapped genes are expressed in the target library, we believe that even a single Prep-I.S.H. microdissection can give a high representation of the cDNA pool hybridized to the chromosome region of interest.

DISCUSSION

The foregoing sections have described Prep-I.S.H. as a technique for the isolation of expressed sequences in a cytogenetically defined region of the human genome and its initial application to discovering candidate genes for a cytogenetically mapped human genetic disease: ataxia telangiectasia at 11q22-23. In this section we will discuss prospects for the advantages of the technique and outline a general strategy for specific gene discovery based on Prep-I.S.H., again using A-T as an example.

Technical Enhancement of Prep-I.S.H

Prep-I.S.H. is a new technique for genome region-specific isolation of expressed sequences that offers several opportunities for significant technical improvement. We have previously presented data (11) related to maximizing representation and specificity of Prep-I.S.H. - derived expressed sequences, starting with fairly standard cDNA preparations as the source of cDNA pools for hybridization to mitotic chromosomes. We believe that significant enhancement of specificity and representation can be derived from optimizing the cDNA source for hybridization. For instance, Hochgeschwender et al (38) have shown that cDNA can be depleted of repetitive sequence elements by prehybridization to genomic DNA covalently bound to cellulose particles. Such repeat-depleted cDNA pools should be simpler to use and are more likely to provide optimal specificity in Prep-I.S.H. than standard libraries, obviating the use of competition hybridization to reduce the effects of genomic repeat elements in cDNAs. Likewise, techniques for normalization of expressed sequences in libraries have been developed (39,40) to enhance the likelihood of detecting rare transcripts in cDNA libraries. Normalization of cDNA pools should be helpful in assuring representation of rare transcripts in Prep-I.S.H.

Short insert cDNA pools (200-800 bp) should be useful because short sequences hybridize to mitotic chromosomes more efficiently than longer sequences (above about 1 kilobase in length). Combining several tissue and/or cell line-specific cDNAs should increase the genome-wide representation of all cDNAs obtained by Prep-I.S.H., which would be useful for those interested in transcription mapping, while subtractive libraries (41) could be used for reducing the complexity of a region-specific Prep-I.S.H., which would be useful when searching for a particular gene of interest. Thus, a short insert, normalized and repeat-depleted library, with expressed sequences spanning multiple tissues, single tissue, or subtractive libraries could all be readily utilized in Prep-I.S.H. strategies for gene discovery.

Prep-I.S.H. and the "Positional Candidate" Approach to Gene Discovery

Ballabio (2) recently gave reasons why positional cloning may not continue to be useful in its present form: genes isolated to date either 1) are relatively common genetic disease genes intensely studied in large pedigrees or 2) often have some special additional associated genetic attribute, such as the occurrence of chromosomal abnormalities, that aids in gene discovery. Therefore, most of the "easy cases" may have been exhausted, leaving more difficult cases with no special useful feature. Single gene disorders with few families to study or with confounding sporadic cases, and multi-gene disorders will be much harder to approach by positional cloning. Ballabio (2) proposes a "positional candidate" approach to gene discovery as an alternative to positional cloning, which is initiated by the assignment of a new disease locus to a small chromosomal region (i.e. a chromosomal band of average 10 Mb). Candidate genes already known to be in that region are inspected from a variety of perspectives (physiological, biochemical, etc.) that can potentially single out a strong candidate as a "match" for a specific disease.

In the current state of the art, there are few undefined, uncharacterized candidate genes in any region of the human genome that can serve as a pool of candidates for such an approach. However, Ballabio points out that through the efforts of the Human Genome Project, new mapping will provide, over time, long lists of disease locations and even longer lists of anonymous gene sequences, such as those reported by Adams and colleagues (6,7) to which the positional candidate gene "matching" strategy can be applied. We agree with this probable evolution in gene discovery but we also believe that Prep-I.S.H. is a way to derive candidate genes for the positional candidate approach, and that Prep-I.S.H. is preferable to random sequencing of cDNAs, because Prep-I.S.H. provides sets of clones with a known mapping location.

Using the A-T gene region on 11q22-23 as an example, we believe the positional candidate approach could find the A-T gene(s). Prep-I.S.H. technology allows the convergence of effort for gene discovery on a chromosome region-specific set of 30-100 candidate genes. Three categories of assays can then be applied to the candidates for A-T gene discovery:

1) Structural assays can be used to look for changes in gene organization and expression. For example, there are numerous A-T cell lines available for Northern analysis of gene expression with candidate gene probes. Here full length candidate gene probes are not essential. When candidate sequence information is available, PCR-based assays, such as single-strand conformational polymorphism (42), can be applied quite efficiently to large numbers of gene probes (43).

2) Functional assays can be applied to candidates when full length genes are available. A battery of functional assays is currently available to rapidly cull candidate cDNAs for A-T genes. For example, a microinjection/*in situ* autoradiography method for measuring de novo DNA synthesis following gamma irradiation (which is specifically affected in A-T homozygous cell lines) is a highly sensitive and rapid method for assessing

candidate cDNA clones for functional complementation purposes. This method exploits digital cytometric imaging microscopy for automated radioautographic quantitation of 3H-thymidine incorporation into the genomic DNA of individual S-phase cells.

3) A category we call "data base assays" is most likely to grow in importance in gene discovery exercises. The ever-increasing availability of genome information, especially encompassing sequence, mapping and gene function information across a wide variety of organisms, will allow a candidate gene-related "query" of the accumulated data for matches with disease gene function as described (2). For instance, with sequence information on numerous candidate genes isolated by Prep-I.S.H. in the 11q22-23 region, the database could be searched for genes of other organisms likely to have a function in DNA metabolism or other AT gene-like features.

Figure 5 shows our view of the difference between a Prep-I.S.H.-related positional candidate approach to gene discovery and positional cloning. Positional cloning emphasizes increasingly fine resolution of the physical location of a disease gene, ignoring other genes in the region and to a large extent their individual "candidacy", while the positional candidate approach "uses" all of the genes in the region of interest in the gene discovery process. A shift to a positional candidate approach using Prep-I.S.H.-generated anonymous expressed sequences would have a number of effects on gene discovery strategies including a reduction in reliance on high resolution physical and genetic linkage mapping, increased use of *in vitro* and *in vivo* physiological model systems and an increased use of mutational and polymorphism analyses applied directly to expressed sequences (4). The Prep-I.S.H.-generated positional candidate approach also changes the nature of collaborations used in disease gene discovery by involving at an early stage those who study the physiology of the disease.

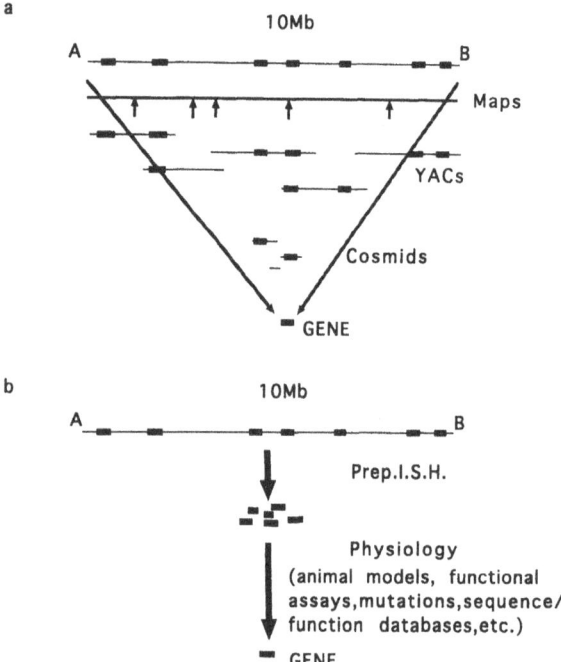

Figure 5. Major distinctions between positional cloning and positional candidate approaches to gene discovery. The top vertical lines represent the same region of 10 Mb between genetic markers A and B as shown in Figure 1 with genes shown as heavy lines. Positional cloning (a) proceeds toward the gene of interest, ignoring all adjacent genes during the cloning process. The positional candidate strategy (b) isolates all of the genes in the region first, and then subjects them to physiologic analysis to sort through them for strong candidates.

In our view Prep-I.S.H., by providing candidate genes at an early stage of gene search,and the Genome Project, by supplying more or less random gene sequence and other data-base information, can be combined to efficiently find disease genes.

ACKNOWLEDGEMENTS

Funded in part by Alberta Cancer Board, Alberta Heritage Foundation for Medical Research, and National Cancer Institute of Canada (to M.P.) and N.I.H. grant EY09478 (to L.M.D.).

REFERENCES

1. F.S. Collins, Positional cloning: Let's not call it reverse anymore, *Nature Genetics* 1:3 (1992).
2. A. Ballabio, The rise and fall of positional cloning?, *Nature Genetics* 3:277 (1993).
3. Human Genome, U.S. Department of Energy, Washington, D.C. (1992).
4. R.C. Levitt, Polymorphisms in the transcribed 3' untranslated region of eukaryotic genes, *Genomics* 11:484 (1991).
5. M. Olson, L. Hood, C. Cantor and D. Botstein, A common language for physical mapping of the human genome, *Science* 245:1434 (1989).
6. M.D. Adams, J.M. Kelley, J.D. Gocayne, M. Dubnick, M.H. Polymeropoulos, H. Xiao, C.R. Merrill, A. Wu, B, Olde, R.F. Moreno, A.R. Kerlavage, W.R. McCombie and J.C. Venter, Complementary DNA sequencing: Expressed sequence tags and human genome project, *Science* 252:1651 (1991).
7. M.D. Adams, M. Dubnick, A.R. Kerlavage, R. Moreno, J.M. Kelley, T.R. Utterback, J.W. Nagle, C. Fields and J.C. Venter, Sequence identification of 2,375 human brain genes, *Nature* 355:632 (1992).
8. J.R. Sawyer and J.C. Hozier, High resolution of mouse chromosomes: Banding homology between man and mouse, *Science* 232:1632 (1986).
9. J.R. Sawyer, Highly conserved segments in mammalian chromosomes, *J. Heredity* 82:128 (1991).
10. J.C. Hozier and L.M. Davis, Cytogenetic approaches to genome mapping, *Anal. Biochem.* 200:205 (1992).
11. J.C. Hozier, R. Graham, T. Westfall, P. Siebert and L.M. Davis, Preparative in situ hybridization: Selection of chromosome region-specific libraries on mitotic chromosomes, In Press: *Genomics* (1994).
12. G. Ruano, W. Fenton and K.K. Kidd, Biphasic amplification of very dilute DNA samples via "booster" PCR., *Nucl. Acids Res.* 17:5407 (1989).
13. G. Ruano, D.E. Brash and K.K. Kidd, PCR: the first few cycles, *Amplifications* 7:1-4 (1991).
14. R.A. Gatti, E. Boder, H.V. Vinters, *et al.*, Ataxia-telangiectasia: An interdisciplinary approach to pathogenesis, *Medicine* 70:99 (1991).
15. R.P. Sedgwick and E. Boder, Ataxia-telangiectasia, *Elsevier Science Publ.*, pp. 347 (1991).
16. M.C. Paterson and P.J. Smith, Ataxia telangiectasia: An inherited human disorder involving hypersensitivity to ionizing radiation and related DNA-damaging chemicals, *Ann. Rev. Genet.* 13:291 (1979).
17. R.A. Gatti, Localizing the genes for ataxia-telangiectasia: a human model for inherited cancer susceptibility, *Adv. Cancer Res.* 56:77 (1991).
18. P.J. McKinnon, Ataxia-telangiectasia: an inherited disorder of ionizing-radiation sensitivity in man, *Hum. Genet.* 75:197 (1987).
19. N.G.J. Jaspers, R.A. Gatti and C. Baan, *et al.*, Genetic complementation analysis of ataxia-telangiectasia and Nijmegen breakage syndrome: a survey of 50 patients, *Cytogenet. Cell Genet.* 49:259 (1988).
20. R.A. Gatti, I. Berkel, and E. Boder, *et al.*, Localization of an ataxia-telangiectasia gene to chromosome 11q22-23, *Nature* 336:577 (1988).

21. C.M. McConville, C.J. Formstone and D. Hernandez, *et al.*, Fine mapping of the chromosome 11q2 2-23 region using PFGE, linkage and haplotype analysis: Localization of the gene for ataxia-telangiectasia to a 5 cM region flanked by NCAM/DRD2 and STMY/CJ52.75, ph2.22, *Nucl. Acids Res.* 18:4335 (1990).

22. T. Foround, S. Wei and Y. Ziv, *et al.*, Localization of an ataxia-telangiectasia locus to a 3-cM interval on chromosome 11q23: linkage analysis of 111 families by an international consortium, *Am. J. Hum. Genet.* 49:1263 (1991).

23. Y. Ziv, G. Rotman and M. Frydman, *et al.*, The ATC (ataxia-telangiectasia complementation group C) locus localizes to chromosome 11q22-q23, *Genomics* 9:373 (1991).

24. K. Komatsu, S. Kodama and Y. Okumura, *et al.*, Restoration of radiation resistance in ataxia-telangiectasia cells by the introduction of normal human chromosome 11., *Mutat. Res.* 235:59 (1990).

25. E. Sobel, E. Lange and N.G.J. Jaspers, *et al.*, Ataxia-telangiectasia: Linkage evidence for genetic heterogeneity, *Am. J. Hum. Genet.* 50:1343 (1992).

26. F. Cornelis, D. Cherif and M. James, *et al.*, Precise localization of a gene responsible for ataxia-telangiectasia on chromosome 11q., *Cytogenet. Cell Genet.* 58:111 (1991).

27. C. Lambert, R.A. Schultz and M. Smith, *et al.*, Functional complementation of ataxia-telangiectasia group D (AT-D) cells by microcell-mediated chromosome transfer and mapping of the AT-D locus to the region 11q22-23, *Proc. Natl. Acad. Sci. USA* 88:5907 (1991).

28. L.N. Kapp, R.B. Painter and L.-C. Yu, *et al.*, Cloning of a candidate gene for ataxia-telangiectasia group D, *Amer. J. Hum. Genet.* 51:45 (1992).

29. NIH/CEPH Collaborative Mapping Group, A comprehensive genetic linkage map of the human genome, *Science* 258:67 (1992).

30. M.W. Smith, S.P. Clark, J.S. Hutchinson, Y.H. Wei, A.C. Churukian, L.B. Daniels, K.L. Diggle, M.W. Gen, A. J. Romo, Y. Lin, L. Selleri, D. L. McElligott and G.A. Evans, A sequence-tagged site map of human chromosome 11, *Genomics* 17:699 (1993).

31. A. Tunnacliffe, H. Perry, P. Radice, M. Budarf and B.S. Emanuel, A panel of sequence tagged sites for chromosome band 11q23, *Genomics* 17:744 (1993).

32. C.W. Richard III, D.R. Cox, L. Kapp, J. Murnane, F. Cornelis, C. Julier, G.M. Lathrop and M.R. James, A radiation hybrid map of human chromosome 11q22-q23 containing the ataxia-telangiectasia disease locus, *Genomics* 17:1 (1993).

33. E. Lange, P. Concannon, N. Uhrhammer, Y. Nakamura, R.A. Gatti, Localization of polymorphic probes in the region of ataxia-telangiectasia at chromosome 11q22, *Am. J. Hum. Genet.* 53:A1028 (1993).

34. R.A. Gatti, K. Peterson, J. Novak, X. Chen, L. Yang-Chen, T. Liang, E. Lange and K. Lange, Prenatal diagnosis of ataxia-telangiectasia by genotyping, *Am. J. Hum. Genet.* 53:A1409 (1993).

35. R. Oskato, A. Bar-Shira, L. Vanagaite, Y. Ziv, S. Ehrlich, G. Rotman, C.M. McConville, A. Chakravartiz and Y. Shiloh, Ataxia-telangiectasia: Allelic association with 11q22-23 markers in Moroccan-Jewish patients, *Am. J. Hum. Genet.* 53:A1055 (1993).

36. C.M. McConville, P.J. Byrd, H. Ambrose, T. Stankovic, Y. Shiloh, J.O. McNamara, T. Kuwahara and A.M.R. Taylor, Mapping of newly identified polymorphic loci together with genes for mitochondrial acetoacetyl-coenzyme A thiolase (ACAT), glutamate receptor 4 (GLUR4), and interleukin 1ß converting enzyme (IL1ßCE) close to the locus for ataxia telangiectasia on chromosome 11q22-23, *Am J. Hum. Genet.* 53:A1043 (1993).

37. K.E. Davies and A.P. Read, Molecular basis of inherited disease, *IRL Press*, Second Edition, p.37.

38. U. Hochgeschwender, J.G. Sutcliffe and M.B. Brennan, Construction and screening of a genomic library specific for mouse chromosome 16, *Proc. Natl. Acad. Sci. USA* 86:8482 (1989).

39. M.S.H. Ko, An 'equalized cDNA library' by the reassociation of short double-stranded cDNAs, *Nucl. Acids Res.* 18:5705 (1990)

40. S.R. Patanjali, S. Parimoo and S.M. Weissman, Construction of a uniform-abundance (normalized) cDNA library, *Proc. Natl. Acad. Sci. USA* 88:1943 (1991).

41. R.M. Myers, The pluses of subtraction, *Science* 259:942 (1993).

42. M. Orita, Y. Suzuki, T. Sekiya and T. Hayashi, Rapid and sensitive detection of point mutations and DNA polymorphisms using the polymerase chain reaction, *Genomics* 5:874 (1989).

43. S.E. Poduslo, M. Dean , U. Kolch and S.J. O'Brien, Detecting high-resolution polymorphisms in human coding loci by combining PCR and single-strand conformation polymorphism (SSCP) analysis, *Am. J. Hum. Genet.* 49:106 (1991).

DIRECT cDNA SCREENING OF GENOMIC REFERENCE LIBRARIES - A RAPID METHOD FOR THE IDENTIFICATION OF TRANSCRIBED SEQUENCES IN LARGE GENOMIC REGIONS

Wolfgang Schwabe, Brenda J. Lawrence, Adelaide S. Robb, Rene M. Hopfinger, Ute Hochgeschwender and Miles B. Brennan

Unit on Genomics, NIMH, 9000 Rockville Pike, Bethesda, MD 20892

ABSTRACT

In order to identify transcribed sequences from large genomic regions we developed and applied direct screening of genomic reference libraries. Briefly, reference libraries in lambda or cosmid vectors are screened with cDNA probes depleted of highly repeated sequence elements. The cDNA probes can be made from the same or related species. Further, by repeating the screenings with cDNA probes made from different tissues, the pattern of expression of sequences in each clone can be inferred. To derive sequence from positive genomic clones, DNA from these are Southern blotted and hybridized with the depleted cDNA probe. Small fragments identified in these blots are then sequenced and these sequences analyzed by GRAIL (1), a neural network program which detects coding exons from genomic sequence. We demonstrate the efficient isolation of expressed sequences from reference libraries of whole chromosomes, chromosomal regions and YAC clones, using cDNA probes from the same and related species.

INTRODUCTION

The goal of transcribed sequence identification is to fill in the genetic and physical map of the human genome. Such an integrated map would contain the sequences, locations and expression patterns of all genes in a given interval. Its significance for disease gene isolation and for the study of genome structure is obvious.

As large genomic regions were made accessible by physical mapping, established techniques, e. g. Zoo blots, Northerns, were found inadequate and several new procedures for identifying expressed sequences from genomic clones were developed. These can be divided into computational, functional and hybridization-based approaches.

The computational approach is based on large scale genomic sequencing. Systematic differences between nucleotide sequences of protein coding exons and non-coding sequences can be recognized by a number of algorithms (1). The functional approaches rely on the ability of splicing and/or polyadenylation sequences associated with expressed sequences to function in heterologous context, e.g. in an exon trap vector (2). Hybridization-based approaches, including direct cDNA screening and direct cDNA selection, are based on the specific hybridization of cDNA to its genomic cognate. In direct cDNA screening, genomic clones carrying transcribed sequences are identified by hybridization with cDNA probe (3); in direct cDNA selection, the cDNAs are eluted and cloned (4-6).

We describe here the identification of genomic clones containing transcribed sequences by direct cDNA screening of reference and YAC derived libraries. Further, we show how coding sequences can rapidly be derived from such genomic clones. The relative advantages of this method are discussed.

METHODS

Genomic Libraries

The high density filters of the ICRF human chromosome 21 reference library (7) were provided by Dr. H. Lehrach, London. High density filters, plated using the same machine, of an Xq28 cosmid library (8) were provided by Dr. A. Poustka, Heidelberg.

A genomic library was constructed in lambda DASH (Stratagene) with human genomic inserts from a YAC containing the ERG locus of chromosome 21 (YAC A125B12, Joint YAC Screening Effort). Yeast DNA from the YAC containing strain was isolated (9), and used to construct a library as described (3). Positive clones (208 clones), giving an estimated 10-fold coverage of the YAC insert were arrayed in microtiter trays and stored and plated as described previously (3).

cDNA Probes

RNA was isolated from tissues or cells by extraction with guanidinium isothiocyanate and centrifugation through a cesium chloride cushion (10). The polyA+ fraction was isolated by oligo(dT) chromatography (11). 2 μg of polyA+ RNA is reverse transcribed using oligo(dT) primers in the presence of [α-32P]dATP traces. After 1 h, the reaction is stopped, and the polynucleotides are separated from the oligo- and mononucleotides by chromatography on a Sephadex G50 column. After ethanol precipitation, this cDNA is used as template in a standard random hexamer labelling reaction (12), to yield a cDNA probe with a specific activity of at least 1 x 10^9 cpm/μg.

Depletion of Highly Repeated Sequences

High molecular weight mammalian genomic DNAs were isolated and coupled to m-aminobenzyloxymethyl cellulose (ABM cellulose; Sigma, cat. # A-3544) as described (3).

The cDNA was denatured by boiling and mixed with a 1,000-fold excess of genomic DNA immobilized on cellulose; the hybridization volume was adjusted to give a concentration of 100 μg of immobilized DNA in 0.8 ml hybridization solution [50 % formamide, 0.75 M NaCL, 50 mM phosphate buffer (pH 6.8), 5 mM EDTA, 0.1 % SDS, 1 x Denhardt's solution (13), 100 μg/ml sonicated salmon sperm DNA]. This mixture was heated to 80°C for 2 min and then incubated for 72 h at 37°C on a rocking platform. About every 12 h (8-14 h), the cellulose was pelleted by centrifugation at medium speed in a clinical centrifuge, the supernatant decanted, heated at 80°C for 10 min, cooled to 37°C, and remixed with the DNA-cellulose. At the end of the depletion, the DNA-cellulose with the annealed highly-repeated sequences from the cDNA probe was pelleted by centrifugation, and the supernatant containing low-repetition and unique sequences from the labelled cDNA was decanted. The supernatant was then used directly as a probe to detect expressed sequences. The DNA cellulose can be stripped and stored for reuse as described (3).

Hybridization

Phage and cosmid filters were prehybridized overnight at 42°C in 5 x SSPE/5 x Denhardt's solution/1 % SDS/100 μg/ml sonicated salmon sperm DNA/50 % formamide. Filters were then hybridized overnight at 42°C (same species) or at 37°C (cross-species) in the same solution supplemented with 10^7 cpm/ml depleted cDNA probe. Filters were initially washed twice for 10 min in 2XSSC/0.1 % SDS at room temperature and then for one h at a higher temperature in 2XSSC/0.1 % SDS, followed by shaking for 10 min. Subsequent washes at still higher temperatures were carried out similarly. Filters were autoradiographed at -80°C with an intensifying screen on Kodak X-AR film.

Southern and Northern Blots

Southern blots of 1 - 2 μg restriction enzyme digested cosmid and phage DNAs were prehybridized, hybridized and washed as described above for colony and phage lifts.

For Northern blots, 3 μg of polyA+ mRNA was electrophoresed in 1.5 % agarose/formaldehyde denaturing gels and transferred to Nytran filters according to the suppliers protocol. Prehybridization and hybridization conditions were as described above.

Sequencing

DNA fragments to be sequenced were subcloned into pUC19 (14) and used for dideoxy sequencing according to the manufacturer's recommendations (USB). The 5' and 3' end sequencing was performed using the M13 forward and reverse sequencing primers (New England Biolabs).

RESULTS

Strategy

The main goal of our method is to generate transcriptional maps of large genomic regions. As starting material we use arrayed genomic reference libraries which can be derived from whole chromosomes, parts of chromosomes, or YAC's. Two considerations lead us to base our approach on genomic reference libraries: First, one genomic library is sufficient to identify all genes, regardless of when and where they are expressed. Second, the representation of a genomic clone is independent of its level of expression.

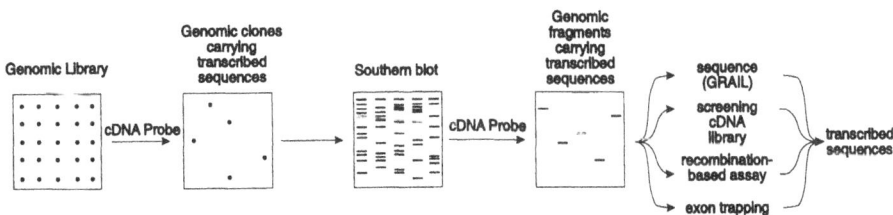

Figure 1. Identification of transcribed sequences by direct cDNA screening: Strategy (see text).

Our strategy is shown in Fig. 1. We hybridize replica filters of a genomic reference library with complex cDNA probes from any source, depleted of highly repeated sequences. Genomic clones carrying transcribed sequences give positive hybridization signals. From these genomic clones, DNA is isolated, digested, electrophoresed, Southern blotted, and reprobed with the same cDNA probe used for the initial screening. This Southern blot hybridization identifies small genomic fragments carrying transcribed sequences. Furthermore, it serves as a rescreening and as a means to eliminate redundant overlapping clones. In the last step, a number of different methods can be used to derive coding sequences from these small genomic fragments. For example the fragments are sequenced and the nucleotide sequence is analyzed by the neural network program GRAIL (1). In a second method this genomic fragment is used as a probe for screening a cDNA library. A third approach applies this genomic fragment in a recombination based assay (15). Fourth, one can identify transcribed sequences using an exon trapping approach (2).

To generate a transcriptional map for a number of tissues, this procedure is repeated using cDNA probes from these tissues (Fig. 2). Note that the repeated cDNA screenings give cumulative data on the expression of sequences from each genomic clone. Further, expressed sequence clones need only be analyzed once, even if they are expressed in many tissues. These characteristics are significant for making comprehensive transcriptional maps (see Conclusion).

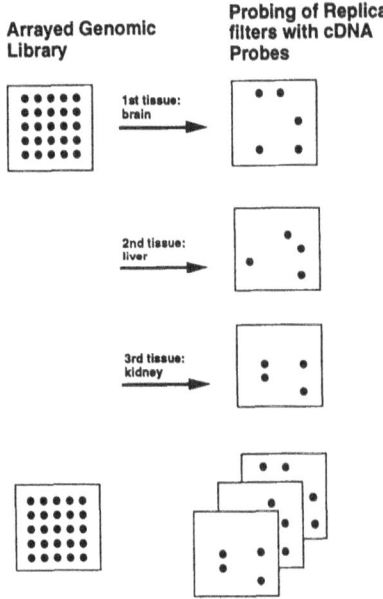

Figure 2. Identification of expression patterns: Arrayed genomic reference libraries can be repeatedly hybridized with cDNA probes derived from various tissues. Each additional hybridization step assembles an expression pattern of the transcribed sequences of genomic clones. These can be expressed either in all tissues or few tissues or in one tissue only. Furthermore, this screening procedure avoids non-redundant analysis: once a genomic clone is identified it need not be reanalyzed.

In probing genomic DNA with cDNA probes we faced a technical problem: around 10 % of mRNAs contain highly repeated sequences. These have to be removed from the cDNA probe. Our method for quantitative removal of repeat sequences from hybridization probes is schematically shown in Fig. 3.

PolyA + RNA is used as a template for a reverse transcription to produce first strand cDNA molecules. During a random hexamer oligo labeling reaction short, second strand molecules with an average length of 500 bp are synthesized. This cDNA probe is a mixture of coding, untranslated, and repeated sequences. To remove the repeated sequences, the cDNA probe is hybridized to an excess of genomic DNA covalently bound to finely divided cellulose. Highly repeated sequences from the probe anneal with the highly repeated sequences in the genomic DNA. The DNA-cellulose with the annealed repetitive sequences from the probe is pelleted by centrifugation, and the unique and low repetitive probe sequences in the supernatant are used directly to hybridize genomic clones.

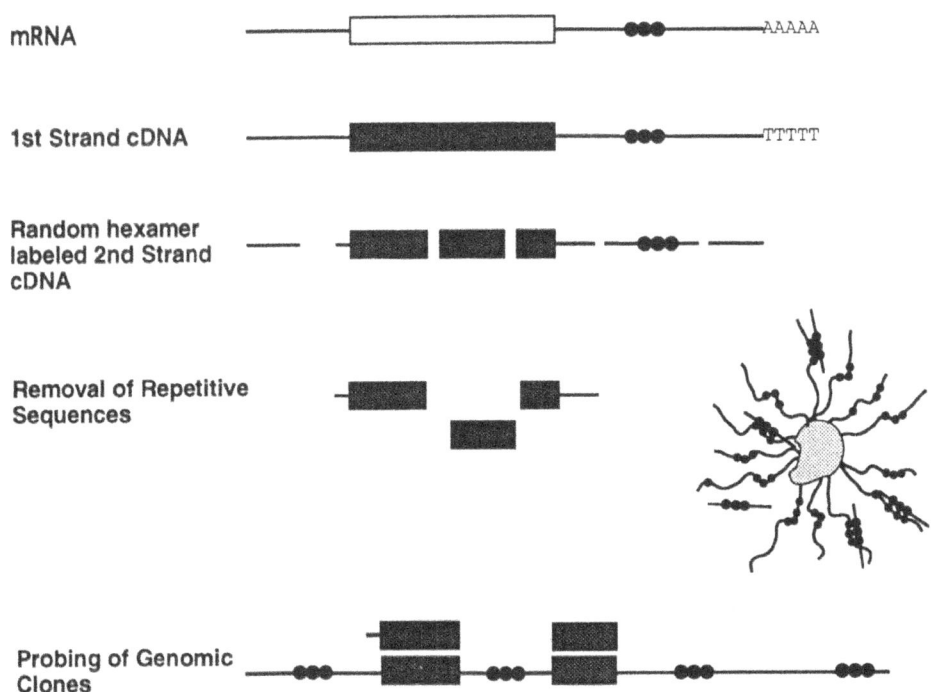

Figure 3. Depletion of highly repeated sequences from cDNA probes by hybridization to DNA-cellulose (see text for details): shaded boxes - coding sequences, black dots - highly repeated sequences.

Identifying Genomic Clones Carrying Transcribed Sequences

To test the robustness of direct cDNA screening, we screened a variety of reference libraries with cDNA probes from different tissues.

Figure 4 shows an example of a screening result of the ICRF Human Chromosome 21 cosmid reference library (7). This library was screened with three cDNA probes: first, with a cDNA probe from a human T-cell line Molt-4, which was depleted of highly repeated sequences, and second with cDNA probes from adult human cerebral cortex either before or after depletion of repeat sequences. The autoradiographs of the filters probed with human cerebral cortex cDNA demonstrate the efficiency of the depletion: Many of the positive hybridization signals from genomic clones hybridized with the undepleted cDNA probe were lost when the cDNA probe was depleted of highly repeated sequences. Comparison of the hybridization signals with the two depleted cDNA probes, from Molt-4 and from cerebral cortex, reveals differences in the patterns of expression of sequences in particular genomic clones. For example, some sequences are expressed in both tissues (Fig. 4, arrow heads), either at the same level or high in one (Molt-4) and low in the other (cerebral cortex). Finally, some appear to be tissue specific, expressed only in Molt-4 (Fig. 4, arrow) or only in cerebral cortex (Fig. 4, triangles).

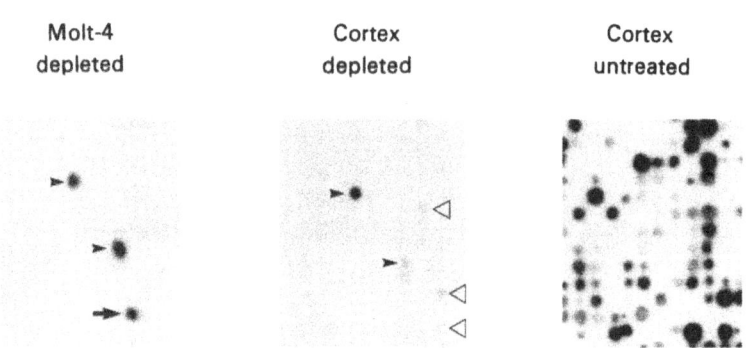

Figure 4. Screening of the ICRF Chromosome 21 cosmid reference library with cDNA probes depleted of highly repeated sequences from human cerebral cortex and from human T-cell line Molt-4 and with an untreated cerebral cortex cDNA probe. Arrow heads show hybridization signal of genomic clones with both cDNA probes, the arrow shows a genomic clone expressed specifically in the T-cell line, and the triangles show cortex specific genomic clones.

Altogether we screened 13,824 cosmid clones of the chromosome 21 reference library. Of these, 202 (1.4%) were positive with Molt-4 cDNA probe only, 205 (1.4%) with the cortex probe only, and 479 (3.5%) gave a signal with both probes. We also screened a part of the Xq28 cosmid reference library (8, data not shown) with human cerebral cortex cDNA probe. Of 300 Xq28 genomic clones, 61 (20%) contained sequences expressed in human cerebral cortex as determined by hybridization with this cDNA probe.

Identifying Genomic Fragments Carrying Transcribed Sequences

Depending upon the particular region and cDNA probe, 5 - 20 % of genomic clones give hybridization signal with a cDNA probe. Unordered reference libraries are highly redundant to ensure coverage; therefore many of the positives will be overlapping clones. To reduce this redundancy and to concentrate our analysis on smaller fragments, we rescreen the positive clones by Southern blot. We will illustrate this in our analysis of the Xq28 library. We selected 19 of the 61 positive clones for further analysis. DNA of these cosmids was digested with Bam HI and Bgl II, electrophoresed on agarose gel, Southern blotted, and hybridized with the cerebral cortex cDNA probe. The nineteen clones resolved into 12 groups: one had four members, two had 3 members, and the rest were singlets. In most clones only one fragment hybridized strongly with the depleted cDNA probe, while two or three hybridizing fragments were seen occasionally. In only one clone no fragment was detected upon rescreening with the depleted cDNA probe. In Fig. 5, the results of this Southern analysis for 2 cosmid clones are shown. Clone 8A4 has three hybridizing fragments with the cDNA probe, clone 12D12 has one. For comparison, hybridizations with undepleted cDNA probe and with total human genomic DNA probes are also shown; these results demonstrate the specificity of the depleted cDNA probe.

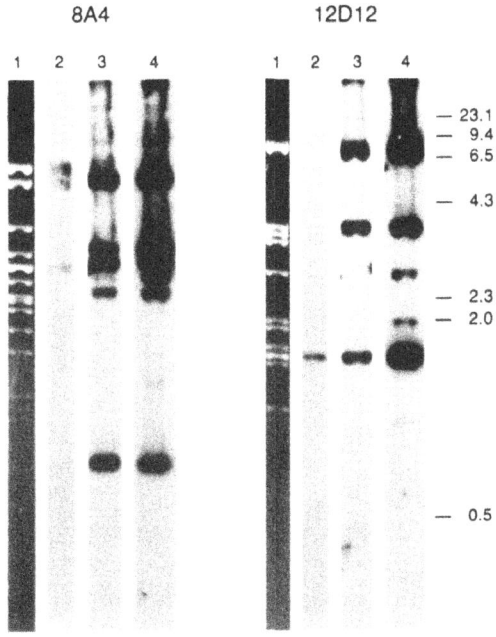

Figure 5. The identification of genomic fragments carrying transcribed sequences of two positive genomic cosmid clones, 8A4 and 12D12, from the Xq28 human cortex screening are shown: lane 1 - restriction digest pattern after a Bam HI/Bgl II digest, blot hybridization of the same gel with a depleted cerebral cortex cDNA probe (lane 2), with an untreated cerebral cortex cDNA probe (lane 3), and with total genomic human DNA (lane 4).

Deriving Exon Sequences

To identify exons from these genomic fragments we tried two different methods: exon trapping, and end sequencing of these fragments followed by GRAIL analysis. First, we subcloned the 1.4 kb Bam HI/Bgl II fragment of the 12D12 cosmid clone into the exon amplification vector pSPL1 (2) in both orientations. In addition, the cosmid clone 12D12 was digested with both Bam HI and Bgl II and the mixture of 15 fragments was shotgun cloned into pSPL1. DNA was prepared in bulk from 8,000 - 24,000 independent recombinants, of which 40 - 70 % had inserts, and was used for transfection into COS-7 cells. Five times more DNA (20 μg) was used for the shotgun cloned material than for the individually subcloned fragments. The individual 1.4 kb fragment gave a novel PCR band after exon amplification, while the shotgunned fragments from the parental 12D12 cosmid showed only the vector derived PCR product. The amplified exon was sequenced and the 217 bp sequence was submitted to GRAIL (Fig. 6).

Xq28 Clone 12D12

length = 217 bp

pos	f-strand	frame	orf				pos	r-strand	frame	orf	
51	0.081	-	- - -	||	161	0.938	1	1 -	217
61	0.926	2	23 - 149	||	151	0.938	1	1 -	217
71	0.935	2	23 - 149	||	141	0.938	1	1 -	217
81	0.881	2	23 - 149	||	131	0.938	1	1 -	217
91	0.938	2	23 - 149	||	121	0.938	1	1 -	217
101	0.938	2	23 - 149	||	111	0.938	1	1 -	217
111	0.938	2	23 - 149	||	101	0.938	1	1 -	217
121	0.103	-	- - -	||	91	0.937	1	1 -	217
131	0.056	-	- - -	||	81	0.937	1	1 -	217
141	0.161	-	- - -	||	71	0.938	1	1 -	217
151	0.085	-	- - -	||	61	0.884	1	1 -	217
161	0.001	-	- - -	||	51	0.921	1	1 -	217

Potential exons are listed in the following

pos	strand	strand_prob	frame	quality	orf	
51 - 161	r	0.78	1	excellent	1 -	217

```
A:   VLRHCVSDKVTVIGAGITVYEALAAADELSK
B:         ::::::|:  |   |:|:|  |||||:|
C:         ::::|::: :|:|  :| |||||:|
D:         :::|:::: |: |: ||:||::|:|
E:         ::::|:|: |  |:|:| ||:||:|
F:         ::::||:: : :| : |:||: |||

A:   QDIFIRVIDLFTIKPLDVATNVSSAKATEGR
B:   ::|    |:|| |::|||::| ::|:: | ||
C:    :|    ||:| ::||| :| |::: :: :
D:   ::|    ::|| |::|||::| ::|:::| ||
E:   ::|    |:|| |::|||::| ::|:: | ||
F:   :::    ||:: ||:|:|::| :::: :::: 

A:   IISVEDHYPQ      Xq28 clone 12D12
B:   | |::           Bacillus subtilis
C:   :::||: |:       Ascaris suum
D:   :: |:: |:       Alcaligenes eutrophus
E:   | |::           Bacillus stearothermophilus
F:   :::||: :||      Homo sapiens
```

B-F: pyruvate dehydrogenases (β subunit)

Figure 6. GRAIL analysis of the 12D12 exon sequence: The upper panel shows the overall results of the GRAIL analysis: The reverse strand was assigned as being an exon with a probability of 93.8 %, the length of an ORF and the quality of this exon were evaluated. In the lower panel the predicted amino acid sequence of the 12D12 exon sequence is shown with the results of the Swiss Prot search - A: Xq28 clone 12D12, trapped exon; B: Bacillus subtilis, pyruvate dehydrogenase, beta subunit (AC #P21882); C: Ascaris suum, pyruvate dehydrogenase, beta subunit (AC #P26269); D: Alcaligenes eutrophus, acetoin:2,6-dichlorophenolindophenol oxidoreductase, beta subunit (AC #P27746); E: Bacillus stearothermophilus, pyruvate dehydrogenase, beta subunit (AC #P21874); F: Homo sapiens, pyruvate dehydrogenase, beta subunit (AC #P11177).

It was assigned a 93.8 % probability of being a coding exon. The program assigned the frame and coding strand (here: frame 1, reverse strand), and evaluated the overall quality of the sequence as excellent. The nucleotide sequence was translated into the amino acid sequence and a search of the Swiss Prot data bank revealed a strong homology to the ß subunit of pyruvate dehydrogenases of different species (see Fig. 6). We used this exon to probe a Northern blot but no signal was detectable. However, from the end sequences of this exon PCR primers were designed, and applied in a RT/PCR using human cortex mRNA (Fig. 7). Here, we were able to find a specific band of 217 bp, the same size as the amplified exon. The amplification of the 12D12 cosmid and of human genomic DNA resulted in a ≈ 320 bp fragment indicating that the trapped exon consists of two fused exons which are separated by an intron of approximately 100 bp in genomic DNA.

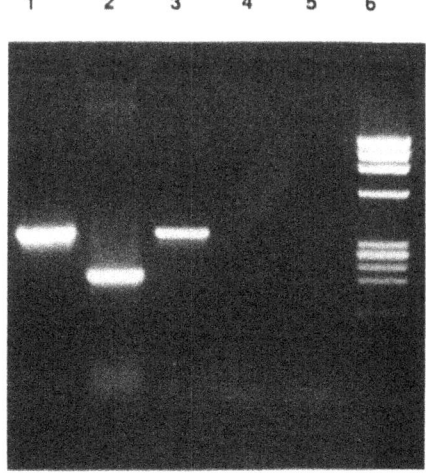

Figure 7. Ethidium bromide stained gel of PCR-products of different templates using primers derived from the exon of the genomic clone 12D12: lane 1: 12D12 genomic cosmid clone; lane 2: a pSPL1 subclone containing the trapped exon 12D12; lane 3: human genomic DNA; lane 4: human cortex cDNA; lane 5: H2O as negative control; lane 6: size marker (PhiX 174 Hae III digested).

Although we found an exon from the cloned 1.4 kb fragment of the cosmid clone 12D12, we were unable to detect this exon using a shotgunned fragment mixture of the same cosmid. We had similar experiences with other clones. Since in our hands exon trapping was low in both reliability and efficiency, we switched to directly sequencing the genomic fragment identified and submitting these sequences to GRAIL.

Xq28 Clone 8A4

```
8a4f16, len = 109

pos   f-strand frame    orf          pos   r-strand frame       orf

 51    0.936      3     1 -  109  ||   51    0.111      -        - - -


       Potential exons are listed in the following


     pos      strand    strand_prob    frame     quality         orf

51 -    51    f          0.98           3        marginal      1 -  109
```

```
                           Phe Leu Val Phe Val Ala Asn Phe Asp Glu
          human rab GDI    TTC CTG GTG TTT GTG GCA AAC TTC GAT GAG
                                        |               |
          bovine rab GDI   --- --- --- --C --- --- --- --T --- ---



                           Asn Asp Pro Lys Thr Phe Glu Gly Val Asp
          human rab GDI    AAT GAC CCC AAG ACC TTT GAG GGC GTT GAC
                            |                           |
          bovine rab GDI   --C --- --- --- --- --- --- --- --C ---



                           Pro Gln Thr Thr Ser Met Arg Asp Val Tyr
          human rab GDI    CCC CAG ACT ACC AGC ATG CGT GAC GTC TAC
                                   ||
          bovine rab GDI   --- --- -AC --- --- --- --- --- --- ---
                                   Asn



                           Arg Lys Phe Asp Leu Gly Gln Asp Val
          human rab GDI    CGG AAG TTT GAT CTG GGC CAG GAT GTC
                                        |   |
          bovine rab GDI   --- --- --- --C T-- --- --- --- ---
```

Figure 8. Exon identification of clone 8A4. A coding strand and frame were assigned as well as the quality of the forward primed sequence (109 bp) was evaluated as marginal (upper panel). The predicted amino acid sequence was submitted to Swiss Prot. The result is shown in the lower panel; the found sequence has 93 % identity to the bovine rab GDI gene.

As an example, both ends of the 6.5 kb fragment of the Xq28 cosmid clone 8A4 were sequenced. The GRAIL analysis of the 109 bp sequence from one end resulted in 98% probability of being a coding sequence with a marginal quality (Fig. 8). The translation into the amino acid sequences and searching the Swiss Prot database revealed this sequence as the human homolog of bovine rab GDI which is a GTP-dissociation inhibitor (16). This 6.5 kb fragment was used to probe a Northern blot of human and mouse cerebral cortex mRNA (Fig. 9). A 2.4 kb transcript in the human RNA and a mouse transcript slightly larger were detected.

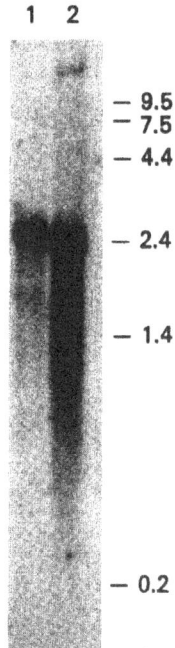

Figure 9. Northern blot hybridization using the 6.5 kb 8A4 genomic fragment as a probe to mouse cerebral cortex mRNA (lane 1) and to human cerebral cortex mRNA (lane 2). Size marker in kb.

Screening for Transcribed Sequences Using Cross-Species cDNA Probes

Since it is frequently difficult to get developmental stage specific tissue from human, we applied the direct cDNA screening approach to cross-species hybridization as well. Figure 10 shows one example of such a cross-species hybridization.

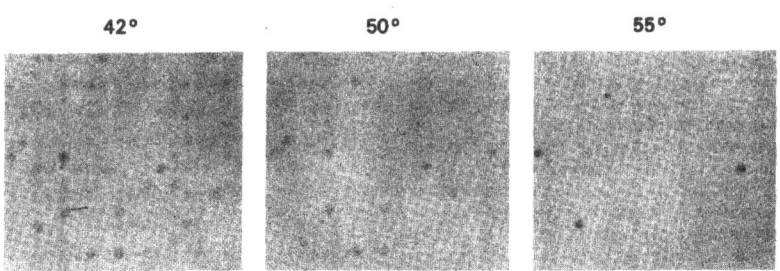

Figure 10. Cross-species direct cDNA screening using P10 cbm mouse cDNA probe depleted of highly repeated mouse sequences against an arrayed, human chromosome 21 sublibrary from a YAC containing the Erg gene. The same filter is shown after washes in 2XSSC/0.1 % SDS at 42°C, at 50°C, and 55°C. Note, the number of genomic clones hybridizing to the cDNA probe declines with rising temperature.

We screened a human chromosome 21 phage sublibrary from a YAC containing the Erg gene with a postnatal day 10 (P10) mouse cerebellum (cbm) cDNA probe depleted of mouse repeated sequences. The filter containing 208 phage clones was hybridized at 42°C and washed successively at increasing stringency by raising the temperature. While most clones gave signal at the initial washing temperature of 42°C, the number of positive clones declined rapidly with increasing temperature: at 50°C we found 22 clones were positive, and at 55°C only 4 clones continued to give positive signal. No hybridization signal could be found after a wash at 60°C.

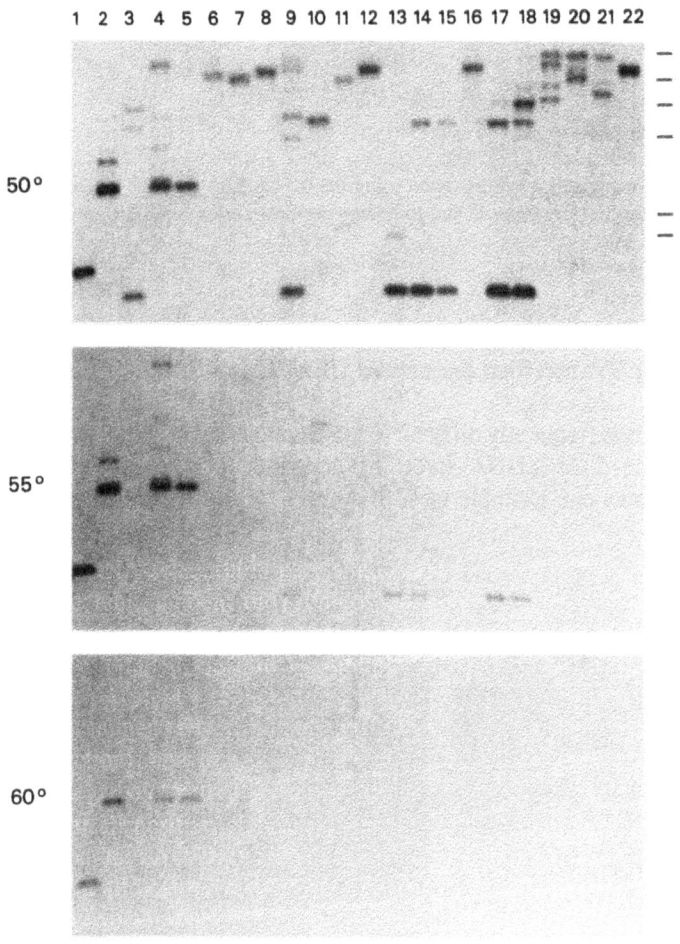

Figure 11. Southern blot analysis of Eco RI digested positive phage clones (see Fig. 10) containing transcribed sequences with a depleted P10 cbm mouse cDNA probe. This Southern filter was successively washed in 2XSSC/0.1 % SDS at 50°C, at 55°C, and at 60°C (see text for details).

DNA of the 22 genomic clones detected at 50°C were digested with Eco RI, electrophoresed, blotted, and once again hybridized with the P10 cbm cDNA probe depleted of repetitive sequences (Fig. 11). The Southern blot was washed at increasing temperatures as well. At 50°C we found many genomic fragments hybridizing to the cDNA probe. As expected the non-specific signal decreased with increasing washing temperature. After washing at 68°C no signal was detectable anymore. We assumed (and later confirmed by hybridization with specific fragments) that the phage clones from the lanes 3, 9, 13, 14, 15, 17, and 18 as well as the clones from the lanes 1, 2, 4, and 5 are overlapping clones due to their similar hybridization behavior including size and Tm of the fragment hybridizing with the cDNA probe. For further analyses we isolated and subcloned the 1.4 kb fragment of the clone 3F3 (lane 1) and the 1.3 kb fragment of the clone 1H5 (lane 17).

Again, the fragments were end sequenced, and the sequence data submitted to GRAIL. Fig. 12 shows the data of the 3F3 search: An exon with a probability of 87 % was detected on the 183 bp long forward primed sequence. A frame and the coding strand were assigned. The translation into an amino acid sequence and a Fasta search of the Swiss Prot database revealed no significant homology to any known proteins.

The GRAIL analyses of the first end sequences of the clone 1H5 did not give any positive results. We subcloned Sau 3A fragments of the 1.3 kb of the 1H5 fragment and end sequenced four of those (Fig. 13). After a GRAIL search, one subclone with a length of 134 bp has a 93.7 % exon probability. Further, a frame and coding strand were assigned. Again, a screen of the Swiss Prot data base with the 1H5 predicted amino acid sequence did not show any significant homology to known proteins.

CONCLUSIONS

Before discussing the relative advantages of direct cDNA screening, it is worthwhile to consider the transcriptional map which is our goal. Every gene in a given interval (location) would be represented by sequence data sufficient for its unequivocal identification and by the pattern of its expression in tissues. As most genes are not expressed in a given tissue, we consider expression data to be of equal importance with sequence data in a transcriptional map. Our approach must be amenable to repeated analyses to identify genes expressed in different tissues. However, pairwise comparison of tissues show that most genes transcribed in either tissue are common to both. Thus a completely independent analysis of each tissue would be extremely redundant.

For these reasons we have based our approach on genomic reference libraries. As mentioned above, these libraries will contain all genes regardless of their pattern or level of expression. The first step then is to derive a reference library in a lambda or cosmid vector. We have presented here and elsewhere (3) the use of whole chromosome libraries, region specific libraries and libraries derived from YAC clones.

The second step is to identify genomic clones carrying transcribed sequences by screening the reference library with cDNA probes. These highly complex probes must be quantitatively depleted of sequences highly repeated in the genome in order that the signal from unique sequences is not obscured; the method of depletion be annealing to immobilized genomic DNA has proven effective while neither preannealling the probe nor blocking the filters is sufficiently quantitative.

Direct screening of genomic libraries with cDNA detects > 50 % of clones

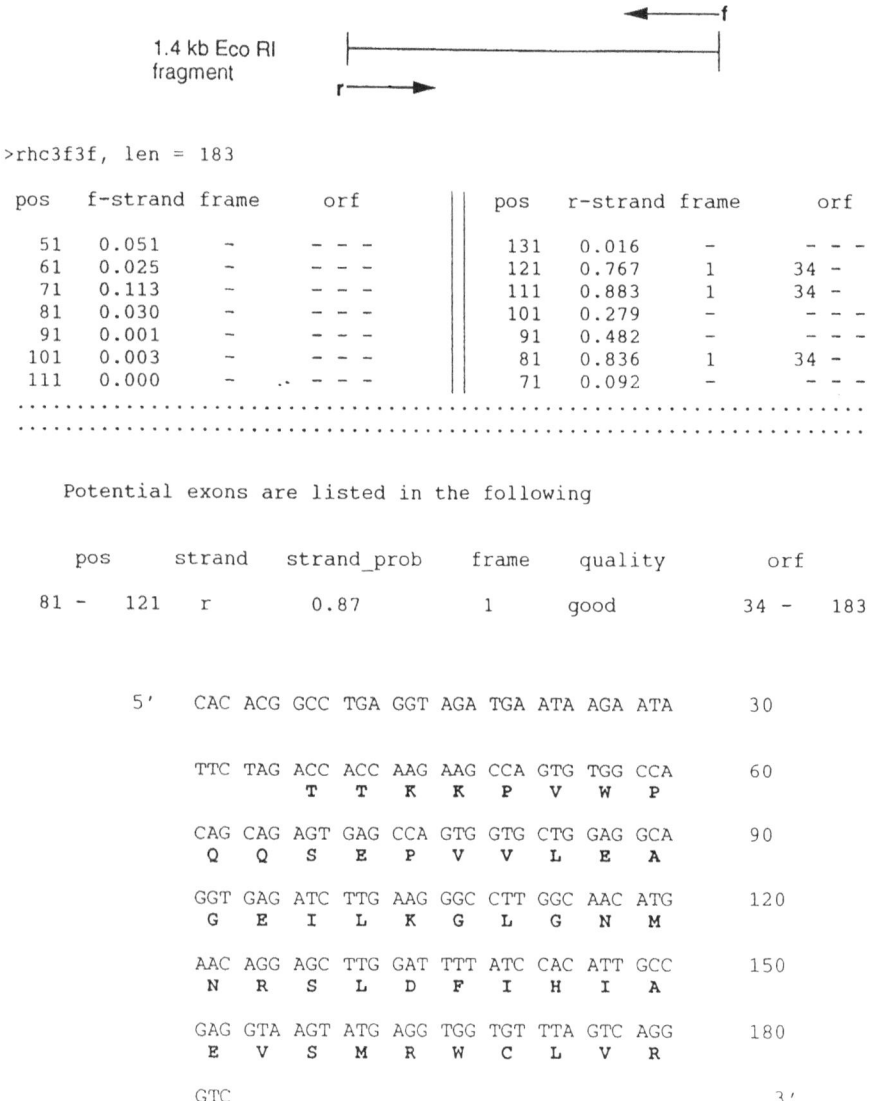

```
                                                              ←──────── f
1.4 kb Eco RI     ├──────────────────────────────────────────────────┤
fragment
                  r ──────►
```

```
>rhc3f3f, len = 183

  pos   f-strand frame      orf      ‖   pos   r-strand frame       orf

   51    0.051     ~       - - -      ‖   131    0.016     ~        - - -
   61    0.025     ~       - - -      ‖   121    0.767     1      34 -   183
   71    0.113     ~       - - -      ‖   111    0.883     1      34 -   183
   81    0.030     ~       - - -      ‖   101    0.279     ~        - - -
   91    0.001     ~       - - -      ‖    91    0.482     ~        - - -
  101    0.003     ~       - - -      ‖    81    0.836     1      34 -   183
  111    0.000     ~    .. - - -      ‖    71    0.092     ~        - - -
...........................................................................
...........................................................................
```

```
     Potential exons are listed in the following

     pos        strand    strand_prob    frame    quality       orf

  81 -  121     r          0.87            1       good       34 -   183
```

```
        5'   CAC ACG GCC TGA GGT AGA TGA ATA AGA ATA        30

             TTC TAG ACC ACC AAG AAG CCA GTG TGG CCA        60
                      T   T   K   K   P   V   W   P

             CAG CAG AGT GAG CCA GTG GTG CTG GAG GCA        90
              Q   Q   S   E   P   V   V   L   E   A

             GGT GAG ATC TTG AAG GGC CTT GGC AAC ATG       120
              G   E   I   L   K   G   L   G   N   M

             AAC AGG AGC TTG GAT TTT ATC CAC ATT GCC       150
              N   R   S   L   D   F   I   H   I   A

             GAG GTA AGT ATG AGG TGG TGT TTA GTC AGG       180
              E   V   S   M   R   W   C   L   V   R

             GTC                                            3'
              V
```

Figure 12. GRAIL analysis of an end sequence of the 1.4 kb subclone 3F3.

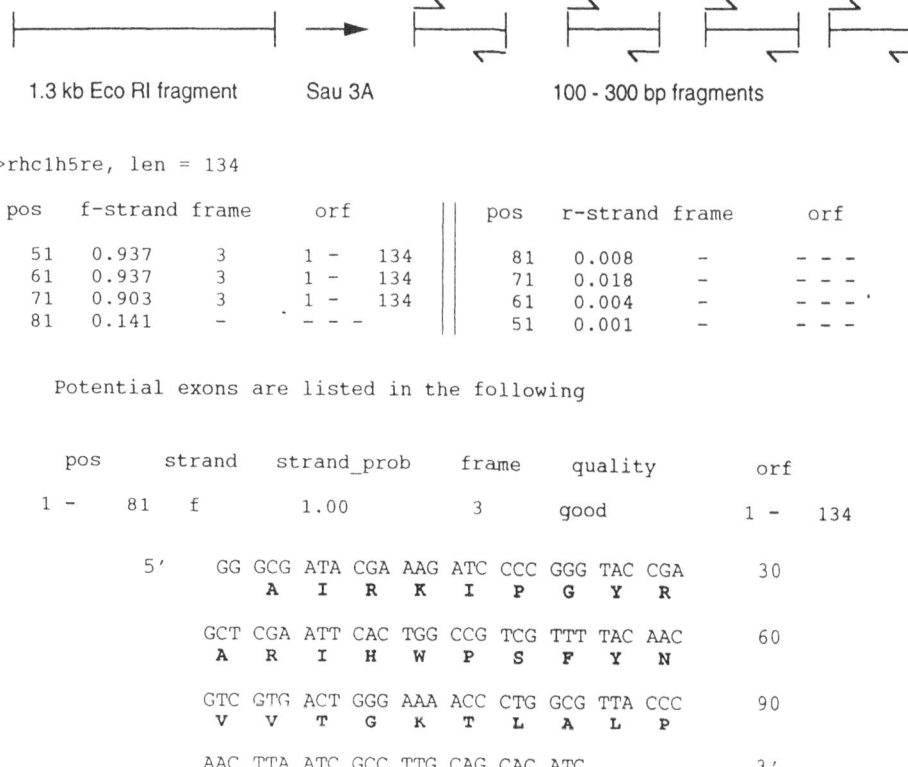

Figure 13. GRAIL analysis of subclones of the 1.3 kb 1H5 fragment. One out of four subclones resulted in a coding strand with a 93.7 % probability being an exon.

expected to carry transcribed sequences. The method is sensitive enough to reliably detect clones carrying sequences expressed > 0.01 % of total mRNA and frequently detects clones carrying sequences expressed at a much lower level. Furthermore, the sensitivity of the cDNA probe can be increased by normalizing the probe (subtracted probe) or by making cDNA probes from individual fractions of size-fractionated mRNA.

The cDNA screenings can be carried out on large scale without loss of sensitivity. Further, as defined developmental stages or tissues from human may not be available, cDNA probes from heterologous mammalian species are also effective.

The last step is to derive sequence data on the expressed sequence. It should be noted that unless the reference library has been ordered into "contigs" of overlapping clones, it will be redundant in order to ensure coverage. Thus, most expressed sequences will be represented by multiple overlapping genomic clones. Both to decrease the size of the fragment to be analyzed and to eliminate redundancies, the positive genomic clones are digested with restriction enzymes, electrophoresed on agarose gels, blotted to filters and reprobed with the cDNA probe. This serves both to eliminate redundant clones and to identify short fragments containing transcribed sequences. We have used both exon trapping and random sequencing followed by GRAIL analysis to identify coding sequences from these small fragments. For our purpose, random sequencing and GRAIL analysis has proven the more efficient.

The sequence data is then used to design oligonucleotides to serve both as PCR primers and as probes for screening cDNA libraries when the full length coding sequence is desired.

The method allows screening of large regions with no decrease in sensitivity; more than 104 clones can be simultaneously screened. For each individual clone the expression pattern can be easily determined. Further, the expression data for each clone is cumulative. Analysis of expressed sequences is also non-redundant: a genomic clone containing sequences expressed in a number of tissues need only be analyzed once. The approach allows use of cross-species probes, which facilitates the study of developmental changes in human gene expression. This method reliably detects genes expressed at 0.01 %, and frequently at much lower abundances. Finally, many steps in this approach are amenable to automation (library plating, filter readings, Southern blots and sequencing) and therefore to large-scale applications.

Two other approaches, exon trapping and direct cDNA selection, have also been used to identify transcribed sequences from genomic clones. The efficiency of trapping an exon from a given gene will depend upon the arrangement of restriction enzyme sites within the introns of that gene; just the inclusion of an exon in a fragment is not sufficient to ensure its presence in the mature mRNA. Further, cryptic "exons", that is sequences which are included in the mature mRNA when cloned in an exon trap vector but which do not correspond to exons in the genome have been isolated. Finally, as exon trapping depends upon PCR amplification of the trapped exon(s), the recovery of exons from pools of clones is subject to skewing from competition in the amplification reaction. This method has been used successfully in the identification of genes from small regions of particular interest, where exhaustive analysis is possible.

Direct cDNA selection involves the specific hybridization of cDNAs to their genomic cognate (cosmid or YAC clones), followed by elution and cloning (4-6). The selected cDNA libraries derived from this procedure have impressive enrichments for expressed sequences mapping to the region. While the recovery is somewhat normalized, the abundance of clones reflects their abundance in the starting library. Analysis of these libraries can be complicated by three factors: the small size of the inserts, the presence of false positive and the redundancy of libraries made from different tissues. Again, this method has been used successfully in small regions, but its utility to large regions is unknown.

ACKNOWLEDGEMENT

We gratefully acknowledge Timothy McPhail for expert help with   the   computer searches.

REFERENCES

1. E.C. Uberbacher and R.J. Mural, Locating protein-coding regions in human DNA sequences by a multiple sensor-neural network approach, *Proc. Natl. Acad. Sci. USA* 88:11261 (1991).
2. A.J. Buckler, D.D. Chang, S.L. Graw, J.D. Brook, D.A. Haber, P.A. Sharp, and D.E. Housman, Exon amplifcation: A strategy to isolate mammalian genes based on RNA splicing, *Proc. Natl. Acad. Sci. USA* 88:4005 (1991).
3. U. Hochgeschwender, J.G. Sutcliffe, and M.B. Brennan, Construction and screening of a genomic library specific for mouse chromosome 16, *Proc. Natl. Acad. Sci. USA* 86:8482 (1989).
4. S. Parimoo, S.R. Pantanjali, H. Shukla, D.D. Chaplin, and S.M. Weissman, cDNA selection: efficient PCR approach for the selection of cDNAs encoded in large chromosomal DNA fragments, *Proc. Natl. Acad. Sci. USA* 88:9623 (1991).
5. M. Lovett, J. Kere, and L.M. Hinton, Direct selection: a method for the isolation of cDNAs encoded by large genomic regions, *Proc. Natl. Acad. Sci. USA* 88:9628 (1991).
6. B. Korn, Z. Sedlacek, A. Manca, P. Kioschis, D. Konecki, H. Lehrach, and A. Poustka, A strategy for the selection of transcribed sequences in the Xq28 region, *Hum. Mol. Genet.* 1:235 (1992).
7. D. Nizetic, G. Zehetner, A.P. Monaco, L. Gellen, B.D. Young, and H. Lehrach, Construction, arraying, and high-density screening of large insert libraries of human chromosomes X and 21: their potential use as reference libraries, *Proc. Natl. Acad. Sci. USA* 88:3233 (1991).
8. P. Kioschis, B. Gross, D. Nizetic, G. Zehetner, and A. Poustka, Molecular analysis of the Xq28 region, *Cytogenet. Cell Genet.* 58:2070 (1991).
9. F.M. Ausubel *et al.* (editors), "Current Protocols in Molecular Biology," Current Protocols, New York (1993), p 6.10.13.
10. J.M. Chirgwin, A.E. Przybyla, R.J. MacDonald, and W.J. Rutter, Isolation of biologically active ribonucleic acid from sources enriched in ribonuclease, *Biochemistry* 18:5294 (1979).
11. H. Aviv and P. Leder, Purification of biologically active globin messenger RNA by chromatography on oligothymidylic acid-cellulose, *Proc. Natl. Acad. Sci. USA* 69: 1408 (1972).
12. A.P. Feinberg and B. Vogelstein, A technique for radiolabeling DNA restriction endonuclease fragments to high specific activity, *Anal. Biochem.* 132:6 (1983).
13. J. Sambrook, E.F. Fritsch, and T. Maniatis, "Molecular Cloning: a Laboratory Manual," Cold Spring Harbor Laboratory Press, New York (1989).
14. C. Yanisch-Perron, J. Vieira, and J. Messing, Improved M13 phage cloning vectors and host strains: nucleotide sequence of the M13mp18 and pUC 19 vectors, *Gene* 33:103 (1985).
15. D.M. Kurnit and B. Seed, Improved genetic selection for screening bacteriophage libraries by homologous recombination in vivo, *Proc. Natl. Acad. Sci. USA* 87: 3166 (1990).
16. T. Sasaki, K. Kaubuchi, A.K. Kabcenell, P.J. Novick, and Y.A. Takai, A mammalian inhibitory GDP/GTP exchange protein (GDP dissociation inhibitor) for smg p25A is active on the yeast SEC4 protein, *Mol. Cell. Biol.* 11:2909 (1991).

IDENTIFICATION OF EXPRESSED SEQUENCES ON HUMAN CHROMOSOME 9q32-34

Jeffrey D. Falk, Hiroshi Usui, and J. Gregor Sutcliffe

Department of Molecular Biology, The Scripps Research Institute, La Jolla, CA 92037

ABSTRACT

Ozelius and colleagues have recently mapped the gene responsible for the neuromuscular disease idiopathic torsion dystonia (DYT1) to human chromosome 9q32-34. Our goal is to identify candidate genes for torsion dystonia, as well as other neurologically important genes localized on 9q32-34. To accomplish this we have identified transcribed sequences within a 3,000 clone collection assembled from a human 9q32-34-specific library. Dystonia candidates are being assessed based on their expression patterns and their localization with respect to the DYT1 locus. cDNA probes from various brain and peripheral tissues were used to screen the clone collection enabling us to elucidate the tissue expression patterns of the genes corresponding to each clone. This resulted in the identification of 143 9q32-34 transcripts, thirty-three of which are expressed in a brain-specific manner and thus are candidates for the dystonia gene. Since none of these transcripts was expressed specifically in the putative dystonic target tissue basal ganglia, the directional tag PCR subtraction method was used to construct a probe enriched in striatal cDNA sequences. Screening of the 9q32-34 clone collection with this subtracted probe identified several additional clones with transcripts expressed preferentially in the striatum.

INTRODUCTION

The gene corresponding to the neuromuscular disease idiopathic torsion dystonia (DYT1) has recently been mapped to human chromosome 9q32-34 by Ozelius and colleagues (1). To identify candidates for DYT1 as well as other neurologically important genes from 9q32-34, we are mapping sequences within this region that are expressed in the putative dystonic target tissues basal ganglia and muscle (2,3).

Several methods have recently emerged for identifying transcribed sequences, each having specific advantages and disadvantages. These include identifying exon sequence tags (4), exon amplification (5), direct cDNA selection schemes (6,7), subtractive hybridization (8), identification of differentially expressed transcripts (9), and identification of expressed sequences in chromosome-specific libraries (10). Of these, the method developed by Hochgeschwender and colleagues (10) for determining the expression patterns of mouse genes identified in arrayed chromosome-specific libraries is best suited for directly identifying genes from a specific chromosomal region and revealing their tissue expression patterns. To identify a set of neurologically important and dystonia candidate genes, we have applied this method to the characterization of transcribed sequences in the human 9q32-34 region. This resulted in the identification of 143 clones of transcribed sequences representing approximately 14-27% of the genes expected to be in this region. One hundred and one of these clones correspond to genes that are expressed ubiquitously in brain and peripheral tissues, whereas, thirty-three of the clones correspond to transcripts that were all expressed specifically in the brain and thus are candidates for the dystonia gene. In addition, we used the directional tag PCR subtraction method of Usui et al. (11) to prepare a subtracted probe enriched in striatal cDNA sequences, which we used to identify genes expressed preferentially in the striatum.

MATERIALS AND METHODS

DNA Probes

Radiolabelled probes for high molecular weight genomic DNA and rat cDNAs were isolated as described by Hochgeschwender et al. (10) with the following modifications. Poly (A)$^+$ RNA was digested with RNase-free DNase (Boehringer Mannheim) to remove any residual DNA prior to first strand cDNA synthesis with MMLV reverse transcriptase (BRL). First strand cDNA was radiolabelled with ^{32}P-dCTP and ^{32}P-dATP (3,000 Ci/mmol) using a random hexamer primed synthesis kit (Boehringer Mannheim).

Directional Tag PCR Subtraction

A rat cDNA probe enriched in striatal sequences was prepared using the directional tag PCR subtraction technique described by Usui et al. (11) in which rat striatal cDNA served as target and rat cerebellum cDNA served as driver for the subtraction.

Cell Lines

The hamster-human hybrid cell line 640-63a12 /CHTG 49, containing human chromosome 9q as its sole human component, and the X ray deletion hybrid E6B which contained only human chromosome 9q32-34 (12) were generously provided by Lisa Henske and David Kwiatkowski.

Library Construction and Screening

Phage libraries were constructed in the λ DASH II vector (Stratagene) with Sau 3A partial digests of genomic DNA from cultured cells, and Gigapak II Gold packaging extracts (Stratagene) using standard procedures (13). Phage lifts and hybridizations were conducted using standard methods (13). Optimal conditions for identifying human repetitive sequences (0.5XSSC, 0.5% SDS, 55° wash) were determined by probing a λ EMBL3 human genomic library with labeled human genomic probe and washing at stringencies ranging from 2XSSC 68° to 0.1XSSC 72°. Cross-hybridization between human and hamster repetitive elements was not observed using the optimized wash conditions described above. A collection of 3,096 human 9q32-34 clones was assembled using labelled human genomic probe to screen 211 150 mm filters containing 3,000 pfu each of the E6B somatic cell hybrid library. Positive human plaques were picked into microtiter

158

plate wells and replicates of the microtiter arrays were gridded on Biotrans (ICN) nylon membranes for subsequent screening procedures.

Screening of the 9q32-34 Collection With Rat cDNA Probes

Prior to screening our human 9q32-34 collection, repetitive sequences were removed from native rat cDNA probes by prehybridization to human genomic DNA-cellulose as described by Hochgeschwender *et al.* (10). Freshly prepared human DNA-cellulose was used for each probe rather than recovering and recycling the cellulose. The clone collection was screened using the hybridization conditions described above and washing conditions of 0.2XSSC, 0.5% SDS at 55°. Identical screening conditions were used to probe the collection with the rat striatum-enriched cDNA probe described above.

RESULTS

Assembling a Collection of Human 9q32-34 Specific Clones

To facilitate the identification of torsion dystonia candidate genes, we modified the method described by Hochgeschwender *et al.* (10) to identify transcribed sequences within a human 9q32-34 -specific library. A phage library was constructed using human genomic DNA isolated from the hamster-human X ray deletion somatic cell hybrid E6B described by Henske and Kwiatowski (12), whose sole human component is duplicate copies of the 9q32-34 region spanning the markers AK1 and DBH. This library was screened with a human genomic DNA probe using conditions that allowed us to identify clones containing human repetitive sequences without detecting the considerable hamster background that was present. These screening conditions (0.5XSSC, 68° wash) were established after conducting control screens in which it was determined that we could: a) identify 99% of the human clones in a λ human genomic library (determined by comparing signals from duplicate filters probed with human genomic DNA or λ DNA); b) eliminate all hamster clones (determined by hybridization of a human genomic DNA probe to a λ hamster genomic DNA library); and c) detect human clones within our λ 9q32-34 E6B genomic library at the frequency (1:200) expected from a somatic cell hybrid containing approximately 10% of a 1.5 x 10^8 human chromosome.

By large scale screening of the λ E6B library, we identified 3,096 human 9q32-34-specific clones that were arrayed into a permanent microtiter collection. Based on arithmetic considerations, we expected that this number of clones would correspond to roughly three genome equivalents of 9q32-34. This was confirmed by measuring the representation of five known 9q32-34 genetic markers within the collection. Three genome equivalents of the 9q32-34 region correlates statistically (Poisson distribution) to a 97% probability that any gene segment is represented at least once within the collection. The 9q32-34 clone collection was subsequently screened with a variety of cDNA probes to identify transcribed genes within the collection and determine the overall tissue expression pattern of each of these genes.

Screening of the Human 9q32-34 E6B Collection With Rat cDNA Probes

Two types of rat cDNA probes were used to screen the arrayed 9q32-34 collection, thereby identifying transcribed sequences: cDNA probes derived from rat brain and peripheral tissues that were depleted of repetitive sequences by prehybridization to human-DBM cellose (10) and a striatum-enriched cDNA probe in which sequences present in both striatal and cerebellar cDNA were removed by subtraction. Figure 1 illustrates the effect of depleting repetitive sequences from rat cDNA probes. A considerable amount of hybridization background was detected when duplicate collection filters were probed with native rat brain or liver cDNA probes (Fig. 1A and 1B). Furthermore, screening with these probes identified only those transcripts expressed at high levels in a tissue-specific/enhanced manner, such as clones 24-1F & 24-5G which exhibited specific or enhanced hybridization to the brain cDNA probe in comparison to the liver cDNA probe. Screening with a rat brain cDNA depleted of repetitive elements (Fig. 1C) significantly reduced the hybridization background levels and increased the hybridization signals such that transcripts expressed in a tissue-specific or ubiquitous manner could be identified.

To ensure that the expression patterns deduced from our cDNA screening experiments accurately reflect *in vivo* expression patterns, DNA from clones 24-1F and 24-5G was used to probe northern blots containing poly (A)$^+$ RNA from rat brain, liver and kidney. A 4 kb transcript was detected in brain but not in liver or kidney with the 24-1F probe (Fig. 2A), and a 0.8 kb transcript found predominantly in the brain but also at comparatively low levels in liver and kidney was detected by the 24-5G probe (Fig. 2B).

Both of these expression patterns are consistent with the cDNA screening results we obtained, confirming the fidelity of our screening procedure. Subsequent screening of the entire 9q32-34 collection was conducted with rat cDNA probes from whole brain, striatum, and liver that were depleted of repetitive elements (Table 1).

Table 1. Screening the E6B human 9q32-34 clone collection with rat cDNA probes depleted of repetitive sequences.

Tissue Expression Pattern	Number of Clones Detected
Brain-Specific (Brain not Liver)	33
Liver-Specific (Liver not Brain)	9
Ubiquitous (Brain and Liver)	101
Caudate-Enhanced (Caudate > Brain)	0
Total Number Of Transcribed Sequences Detected	143

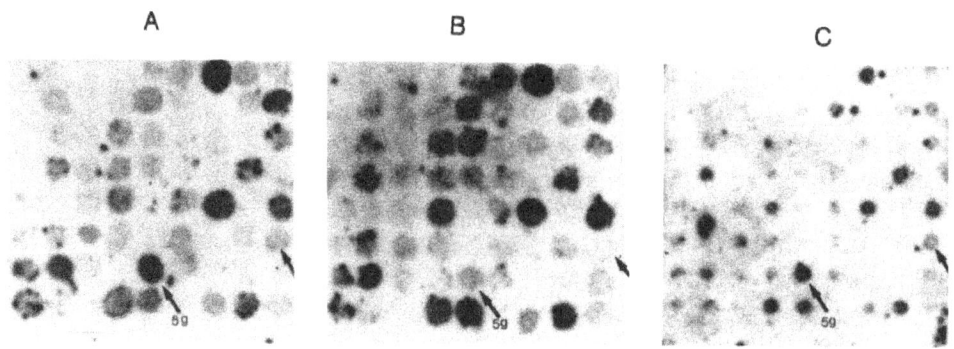

Figure 1. Screening of a human 9q32-34-specific clone collection with rat cDNA probes. Duplicate filters representing one microtiter dish hybridized with native rat cDNA probes from (A) brain and (B) liver, and (C) rat brain cDNA probe that has been depleted of repetitive elements. Arrows designate clones 24-1F and 24-5G that exhibit brain specific/enhanced expression patterns.

160

These probes hybridized to a total of 143 clones: 33 of these were detected with a whole brain probe but not with a liver or kidney probe; 9 were detected with a liver probe but not with a brain probe; and 101 were detected with both brain and liver probes. All of the clones detected with the caudate cDNA probe also exhibited similar levels of hybridization with the whole brain cDNA probe, suggesting that none of these clones corresponded to RNAs expressed in a striatal-enhanced manner. Since our collection represents three genome equivalents of the 9q32-34 region, the 143 transcripts detected by our screening may correspond to the identification of approximately 47 distinct genes.

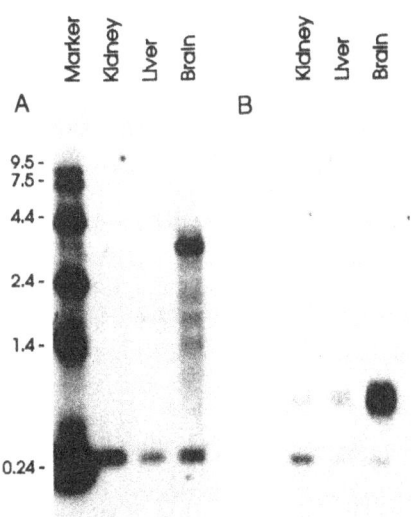

Figure 2. Northern blot analysis of 9q32-34 clones exhibiting brain specific/enhanced expression. Northern blots containing 2 μg/lane of rat poly(A) + RNA from brain, liver and kidney were hybridized with (A) whole phage DNA from clone 24-1F or (B) insert DNA from clone 24-2G.

Further characterization of these transcribed gene sequences is being accomplished through northern blot analysis using RNA from a variety of brain regions and peripheral tissues, and by mapping these expressed sequences with respect to YAC clones that have been localized within 1 Mb of the dystonia locus, DYT1 (14). This has resulted in the identification of three genes located proximal to the DYT1 locus, two of which are expressed in a brain-specific manner. Clone #24-5G encodes an 0.8 kb brain-enriched transcript whose concentration is highest in the cortex region of the brain. Clone # 12-7F specifies a 1.5 kb transcript whose expression was detected in the striatum, cortex, hippocampus, medulla, pons and brainstem, but was not detected in either the cerebellum or the hypothalamus. These clones are under investigation as candidates for the dystonia gene.

Identification of 9q32-34 Genes Exhibiting Enhanced Striatum Expression

Since the dystonic defect is thought to affect the basal ganglia region of the brain and we were unable to detect 9q32-34 clones exhibiting enhanced hybridization to conventional striatal cDNA probes (Table 1), the 9q32-34 collection was screened with a probe that was enriched in striatal-specific sequences. The probe was prepared using the directional tag PCR subtraction method (Fig. 3) described by Usui *et al.* (11). This entailed constructing directional libraries from striatal and cerebellar cDNA for use as sources of target and driver sequences, respectively. T7 polymerase was used to synthesize sense RNA from the target and driver cDNA, followed by the removal of any template DNA by digestion with RNase-free DNase. Tagged antisense striatal cDNA was then synthesized with reverse transcriptase, and the target and driver were hybridized at a 1:20 ratio. Single stranded, tagged, striatal-enriched target cDNA was isolated by hydroxyapatite chromatography.

Several advantages were inherent in the use of the directional tag PCR subtraction procedure. The primer sites present on the subtracted product facilitated PCR amplification of the product, labelling of the product for use as a probe, and cloning the product for library construction. Furthermore, driver contamination of the subtracted product was eliminated in this procedure since the target DNA is selectively amplified by PCR. In addition, this subtraction procedure has been performed successfully with as little as 10 ng, allowing very small quantities of RNA to be used.

To demonstrate that the probe prepared by this subtraction was in fact enriched in striatal sequences, it was extensively characterized prior to its use as probe to screen the 9q32-34 collection. Figure 4 demonstrates a comparison of native striatal and subtracted cDNA probes used to screen a panel of control cDNA clones of mRNAs whose expression patterns comprise two groups: RC3, CPU1 and 1G5 represent genes that are expressed in striatum but not in cerebellum; NSE and cyclophillin represent genes that are expressed in both striatum and cerebellum. The native striatal probe clearly hybridizes to both groups of genes (Fig. 4A), whereas, the subtracted cDNA probe hybridizes only to the genes expressed specifically in striatum but not in cerebellum, and the intensity of the hybridization signal is significantly enhanced (Fig. 4B).

Quantitation of the prevalence of these control genes in a native and subtracted striatum cDNA libraries (Table 2) demonstrated that genes expressed in striatum but not cerebellum were enriched by as much as 12 fold in the subtracted striatal library, while genes present in both striatum and cerebellum were depleted at least 40 fold from the subtracted library. These results are indicative of a successful subtraction in which sequences present in both cerebellum and striatal cDNA have been removed, resulting in a probe that is enriched in striatum-specific sequences.

Final confirmation of the effectiveness of this subtraction was obtained by identifying clones of novel striatum-enriched mRNAs through comparative screening of the subtracted striatal cDNA library with native cerebellum (Fig. 5A) and striatum (Fig. 5B) cDNA probes, as well as a subtracted striatum cDNA probe (Fig. 5C). This screening identified several clones that appeared to hybridize preferentially to the striatum-enriched subtracted probe.

Northern blot analysis with the clones confirmed that most corresponded to RNAs whose expression was restricted to or enriched in the striatum (Fig. 6).

Quantitation of the frequency of these clones in native and subtracted striatal cDNA libraries showed that they were enriched as much as 30 fold in the subtracted library (clone SE6C), and that clones of low abundance mRNAs, such as SE23B which represented less than 0.01% of the cDNA population, could be detected using this subtraction procedure (Table 3).

The striatum-enriched subtracted probe was subsequently used to screen the 9q32-34 collection, resulting in the detection of twelve clones. Of these, three exhibited strong hybridization signals while the remaining nine clones exhibited weak hybridization signals. Six

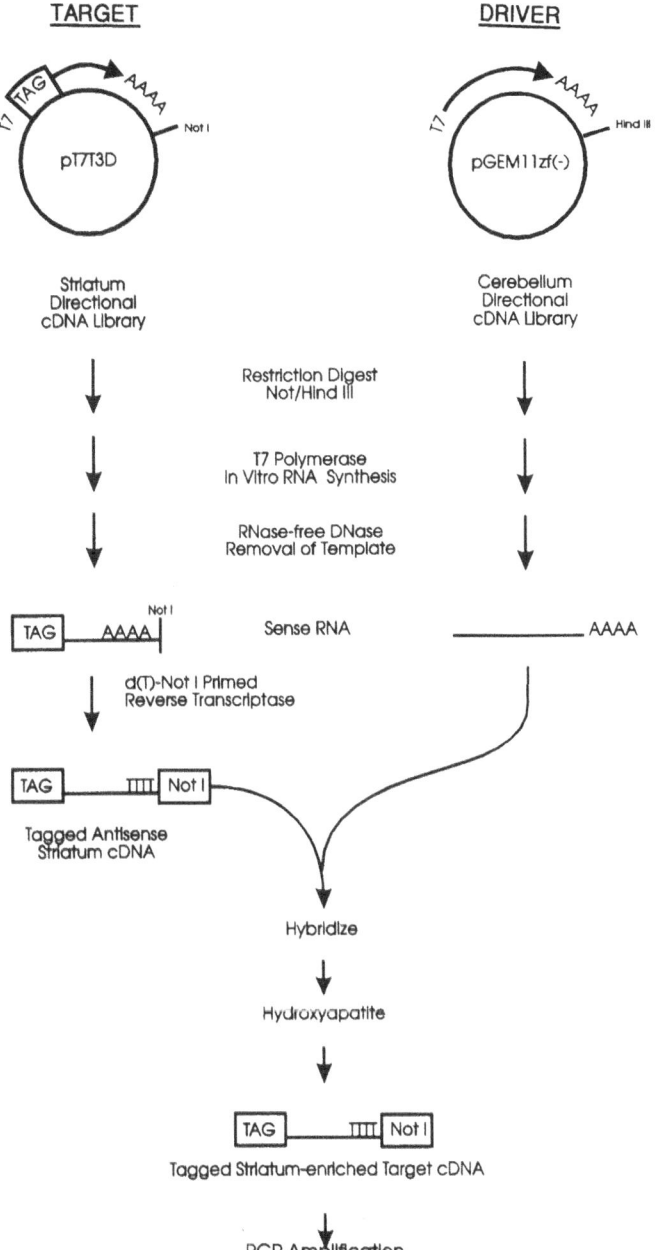

TARGET

DRIVER

T7 — TAG — AAAA
Not I
pT7T3D

T7 — AAAA
Hind III
pGEM11zf(-)

Striatum
Directional
cDNA Library

Cerebellum
Directional
cDNA Library

Restriction Digest
Not/Hind III

T7 Polymerase
In Vitro RNA Synthesis

RNase-free DNase
Removal of Template

Not I
TAG — AAAA   Sense RNA   ———————— AAAA

d(T)-Not I Primed
Reverse Transcriptase

TAG — IIII — Not I

Tagged Antisense
Striatum cDNA

Hybridize

Hydroxyapatite

TAG — IIII — Not I

Tagged Striatum-enriched Target cDNA

PCR Amplification

Figure 3. Summary of the directional tag PCR subtraction method used to isolate a rat striatum-enriched cDNA probe.

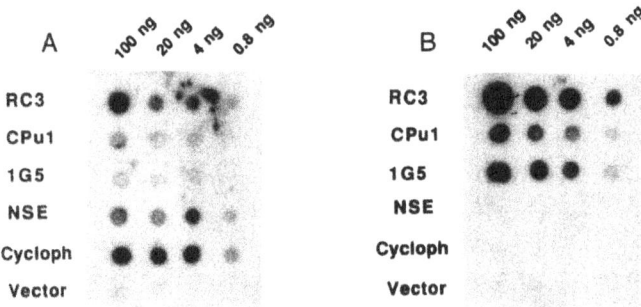

Figure 4. Screening of a panel of control genes with known expression patterns with a native striatal cDNA probe (A), or a cerebellum-minus-striatum subtracted cDNA probe (B).

Table 2. Frequencies of control clones in native striatum cDNA and subtracted striatum cDNA libraries.

Clone	Control cDNA Library		Subtracted cDNA Library		Relative Frequency
RC3	14/43,000	(0.32%)	31/2,860	(1.08%)	3.4-fold Increase
CPU1	2/10,910	(0.018%)	13/5,640	(0.23%)	12.8-fold Increase
IG5	5/65,470	(0.023%)	25/14,110	(0.18%)	7.8-fold Increase
Cyclo-phillin	19/4,300	(0.44%)	0/8,470		>40-fold Decrease

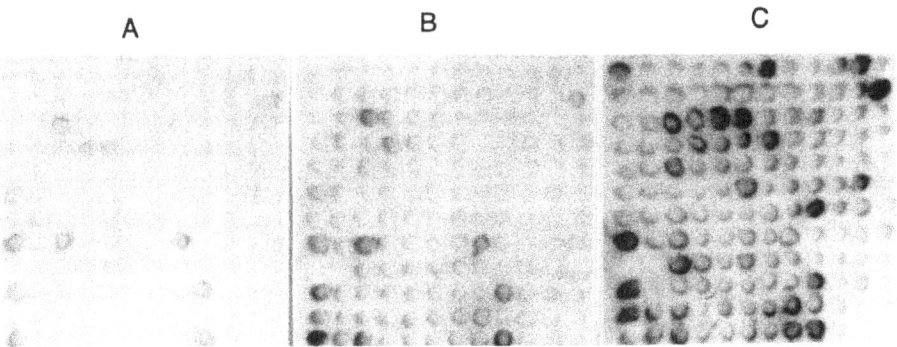

Figure 5. Screening of a gridded subtracted striatum library with a native rat cerebellum cDNA probe (A), a native rat striatum cDNA probe, and a cerebellum-minus-striatum subtracted rat cDNA probe (C).

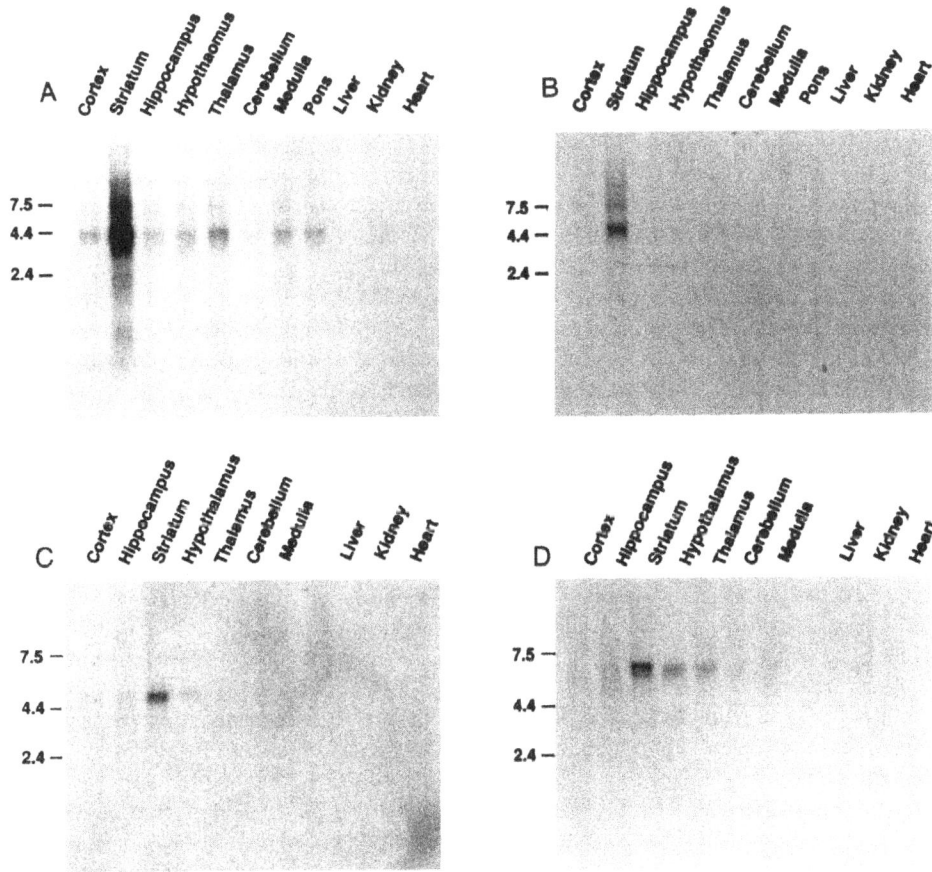

Figure 6. Northern blot analysis of novel rat genes expressed in a striatum-specific/enhanced manner. Northern blots containing 2 μg/lane of rat poly (A) $^+$ RNA from various brain and peripheral tissues were probed with clones SE7D (A), SE6C (B), SE23B (C) and SE7E (D).

Table 3. Frequencies of novel striatum-enriched clones in native striatum cDNA and subtracted striatum cDNA libraries.

Clone	Control cDNA Library	Subtracted cDNA Library	Relative Frequency
SE6C	2/4,900 (0.04%)	19/1,560 (1.22%)	30.5-fold Increase
SE7D	3/4,900 (0.06%)	13/1,560 (0.83%)	13.8-fold Increase
SE7E	1/9,800 (0.01%)	3/10,900 (0.027%)	2.7-fold Increase
SE23B	0/9,800 (<0.01%)	5/10,920 (0.04%)	>4-fold Increase

of the nine weakly hybridizing clones corresponded to clones previously identified by screening the collection with native cDNA probes. Northern blot analysis was used to characterize the expression patterns of the three clones that exhibited strong hybridization to the subtracted probe. The corresponding mRNAs for each clone were expressed preferentially in striatum as compared to cerebellum, in accordance with the parameters used for the subtraction procedure. However, none of the clones was expressed in an absolutely striatum-specific manner that would clearly implicate them as primary candidates for the dystonia gene.

DISCUSSION

We have assembled an arrayed collection of three genome equivalents of human chromosome 9q32-34-specific genomic clones that has been used to identify genes within this region and determine their transcription patterns. Rat cDNA probes from various tissues were used to identify 143 clones within this collection that contained expressed sequences. Thirty-three of these clones corresponded to transcripts that were expressed in whole brain and striatum, but exhibited little or no expression in liver (or kidney), which suggests that they were expressed in a brain-specific/enhanced manner. Transcripts of this nature represent potentially important neurological genes worthy of additional characterization to determine their complete identity and function. They also represent candidates for the gene responsible for idiopathic torsion dystonia and thus are being characterized based on their expression patterns and proximity to the dystonia locus, DYT1. Two of these genes are expressed specifically in the brain and map within 1 Mb of the DYT1 locus.

Since the basal ganglia has been implicated as the primary brain region affected by the dystonic defect, we attempted to isolate a 9q32-34 gene that was expressed specifically in the striatum. However, we have not yet been successful in identifying a striatum-specific gene using either a native striatal cDNA probe or a subtracted striatum-enriched probe. The directional tag PCR subtraction method used to generate the striatum-enriched probe was able to identify several clones that adhered to the basic principle of the striatum-minus-cerebellum subtraction in that they corresponded to RNAs expressed preferentially in striatum as compared to cerebellum, confirming that the subtraction procedure was successful. Characterization of the remaining nine clones that were detected by the subtracted probe may result in the identification of a striatum-specific gene.

The studies described in this paper demonstrate that the method of Hochgeschwender, Sutcliffe and Brennan (10) is applicable to the identification of transcribed sequences within a specific human chromosomal region. The ability to directly determine the chromosomal location and tissue expression pattern of each identified transcript make this an attractive method of gene mapping. It is particularly amenable to the identification of genes with a defined chromosomal location and expression pattern. Disease genes, such as the torsion dystonia gene, are well suited to characterization using this procedure. However, there are limitations inherent in this method such as its inability to detect rare transcripts. We estimate that the 143 transcripts we identified correspond to approximately 47 unique genes (based on the three 9q32-34 genome equivalents present in our clone collection). If there are 50,000-100,000 human genes, and genes are present within the 9q32-34 region at the average density, then we have detected 14-27% of the genes from this region while overlooking the remainder. In order to increase our detection sensitivity, we are exploring the use of PCR-based methods to enrich our cDNA probes for sequences localized in the 9q32-34 region.

ACKNOWLEDGEMENTS

We thank Patria Danielson for her technical assistance in RNA preparations and northern blot analysis. This work was supported by a grant from the Dystonia Medical Research Foundation and in part by NIH HG00332, NS22347 and NS22111.

REFERENCES

1. L. Ozelius, P.L. Kramer, C.B. Moskowitz, D.J. Kwiatkowski, M.F. Brin, S.B. Bressman, D.E. Schuback, C.T. Falk, N. Risch, D. de Leon, R.E. Burke, J. Haines, J.E. Gusella, S. Fahn, and X.O. Breakefield, Human gene for torsion dystonia located on chromosome 9q32- 34, *Neuron* 2:1427 (1989).

2. D.B. Calne and A. Lang, Secondary dystonia, *Adv. Neurol.* 50:9 (1988).

3. C.D. Marsden, J.A. Obeso, J.J. Zarranz, and A.E. Lang, The anatomical basis of symptomatic hemidystonia, *Brain* 106:461 (1985).

4. M.D. Adams, J.M. Kelly, J.D. Gocayne, M. Dubnick, M.H. Polymeropoulos, H. Xaio, C.R. Merril, A. Wu, B. Olde, R.F. Moreno, A.R. Kerlavage, W.R. McCombie, and J.C. Venter, Complementary DNA sequencing: expressed sequence tags and human genome project, *Science* 252:1651 (1991).

5. A.J. Buchler, D.C. Chang, S.L. Graw, J.D. Brook, D.A. Haber, P.A. Sharp, and D.E. Houseman, Exon amplification: A strategy to isolate mammalian genes based on RNA splicing, *Proc. Natl. Acad. Sci. USA* 88:4005 (1991).

6. S. Parimoo, S.R. Payanjali, H. Shukla, D.D. Chaplin, and S.M. Weissman, cDNA selection:Efficient PCR approach for the selection of cDNAs encoded in large chromosomal DNA fragments, *Proc. Natl. Acad. Sci. USA* 88:9623 (1991).

7. M. Lovett, J. Kere, and L.M. Hinton, Direct selection: A method for the isolation of cDNAs encoded by large genomic regions, *Proc. Natl. Acad. Sci. USA* 88:9623 (1991).

8. G.H. Travis, M.B. Brennan, P. Danielson, C.A. Kozak, and J.G. Sutcliffe, Identification of a photoreceptor specific nRNA encoded by the gene responsible for retinal degeneration slow (rds), *Nature* 338:70 (1989).

9. P. Liang and A.B. Pardee, Differential display of eukaryotic messenger RNA by means of the polymerase chain reaction, *Science* 257:967 (1992).

10. U. Hochgeschwender, J.G. Sutcliffe, and M.B. Brennan, Construction and screening of a genomic library specific for mouse chromosome 16, *Proc. Natl. Acad. Sci. USA* 86:8482 (1989).

11. H. Usui, J. D. Falk, A. Dopazo, L. de Lecea, M.G. Erlander, and J.G. Sutcliffe, Isolation of clones of rat striatum-specific mRNAs by diectional tag PCR subtraction, *J. Neurosci.* In press, (1994).

12. E.P. Henske, L. Ozelius, M.A. Anderson, and D.J. Kwiatkowski, A radiation-reduced hybrid cell line containing 5 Mb/17 cM of human DNA from 9q34, *Genomics* 13:841 (1992).

13. J. Sambrook, E.F. Fritsch, and T. Maniatis, Molecular Cloning: A Laboratory Manual, Cold Spring Harbor Laboratory Press, New York (1989).

14. L. Ozelius and X. Breakefield, personal communication.

AN EXON TRAPPING SYSTEM PROVIDING SIZE SELECTION OF SPLICED CLONES AND FACILITATING DIRECT CLONING

Nicole A. Datson[1], Geoffrey M. Duyk[2], Lau A.J. Blonden[1], Gert-Jan B. Van Ommen[1] and Johan T. Den Dunnen[1]

[1]Department of Human Genetics, Medical Genetics Centre South-West Netherlands, Leiden University, Wassenaarseweg 72, 2333 AL LEIDEN, The Netherlands
[2]Department of Genetics, Howard Hughes Medical Institute, Harvard Medical School, Boston MA 02115

ABSTRACT

Exon trapping is a method to functionally clone expressed sequences from genomic DNA. We have developed an exon trapping procedure based on the use of vector pETV-SD2. Cosmid DNA is partially digested and cloned in pETV-SD2. DNA of an entire library of subclones is introduced into COS-1 cells and transiently expressed. RNA is isolated and vector-derived transcripts are amplified by RT-PCR. Cloning of the RT-PCR products, which contain a ColEI-origin of replication and a supF marker, is established by NotI digestion and intramolecular circularisation. Due to their shorter length, spliced clones are preferentially amplified and cloned.

Using this approach, we have been able to trap several exons from test cosmids of the DMD-gene, including the single 176 bp exon 45 present in 40 kb of intronic sequences. We have applied this system in the search for coding sequences in various genomic regions, including the candidate region for the gene involved in facioscapulohumeral muscular dystrophy (FSHD) on 4q35. We have developed a screening procedure which allows more efficient identification of potential exon containing clones, thus discarding the high background of false positives.

INTRODUCTION

In the last decade major advances have been made in cloning technology, allowing the cloning of increasingly large fragments of genomic DNA, as in P1 and YAC clones (1-3). The identification of transcribed sequences in these large stretches of DNA is a limiting factor in the construction of detailed transcription maps of chromosomal regions. Commonly applied indirect methods in the search for genes include the identification of HTF-islands (4), interspecies sequence conservation (5) and the use of computer programs which calculate the probability that a stretch of DNA encodes a protein (6,7). Direct methods identify genes based on their expression by screening cDNA libraries and Northern blots with phage, cosmid or YAC DNA (8) or by hybridisation selection in which cDNAs homologous to a specific cosmid or YAC clone are selected in multiple rounds of enrichment by hybridisation (9,10).

With many of these techniques, subcloning of the DNA into smaller units is necessary and only small numbers of sequences can be analysed simultaneously. Consequently, the screening of large genomic regions is not technically feasible. Hybridisation based methods do allow large genomic regions to be screened, but as with all expression-based techniques, their success depends on the availability of RNA from a tissue expressing the gene of interest. Furthermore, there is the intrinsic risk of cloning homologous genes which reside on other chromosomes.

Exon trapping is a gene identification technique that is independent of gene expression since it selects genes at the DNA level by the functional detection of sequences involved in RNA splicing. Several variants have been described (11-14), allowing the relatively easy generation of probes containing potential coding sequences, that can be used to probe cDNA libraries and Northern blots or for sequence analysis. Recently, exon trapping has led to the successful cloning of a number of genes, including those responsible for Huntington's Disease (15), Neurofibromatosis type II (16), X-linked glycerol kinase deficiency (17) and α-adducin (18).

In most exon trapping methods DNA of interest is cloned between a splice donor (SD) and splice acceptor (SA) site (12-14). Retention of exons from in the cloned DNA between the vector-derived exons will result in a transcript of increased length. Due to the abundance and shorter length of the unspliced product; however, it will be favoured during PCR amplification above the larger, less abundant products containing trapped exons. In this paper we describe a variant of exon trapping based on the method of Duyk et al. (11). We use an exon trap cassette which contains a SD site only and an internal LacZα gene. Spliced transcripts will not contain the LacZα-region and are smaller than their unspliced counterparts. Consequently they will have a white clone phenotype and an amplification advantage in the subsequent PCR step. Furthermore, an amplifiable cassette with a dominant selectable marker (supF gene), an origin of replication and primer-derived NotI sites, allow efficient cloning of the RT-PCR products, by digestion and circularisation.

MATERIALS AND METHODS

Bacteria and Cell Lines

Vector pETV-SD2 and the cosmid subclone libraries were propagated in *Escherichia coli* DH10BP3, a strain derived from *E. coli* DH10 containing a P3 plasmid (kan^R tet(am) amp(am)) (19) obtained from BRL. The P3 plasmid was isolated from MC1061/P3 by an alkaline lysis mini-prep procedure. COS-1 cells were cultured in DMEM with 10% inactivated fetal calf serum (GIBCO BRL).

Vector Construction

pETV-SD2 is described in Fig. 1A. pETV-SD2 is a derivative of pETV-SD (11) and was constructed using standard methods. Briefly, the SV40 Ori-Tn5 Neo cassette was deleted and a unique EcoRI site was inserted in its place. The exon trap cassette was modified by the insertion of a synthetic supF gene immediately 5' to the human betaglobin exon 1 domain. The supF gene was subcloned from CDM8 (pos. 1181-1395) by PCR amplification using the synthetic oligonucleotide primers ODC 114 (5' GGGGAGATCTCTAGAGCGGTCTTTCTC) and ODC 115

(GGCTGGATCCGGGAGCAGATTCTTTCGGAC).   The SV40 ori incorporated into the backbone of pETV-SD2 is identical to the SV40 ori present in pETV-SD (position 583-927) and is transcribed in a clockwise direction.

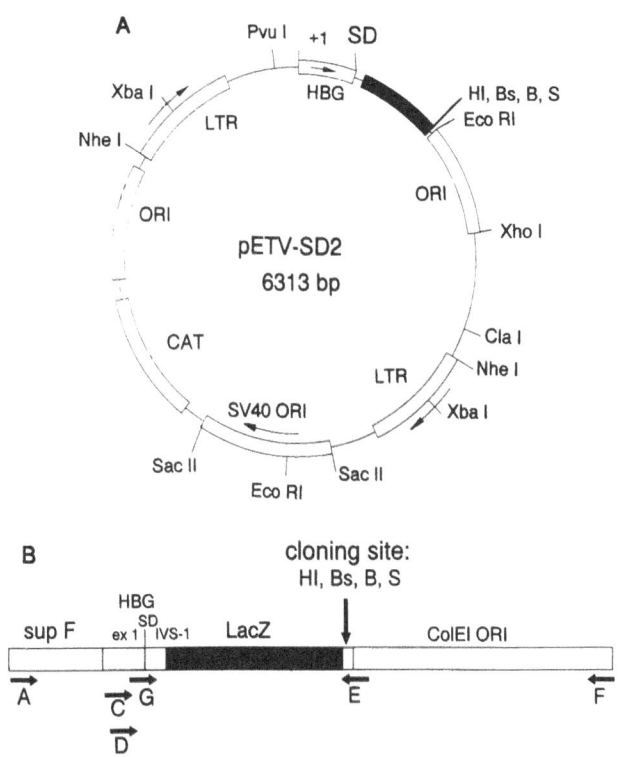

Figure 1. A) Physical map of pETV-SD2. **B)** Exon trap cassette. The cassette contains exon 1 of the human ß-globin gene (HBG, positions 54-142) and 50 bp of intervening sequence 1 (IVS-1, positions 143-192). A 404-bp fragment encoding the LacZα gene is followed by unique BamHI, BstXI, BclI and SalI cloning sites. The supF-marker allows selection for ampicillin and tetracycline resistance in DH10BP3. Circularisation of RT-PCR product A/F creates a functional plasmid.

Oligonucleotides and Hybridisation Probes

The positions of the various oligonucleotides are shown in Fig. 1B. The sequences are as follows (5'→3'):
A=TCAGCGGCCGCAGATCTCTAGAGCGGTCTTTCTCAACGTAAC,
C=GGAGAAGTCTGCCGTTACTG, D=GGTGAACGTGGATGAAGTT,
E=GCGGAAT TCGTCGACTCTAGCCATGATC,
F=ACTGCGGCCGCTCGAGTGCTGGCGTTTTTCC ATAGGC and
G=CCTGGGCAGGTTGGTATCA.   cDNA (5b-7) and (10b-11) are subclones of the DMD cDNA (20) and were used to screen for exon containing clones. cDNA (5b-7) contains exons 32 to 47 and cDNA (10b-11) exons 65 to 76. PCR products were purified by excising the band from an agarose gel and spinning through glasswool.

Cosmids

cPT1 (21) contains exon 45 (176 bp) and cos3 (A.P. Monaco, unpublished) contains exons 68-74 (167, 111, 136, 39, 66, 66 and 159 bp, respectively) (22) of the DMD-gene. cAL24 (21) is located within the P20 intron of the DMD-gene. cy13 and cy34 (T.J. Wright, unpublished) are cosmids in the FSHD candidate region.

Construction of Cosmid Subclone (input) Libraries

Cosmids were subcloned in pETV-SD2 after partial digestion with Sau3AI. The partial digest with most fragments larger than 2 kb was selected for cloning into pETV-SD2, linearised with BamHI and dephosphorylated with Shrimp Alkaline Phosphatase (United States Biochemical Corporation). The ligated products were introduced into DH10B/P3 by electroporation (Gene Pulser, BioRad) using standard *E. coli* electroporation parameters. A 1 litre culture of *E. coli* cells was prepared for electroporation by spinning down and resuspending the cell pellet in 1 litre of a cold 10% glycerol solution. This was repeated, lowering the volume of 10% glycerol solution from 1 liter via 500 ml and 20 ml to finally 2 ml. Subsequently, the cells were aliquotted into 40 μl portions and frozen at -80°C until required. After electroporation the libraries were plated on LB containing ampicillin (30 μg/ml), tetracycline (25 μg/ml) and chloramphenicol (25 μg/ml). DNA of the total library was isolated by transferring agar from a plate with at least 200 colonies to a large flask containing Terrific Broth (23) and antibiotics. After overnight growth, agar was removed from the cultures by filtering through cheese cloth. Plasmid DNA was isolated using the alkali lysis method followed by cesium-chloride purification (23).

Introduction of DNA into COS-1 Cells

10 μg of DNA of the input libraries was co-transfected with 1 μg of the human growth hormone construct pXGH5 (Nichols Institute, San Juan Capistrano, USA) to establish transfection efficiency. This was assessed after harvesting by measuring the concentration of human growth hormone (hGH) secreted into the COS-1 cell culture medium (see below). DNA was introduced into the COS-1 cells by electroporation: 10^7 cells were collected, washed extensively in cold PBS, resuspended in 0.5 ml cold PBS and placed in a prechilled electroporation cuvette (0.4 cm chamber, BioRad). 100 μl of PBS, containing 11 μg of DNA, was added to the cells. After 5 min on ice, the cells were electroporated in a Biorad Gene Pulser (300 V [750 V/cm]; 960 μF), and placed on ice again. After 5 min, the cells were transferred gently to a 100 mm tissue culture dish containing 10 ml of prewarmed, equilibrated DMEM + 10% FCS.

As an alternative approach, 1 μl of ligation mixture of the input libraries was transfected directly into COS-1 cells with 10 μg carrier DNA (human placenta).

RNA Isolation and RT-PCR

48-72 h after transfection the cells were harvested and RNA was isolated as described (24). The transfection efficiency was tested in 100 μl of the cell culture medium by assaying the HGH-concentration using the Allégro HGH Transient Gene Assay Kit (Nichols Institute, San Juan Capistrano, USA). First-strand cDNA synthesis was performed by adding 1 μg of random primers (Promega) to 10 μg total RNA in a volume of 32 μl. The mixture was incubated at 65°C for 10 min and chilled on ice. 28 μl of a mix containing 6 μl 0.1 M DTT, 6 μl 10 mM dNTPs, 1 μl RNasin (40 units/μl) (Promega), 12 μl 5x reaction buffer (250 mM Tris-HCl pH 8.3, 375 mM KCl, 15 mM MgCl$_2$; GIBCO BRL) and 600 units Moloney Murine Leukemia Virus H$^-$ Reverse Transcriptase (GIBCO BRL) were added to a final volume of 60 μl, and incubated at 42°C for 1 h. Subsequently, the solution was heated to 95°C for 5 min and chilled on ice. 4.5 units of RNaseH (Promega) were added and the solution was incubated at 37°C for 20 min. 10 μl of the solution was used in a PCR reaction containing 12.5 pmol each of primers A and F, 50 mM KCl, 1.5 mM MgCl$_2$, 10 mM Tris.HCl pH 8.0, 0.2 mM dNTPs, 0.2 mg/ml BSA and 0.25 units SuperTaq (HT Biotechnology Ltd) in a reaction volume of 50 μl, followed by an initial denaturation step of 10' at 94°C, 30 cycles of amplification (1 min at 94°C, 1 min at 60°C and 3.5 min at 72°C) and a final extension of 10' at 72°C.

Cloning of the RT-PCR Products by Intramolecular Circularisation

Products of four parallel reactions were pooled, extracted with phenol and phenol-chloroform and precipitated with ethanol. The products were digested with NotI and purified on a Sephadex G-50 column. The fraction containing the PCR material was ethanol precipitated, using 50 μg carrier tRNA. The pellet was dissolved in 20 μl of H$_2$O. 4 μl was used in a ligation, containing 0.1 mg/ml BSA, 1 μl 10 x ligation buffer (Promega) and 0.6 units T4 DNA ligase (Promega) in a total volume of 10 μl. After overnight incubation at 15°C, 1 μl of ligation mix was transformed by electroporation into *E. coli* DH10B/P3. Clones were plated on LB containing ampicillin (30 μg/ml), tetracycline (25 μg/ml), X-gal (30 μg/ml) and IPTG (30 μg/ml). After 24-48 h of incubation at 37°C, white clones were selected for further analysis.

Sequencing of PCR Products

Direct sequencing was performed on PCR products synthesized with primers C and E. Biotinylated primer E allowed purification of the single stranded DNA using avidin-coated magnetic beads (Dynal). An internal primer, D, was used for sequencing.

Southern Blot Analysis and Filter Lifting

Restriction digestion, electrophoresis, blotting, labelling, hybridisation and filter lifting were performed using standard procedures (21,23).

RESULTS

Outline of Exon Trapping Procedure

The procedure for exon trapping with vector pETV-SD2 is summarised in Fig. 2. Cosmid DNA is subcloned in pETV-SD2 and introduced into COS-1 cells by electroporation. Transcription of the introduced DNA is driven by a SV40 promotor present in the vector. The cis-acting sequences on the ß-globin exon-intron boundary serve as a functional splice donor site (SD) for splice acceptor (SA) sequences cloned downstream. RNA is isolated and vector-derived transcripts are amplified by reverse transcriptase PCR (RT-PCR) with vector-specific primers, generating products that give rise to a functional plasmid after circularisation.

The exon trap cassette of pETV-SD2 is depicted in Fig. 1B. pETV-SD2 contains exon 1 of the human ß-globin gene, followed by part of intron 1 (IVS-1). The α-complementing factor of the LacZ gene is located between the SD and the multiple cloning site. A supF marker and a ColEI replication origin are located at either end of the cassette. Direct cloning is achieved by transformation of circularised RT-PCR products to *E. coli* DH10BP3. Spliced clones can be distinguished in a blue/white screening based on splicing of the LacZα gene.

Functional Cloning of a Known Exon

To test the system, cosmid cPT1, known to contain exon 45 of the human DMD-gene (21), was subcloned in pETV-SD2. To test if specific cosmid regions were enriched, RT-PCR material of the cPT1 trapping was used directly to probe HindIII digested cPT1. The 0.5 kb HindIII fragment known to contain exon 45 (21) was the only hybridising band (Fig. 3). This indicates that a hybridisation test can be used for an initial analysis of the success of the exon trapping protocol, i.e. to check if specific sequences have been enriched. Furthermore, without further analysis, this probe can be used to directly screen cDNA libraries or zooblots.

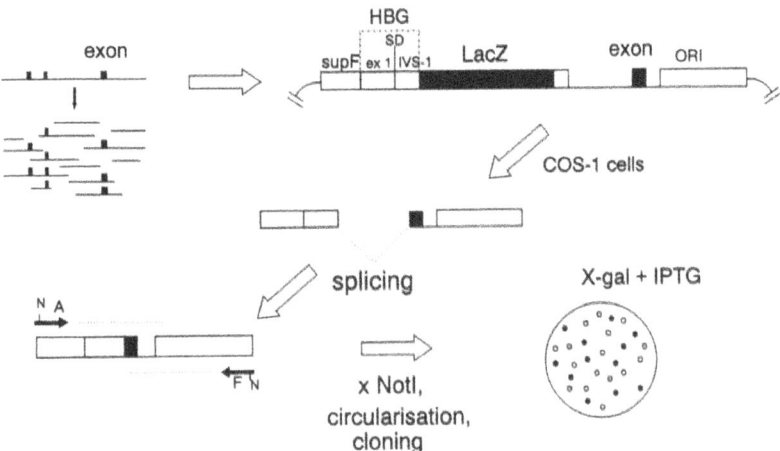

Figure 2. Exon trapping procedure. Input DNA is partially digested with Sau3AI and fragments of 1-2 kb are cloned in pETV-SD2. DNA of the pooled clones is introduced into COS-1 cells, with vector sequences allowing plasmid propagation and expression. Posttransfection, total RNA is isolated and a first-strand random primed cDNA synthesis is performed. cDNA is PCR-amplified using vector-derived primers containing NotI-sites (N). The RT-PCR products are digested with NotI, and circularised, creating a functional plasmid. Clones are colour-selected when loss of the LacZα gene, i.e. splicing, has occurred. White clones are analysed further.

White colonies were picked onto a fresh plate and after overnight incubation at 37°C a filter lift of the colonies was made. Hybridization with cDNA (5b-7) showed that 12 out of 78 (15%) of the tested cPT1-derived white clones were positive. Colony PCR was performed on the positive clones with primer pair C/E (Fig. 1B), resulting in a PCR product of 375 bp. This PCR product was hybridised to HindIII digested cPT1 and mapped to the same band 0.5 kb fragment that was positive in the hybridisation with the total RT-PCR material (Fig. 3).

Unequivocal proof of correct splicing of the clones was obtained by sequence analysis. This showed that the ß-globin exon had been spliced correctly to exon 45 of the DMD-gene. The trapped sequence differed from the genomic sequence at the 3' end of exon 45, indicating a second splicing event at its splice donor site (Fig. 4).

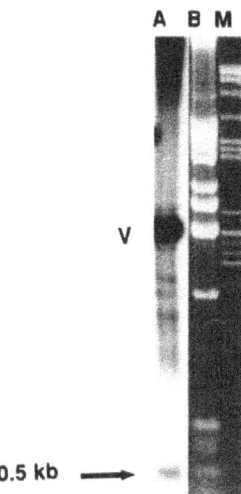

Figure 3. Hybridisation of the RT-PCR products derived from exon trapping of cosmid cPT1 against a HindIII-digest of cPT1. A: RT-PCR products of cPT1 hybridised against HindIII-digested cPT1; B: ethidium-bromide stained gel of HindIII-digested cPT1; M: marker (λ HindIII/PstI); V: vector fragments. The hybridising 0.5 kb HindIII fragment contains exon 45 of the DMD-gene. The other hybridising bands are probably due to partial digestion of cPT1.

```
                      ┌─── exon 45 DMD-gene
         intron 44    │                        ┌── intron 45
         tatcttacagGAACT...      ...AAGAGgtagggcgac
                   | | | | |              | | | | |
cPT1.5   CCCTGGGCAGGAACT...      ...AAGAGGTTTCCACCT
         | | | | | | | | |
         CCCTGGGCAGgttgg...
         HBG exon 1  IVS-1
         └─────────┘


                      ┌─── exon 68 DMD-gene
         intron 67    │                        ┌── intron 68
         gtctttgcagGCTAA...      ...TTCAGgtattaggaa
                   | | | | |              | | | | |
cos3.23  CCCTGGGCAGGCTAA...      ...TTCAGTTGCTGTTAG
         | | | | | | | | |
         CCCTGGGCAGgttgg...
         HBG exon 1  IVS-1
         └─────────┘
```

Figure 4. Sequence of exon trap clones cPT1.5 and cos3.23 containing trapped exons 45 and 68 of the DMD-gene respectively. Intronic sequences are in lower case, exonic sequences in upper case. Both clones have undergone a second splicing event at their SD site.

Elimination of False Positives

Since only 15% of the gridded white clones contained a correctly spliced exon 45, this, while encouraging, suggested a significant source of false positives amongst the white colonies, which needed to be identified and discarded. Sequence analysis of white clones obtained from several trapping experiments revealed two different classes of false positives: i) clones that were spliced to a cryptic SA in the vector downstream of the cloning site and ii) clones with mutations resulting in a non-functional LacZα gene: internal deletions or translation terminating point mutations. The cryptically spliced clones can be easily recognized by a PCR assay using a reverse primer upstream of the major cryptic splice site (primer E); only clones that possess the region between the cloning site and the cryptic SA-site will yield a product. Of 38 white cPT1 exon trap clones screened with primer pair C/E, 28 (74%) gave no product and could thus be immediately discarded, improving the yield of potentially positive clones four-fold. Alternatively, hybridisation with oligo E can be performed, allowing identification of cryptically spliced clones which are negative.

Hybridisation with oligo G, which straddles the ß-globin exon/intron junction and should be absent in bona fide spliced clones, allows identification of clones with a defective LacZα gene that still contain the IVS-1 region.

To screen out false positives we now routinely pick white clones into microtiter dishes and prepare high-density grids using a Biomek Automated Workstation (Beckmann). Based on the hybridisation results with oligo's E and G, most of the false positives are discarded, i.e. cryptically spliced clones (E-/G-) and deletion clones (E+/G+). Finally, the E+/G- clones are further processed (Fig. 5) and checked for size in a PCR with primer sets A/F and C/E. Only clones that are large enough to contain a spliced exon (over 937 bp with C/E) and that manifest the correct size difference between PCR products A/F and C/E (102 bp), are subjected to sequence analysis to confirm splicing of a novel sequence to exon 1 of ß-globin.

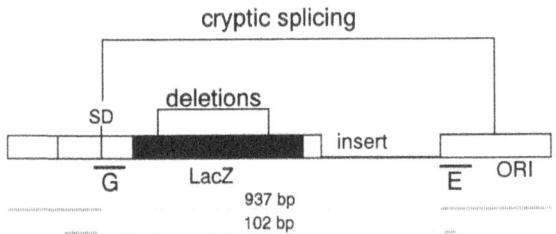

Figure 5. The screening procedure applied to distinguish clones containing trapped exons from the two classes of false positive clones, i.e. cryptically spliced clones and clones with a defective LacZα gene.

Trapping in Other Genomic Regions

Having shown that the procedure was capable of detecting a single 176 bp exon in 40 kb of surrounding DNA, we proceeded to use the system to trap other regions, including regions exhibiting differential splicing and regions with unknown gene content.

Cosmid cos3 spans a region of the DMD-gene which is involved in differential splicing (22,25). Our interest in this region was fundamental, i.e. to see how pETV-SD2 performed in trapping constitutively expressed exons versus exons that are frequently skipped in tissue-specific transcripts. We obtained several correctly spliced clones corresponding to exon 68 (Table 1). The other six exons present in the cosmid were not trapped in the set of clones analysed.

Exon trapping was also performed in 2 genomic regions with unknown gene content: the P20 intron of the DMD-gene on Xp21 and the candidate region for FSHD on 4q35. Cosmid cAL24 from the deletion-sensitive intron 44 of the DMD-gene, which has been hypothesised to contain another gene residing within the DMD-gene (26), was used in the exon trapping procedure. Computational analysis with the program GRAIL predicts the presence of 3 excellent coding regions within cAL24. One correctly spliced clone was obtained among 6 potential exon containing products (Table 1).

Recently chromosome 4 specific DNA rearrangements were observed with probe p13E-11 in FSHD patients (27). A shorter EcoRI fragment was observed in patients than in normal individuals caused by deletions of integral copies of a 3.2 kb tandemly repeated unit (28). Exon trapping was performed on two cosmids (cy13, cy34) in the vicinity of the FSHD-specific DNA rearrangements to identify a possible candidate gene involved in the pathogenesis of this muscle disorder. Cosmid cy13 encompasses the entire EcoRI fragment containing the tandem repeat and extends in proximal direction. Cosmid cy34 shows a slight overlap with cy13 and is located proximal to cy13. One correctly spliced clone was derived from cy13. Cosmid cy34 did not yield any correctly spliced products. Table 1 summarises the trapping data.

DISCUSSION

We have designed an exon trapping system, using the vector pETV-SD2, which distinguishes itself most prominently from other systems (12-14) by containing a SD site only. Since exon containing clones lose the LacZα-region, spliced transcripts are substantially smaller than unspliced transcripts derived from exon-less clones, resulting in a more efficient amplification. Other methods, using both a SD and SA (12-14), give increased transcript lengths when exons are trapped, leading to less efficient amplification. The pETV-SD2 system also provides direct cloning of RT-PCR products by NotI digestion and intramolecular circularisation. Simultaneously this is a selective step, since only full-length vector-derived transcripts with the SupF and ColEI-ori will give rise to a functional plasmid.

The flexibility of the pETV-SD2 system, using only a SD, allows the trapping of exons which are not cloned entirely, e.g. exons which contain an internal Sau3AI-site. Furthermore, with a few minor modifications, pETV-SD2 can be used to clone 3' exons simply by performing 3' RACE, i.e. RT-PCR with a forward vector primer and a reverse tagged oligo-dT primer (29). This can be performed on the same RNA used for trapping of internal exons, requiring construction of one subclone library only. The advantage of trapping 3' exons (30) is that they are multiply selected for a SA site and polyadenylation, which can be easily checked later. Furthermore, they are larger than internal exons, thus giving better probes for screening of cDNA libraries. Finally, since most genes only contain a single 3' exon, they are excellent landmarks of genes.

We have performed 3' exon trapping on a cosmid containing 4 entire genes of the α-globin gene cluster (α-1, α-2, zeta-1 and theta-1). Gel electrophoresis, blotting and hybridisation of the 3' RACE products with a probe directed against one of the α-globin cluster genes and a probe for ß-globin exon 1, allowed the detection of discrete bands which were positive for both probes. The length of the fragments corresponded to splicing of ß-globin exon 1 to the 3' terminal exon of the globin genes. Sequence analysis should provide the definite proof that the visible bands represent trapped 3' exons (Datson et al., manuscript in preparation).

Improving the Procedure

There are several possibilities to improve the exon trapping protocol described. The high background of false positives is a problem, but fortunately the majority can be easily identified. Eliminating the major cryptic SA downstream of the cloning site was not attempted since it might activate a new cryptic splice site that is not as easily identifiable.

The cloning strategy using Sau3AI partials prevents exons that do not have a Sau3AI site in their vicinity from being picked up. First, the cloning step favours smaller insert clones, and second, the RT-PCR step selects against exons where Sau3AI sites reside far downstream of the SA-site. Cloning fragments derived from sonicated DNA or random PCR-amplified DNA (31) by linker/adaptor addition would solve this problem by yielding a more randomised clone library. The latter possibility is also useful to obtain a sufficient amount of YAC-derived DNA for cloning in pETV-SD2.

Finally, experimental time could be saved if the propagation of a subclone library preceding the COS-1 cell transformation would not be necessary. Preliminary experiments show that direct transfection of the ligation products yields exon trap clones in the same

cosmid	white (n)	hyb				PCR			e.c.p. (n)	g (n)	spliced (n)
		cDNA+	E+ (n)	G- (n)	E+/G- (n)		A/F (n)	C/E (n)			
cPT1	78	12/78 (15%)	nt	nt	nt	>\<\NP	18/38 (47%) 12/38 (32%) 8/38 (21%)	9/37 (24%) 6/37 (16%) 22/37 (59%)	9	4	2
cos3	94	6/94 (6%)	nt	nt	nt	>	nt	6*/6 (100%)	6	2	1
cAL24	48	nt	nt	nt	nt	>\<\NP	nt	9/48 (19%) 0 (0%) 39/48 (81%)	9	6	1
cy13	96	nt	20/96 (21%)	59/96 (61%)	6/96 (6%)	>\<\NP	19/96 (20%) 76/96 (79%) 1/96 (1%)	36/96 (38%) 2/96 (2%) 58/96 (60%)	1	1	0
cy34	96	nt	13/96 (14%)	58/96 (60%)	10/96 (10%)	>\<\NP	22/96 (23%) 69/96 (72%) 5/96 (5%)	28/96 (29%) 2/96 (2%) 66/96 (69%)	7	6	2
cy34 (lig.mix)	96	nt	34/96 (35%)	23/96 (24%)	7/96 (7%)	>\<\NP	36/96 (38%) 54/96 (56%) 6/96 (6%)	25/96 (26%) 6/96 (6%) 65/96 (68%)	0	0	0

quantity and with the same blue-white ratio as transfection of DNA that has undergone an amplification step in bacteria (Table 1). A higher complexity of the input library will be obtained using this procedure.

Differential Splicing

Exon trapping has been described to discriminate between constitutive and differentially spliced exons (32) by specific trapping of constitutive exons only. We have tested this in pETV-SD2 by exon trapping a cosmid containing exons 68-74 of the DMD-gene, a region which exhibits extensive differential splicing. Only exon 68 was detectable as a trapped product. Although the SA of exon 68 scores best for adherence to consensus splice sites, alternative splicing of this exon has been observed in smooth muscle (25). In contrast, the two constitutive exons 69 and 70 were not detectable as trapped products in the set of 94 clones we have analysed. Perhaps the parallel use of a SD based and a SD/SA based exon trapping system will allow trapping of more exons in differentially spliced regions, but a major influence of the SA on the use of the SD cannot be excluded.

P20 Intron of the DMD-Gene

Exon trapping of a cosmid from this deletion-sensitive hotspot in the DMD-gene yielded one correctly spliced clone. Computational analysis of the sequence of this cosmid with the program GRAIL predicted three regions with excellent coding potential. The trapped sequence from this cosmid shows overlap with one of the predicted regions, although on the opposite strand. The trapped sequence itself is also predicted to contain a good exon by GRAIL. Future experiments will show if it is part of a putative gene within this region.

FSHD Candidate Region

Trapping of two cosmids within this region yielded one correctly spliced clone. The FSHD candidate region on 4qter is subtelomeric and full of repetitive sequences, and so is a difficult region from which to isolate coding sequences by hybridisation based methods. Direct cDNA selection experiments (33) with the same two cosmids showed a clustering of clones on a specific EcoRI fragment of cosmid cy34. Many of these clones are derived from homologous regions (R.R. Frants, personal communication). The sequence trapped with pETV-SD2 maps to the same fragment, confirming the results obtained by gene tracking. Using an oligo designed to the trapped sequence we have amplified inserts from various cDNA libraries. Currently we are analysing these sequences in more detail.

Problems Inherent to Exon Trapping

On average, in a randomly chosen genomic region one would expect less than one third of the cosmids to contain coding sequences. So far we have detected at least one

---

Table 1. Summary of exon trap data. (See facing page.) **NP**: no product; **nt**: not tested; >,< 937 bp (A/F); >,< 102 bp (C/E); **e.c.p.**: potential exon containing products; for cosmids cPT1, cos3 and cAL24, all clones that yielded a PCR product with primer set C/E were considered e.c.p.'s; for cosmids cy13 and cy34 more stringent criteria were used: E+/G- clones with a minimum PCR product of 937 bp with primer set C/E and manifesting the correct size difference of 102 bp between PCR products C/E and A/F; **g**: different groups of PCR products based on difference in length of PCR products; **spliced**: number of correctly spliced groups *: only the clones positive with the DMD cDNA were tested.

correctly spliced sequence per cosmid, testing regions that are known to contain a gene. The most difficult part of exon trapping is to conclude that a region is devoid of exons. One may always get clones, but not all trapped clones will represent bona fide spliced exons. Within genes, cryptic SA sites may be activated when they are taken out of their genomic context on a DNA fragment containing no other splice signals. They remain silent *in vivo* either because they are not followed by a proper SD, i.e. do not constitute an exon) or reside on a non-coding strand. Other sequences representing potential exons may be trapped from regions which are never transcribed, for instance intergenic regions.

In summary, while the use of a vector with only a SD may reduce some of the specificity and sensitivity, our system provides a positive size selection for spliced products as well as facilitating cloning of trapped products. Its flexibility allows trapping of exons which are not entirely cloned as well as of differentially spliced exons. Moreover, with a few minor modifications the system can be applied to trap 3' terminal exons.

ACKNOWLEDGEMENTS

We thank Dr. A. Monaco for providing cos3, Dr. T. Wright and Dr. C. Wijmenga for contributing cy13 and cy34 and the clone maps, and Dr. D. Cox and Dr. R. Myers for their contribution to the development of the exon trap vector. This work was supported by a grant from the Medical Genetics Centre-South West Netherlands (MGC-ZWN).

REFERENCES

1. D.T. Burke, G.R. Carle and M.V. Olson, Cloning of large segments of exogenous DNA into yeast by means of artificial chromosome vectors, *Science* 236:806 (1987).
2. R. Anand, A. Villasante and C. Tyler-Smith, Construction of yeast artificial chromosome libraries with large inserts using fractionation by pulsed-field gel electrophoresis, *Nucl. Acids Res.* 17:3425 (1989).
3. J.C. Pierce, B. Sauer and N.L. Sternberg, A positive selection vector for cloning high molecular weight DNA by the bacteriophage P1 system: improved cloning efficiency, *Proc. Natl. Acad. Sci. USA* 89:2056 (1992).
4. A.P. Bird, CpG-rich islands and the function of DNA methylation, *Nature* 321:209 (1986).
5. A.P. Monaco, R.L. Neve, C. Colletti-Feener, C.J. Bertelson, D.M. Kurnit and L.M. Kunkel, Isolation of candidate cDNAs for portions of the Duchenne muscular dystrophy gene, *Nature* 323:646 (1986).
6. S. Dear and R. Staden, A sequence assembly and editing program for efficient management of large projects, *Nucl. Acids Res.* 19:3907 (1991).
7. E.C. Uberbacher and R.J. Mural, Locating protein-coding regions in human DNA sequences by a multiple sensor-neural network approach, *Proc. Natl. Acad. Sci. USA* 88:11261 (1991).
8. P. Elvin, G. Slynn, D. Black, A. Graham, R. Butler, J. Riley, R. Anand and A.F. Markham, Isolation of cDNA clones using yeast artificial chromosome probes, *Nucl. Acids Res.* 18:3913 (1990).
9. S. Parimoo, S.R. Pantanjali, H. Shukla, D.D. Chaplin, and S.M. Weissman, cDNA selection: efficient PCR approach for the selection of cDNAs encoded in large chromosomal DNA fragments, *Proc. Natl. Acad. Sci. USA* 88:9623 (1991).
10. J.G. Morgan, G.M. Dolganov, S.E. Robbins, L.M. Hinton and M. Lovett, The selective isolation of novel cDNAs encoded by the regions surrounding the human interleukin 4 and 5 genes, *Nucl. Acids Res.* 20:5173 (1992).
11. G.M. Duyk, S. Kim, R.M. Myers and D.R. Cox, Exon trapping: A genetic screen to identify candidate transcribed sequences in cloned mammalian genomic DNA, *Proc. Natl. Acad. Sci. USA* 87:8995 (1990).
12. A.J. Buckler, D.D. Chang, S.L. Graw, J.D. Brook, D.A. Haber, P.A. Sharp and D.E. Housman, Exon amplification: a strategy to isolate mammalian gene based on RNA splicing, *Proc. Natl. Acad. Sci. USA* 88:4005 (1991).

13. D. Auch and M. Reth, Exon trap cloning: using PCR to rapidly detect and clone exons from genomic DNA fragments, *Nucl. Acids Res.* 18:6743 (1990).

14. M. Hamaguchi, H. Sakamoto, H. Tsuruta, H. Sasaki, T. Muto, T. Sugimura and M. Terada, Establishment of a highly sensitive and specific exon-trapping system, *Proc. Natl. Acad. Sci. USA* 89:9779 (1992).

15. The Huntington's Disease Collaborative Research Group, A novel gene containing a trinucleotide repeat that is expanded and unstable on Huntington's Disease chromosomes, *Cell* 72:971 (1993).

16. J.A. Trofatter, M.M. MacCollin, J.L. Rutter, J.R. Murrell, M.P. Duyao, D.M. Parry, R. Eldridge, N. Kley, A.G. Menon, K. Pulaski, V.H. Haase, C.M. Ambrose, D. Munroe, C. Bove, J.L. Haines, R.L. Martuza, M.E. MacDonald, B.R. Seizinger, M.P. Short, A.J. Buckler and J.F. Gusella, A novel moesin-, ezrin-, radixin-like gene is a candidate for the neurofibromatosis 2 tumor suppressor, *Cell* 72:791 (1993).

17. A.P. Walker, F. Muscatelli and A.P. Monaco, Isolation of the human Xp21 glycerol kinase gene by positional cloning, *Hum. Mol. Genet.* 2:107 (1993).

18. S.A.M. Taylor, R.G. Snell, A. Buckler, D. Church, C.S. Lin, M. Alther, N. Groot, G. Barnes, D.J. Shaw, J.J. Wasmuth, P.S. Harper, D.E. Housman, M.E. MacDonald and J.F. Gusella, *Nature Genetics* 2:223 (1993).

19. B. Seed, Purification of genomic sequences from bacteriophage libraries by recombination and selection in vivo, *Nucl. Acids Res.* 11:2427 (1983).

20. M. Koenig, E.P. Hoffman, C.J. Bertelson, A.P. Monaco, C.A. Feener and L.M. Kunkel, Complete cloning of the Duchenne muscular dystrophy (DMD) cDNA and preliminary genomic organization of the DMD gene in normal and affected individuals, *Cell* 50:509 (1987).

21. L.A.J. Blonden, J.T. Den Dunnen, H.M.B. Van Paassen, M.C. Wapenaar, P.M. Grootscholten, H.B. Ginjaar, E. Bakker, P.L. Pearson and G.J.B. Van Ommen, High resolution deletion breakpoint mapping in the DMD-gene by whole cosmid hybridization, *Nucl. Acids Res.* 17:5611 (1989).

22. R.G. Roberts, A.J. Coffey, M. Bobrow and D.R. Bentley, Determination of the exon structure of the distal portion of the dystrophin gene by vectorette PCR, *Genomics* 13:942 (1992).

23. T. Maniatis, E.F. Fritsch, and J. Sambrook, J, "Molecular Cloning (A Laboratory Manual)", New York, Cold Spring Harbor Laboratory, 1982.

24. R.G. Roberts, D.R. Bentley, T.F.M. Barby, E. Manners and M. Bobrow, Direct diagnosis of carriers of Duchenne and Becker muscular dystrophy by amplification of lymphocyte RNA, *Lancet* 336:1523 (1990).

25. C.A. Feener, M. Koenig and L.M. Kunkel, Alternative splicing of human dystrophin mRNA generates isoforms at the carboxy terminus, *Nature*:338:509 (1989).

26. L.A.J. Blonden, P.M. Grootscholten, J.T. Den Dunnen, E. Bakker, S. Abbs, M. Bobrow, C. Boehm, C. Van Broeckhoven, L. Baumbach, J. Chamberlain, C.T. Caskey, M. Denton, L. Felicetti, G. Galluzzi, K.H. Fischbeck, U. Francke, B. Darras, H. Gilgenkrantz, J.-C. Kaplan, F.H. Herrmann, C. Junien, C. Boileau, S. Liechti-Gallati, M. Lindlof, T. Matsumoto, N. Niikawa, C. Muller, J.E. Poncin, S. Malcolm, E. Robertson, G. Romeo, A.E. Covone, H. Scheffer, E. Schroder, M. Schwartz, C. Verellen, A. Walker, R. Worton, E. Gillard and G.J.B. Van Ommen, 242 breakpoints in the 200-kb deletion-prone P20 region of the DMD-gene are widely spread, *Genomics* 10:631 (1991).

27. C. Wijmenga, J.E. Hewitt, L.A. Sandkuijl, L.N. Clark, T.J. Wright, J.G. Dauwerse, A.M. Gruter, M.H. Hofker, P. Moerer, R. Williamson, G.J.B. Van Ommen, G.W. Padberg and R.R. Frants, Chromosome 4q DNA rearrangements associated with facioscapulohumeral muscular dystrophy, *Nature Genetics* 2:26 (1992).

28. J.C.T. van Deutekom, C. Wijmenga, E.A.E. van Tienhoven, A-M. Gruter, J.E. Hewitt, G.W. Padberg, G.J.B. Van Ommen, M.H. Hofker and R.R. Frants, FSHD associated DNA rearrangements are due to deletions of integral copies of a 3.2 kb tandemly repeated unit, *Hum. Mol. Genet.* (1993).

29. M.A. Frohman, M.K. Dush and G.R. Martin, Rapid production of full-length cDNAs from rare transcripts: amplification using a single gene-specific oligonucleotide primer, *Proc. Natl. Acad. Sci. USA* 85:8998 (1988).

30. D.B. Krizman and S.M. Berget, 3'-terminal exon trapping, *in*: "Genome Mapping and Sequencing," Cold Spring Harbor Press, NY, USA:p202 (1993).(Abstract)

31. L. Zhang, X. Cui, K. Schmitt, R. Hubert, W. Navidi and N. Arnheim, Whole genome amplification from a single cell: implications for genetic analysis, *Proc. Natl. Acad. Sci. USA* 89:5847 (1992).

32. A. Andreadis, P.E. Nisson, K.S. Kosik and P.C. Watkins, The exon trapping assay partly discriminates against alternatively spliced exons, *Nucl. Acids Res.* 21:2217 (1993).

ISOLATION OF GENE SEQUENCES FROM THE BRCA1 REGION OF CHROMOSOME 17q21 BY EXON AMPLIFICATION

Kenneth J. Abel[1], Lucio H. Castilla[2], Alan J. Buckler[5], Fergus J. Couch[1], Peggy Ho[1], Ida Schaefer[4], Settara C. Chandrasekharappa[6], Francis S. Collins[1,3,6], and Barbara L. Weber[1]

Departments of [1]Internal Medicine, [2]Biology, and [3]Human Genetics, and [4]Michigan Human Genome Center and the Biomedical Research Core Facilities, University of Michigan Medical School, Ann Arbor, Michigan 48109. [5]Massachusetts General Hospital and Department of Neurology, Harvard Medical School, Boston, Massachusetts 02139. [6]Laboratory of Gene Transfer, National Center for Human Genome Research, National Institutes of Health, Bethesda, Maryland 20892.

ABSTRACT

A variety of techniques now exist for the identification and isolation of gene sequences from cloned genomic DNA. We report the use of exon amplification to isolate candidate exons of genes in the chromosome 17q21 region associated with familial breast and ovarian cancer. We have used the second generation splicing vector pSPL3, which provides greater flexibility for cloning genomic fragments and which reduces the frequency of the major classes of false positive clones. In two experiments, exon amplification was performed using DNAs of approximately 170 cosmids spanning 1-2 Mb of this region. Cosmid DNAs were pooled in groups of 6-10 each. More than 2000 candidate exon clones from these experiments have been arrayed in microtiter dishes. The average size determined for nearly 400 cloned inserts was approximately 200 base pairs. Ongoing efforts to identify and eliminate clone redundancy have thus far yielded more than 100 unique exon clones. Less than 10% of the clones were found to be repetitive or to be artifacts resulting from cryptic splicing involving sequences present in the splicing vector. Thus the great majority of clones were found to be single copy and to derive from the correct chromosomal location. These exons have been used as hybridization probes to isolate cDNA clones derived from normal breast tissue. The cloned exons and corresponding cDNAs are being localized within developing cosmid contigs in order to assemble a transcription map of the region, and to position transcribed sequences with respect to critical recombinants in breast/ovarian cancer families. While database searches suggest that many of the exon sequences are unique, these searches have also identified several genes which were either mapped previously to proximal 17q or which appear to be homologs of genes in other species. Exon amplification represents a rapid and efficient means for isolating candidate gene sequences from genomic clones, facilitating efforts to identify specific genes associated with disease using positional cloning strategies. Utilization of this technique to survey large genomic regions will also assist in efforts to construct transcription maps of chromosomes.

INTRODUCTION

Efforts to describe the complete coding potential of complex mammalian genomes require efficient strategies to isolate genes or portions of genes from cloned genomic DNA. A variety of transcript identification techniques now exist which rely upon large insert genomic clones such as cosmids or yeast artificial chromosomes (YACs), frequently in conjunction with libraries of cDNA clones (1-5). These techniques have been used successfully in positional cloning approaches to identify genes associated with specific diseases (6-8). The same strategies, however, can also be used to survey large genomic regions for coding sequences as part of efforts to develop long-range transcription maps (9-11).

Exon amplification (4) is a powerful technique for isolating gene sequences from genomic clones. This technique relies upon the normal processing of hnRNA transcripts, whereby noncontiguous sequences destined to make up the mature messenger RNA (exons) are accurately spliced together and the intervening sequences (introns) are excised. The exons are flanked by specific sequences which serve as signposts for the cellular splicing machinery, identifying the precise sites where cutting and joining of the exons occurs. Exon amplification permits the selective recovery of sequences present in cloned genomic DNA which are flanked by functional splice sites, yielding clone libraries highly enriched for gene sequences. A number of versions have been reported based upon utilization of either one or two splice sites (4, 12-14), or upon recovery of the last exons of genes (15,16). Since the technique relies only upon genomic DNA, it is particularly well suited for isolating candidate gene sequences when little or no information is available regarding levels, source, or developmental stage of expression. Exon amplification has been used to search for genes both in defined chromosomal regions (17,18) and entire chromosomes (19), to characterize the exon/intron structure of individual genes (20), and in studies of alternative splicing (21). It has also proved useful in positional cloning efforts to isolate genes defective in Huntington's disease (8), Menkes disease (22), and neurofibromatosis type 2 (23).

Among the versions of exon amplification which have been described, the technique described by Buckler et al. (4) is particularly attractive because the requirement for two functional splice sites improves the stringency of the technique for isolating "true" exons. Moreover, the types of most artifactual clones which can arise have been well documented (4,18,19). In the original system, genomic restriction fragments are ligated into a plasmid splicing vector (pSPL1) containing the HIV tat intron flanked by functional donor and acceptor splice sites. Upon transfection into COS-7 cells, high level transcription across the cloned genomic DNA is driven by upstream SV40 sequences. Genomic exons included in these primary transcripts are efficiently joined with the vector splice sites, resulting in the "trapping" of exons in the processed mRNAs. Following reverse transcription and PCR amplification, the candidate exons can be cloned into commonly used plasmid vectors.

Although this method was effective, two classes of false positives led to improvements in the technique. The most commonly observed artifacts were those resulting from splicing events involving only vector donor and acceptor sites, or from cryptic splicing involving sites present in the vector intron. Recently, pSPL3, a modified version of the original splicing vector has been developed (24) which is designed to exclude these artifacts from the cloning process. The modified vector contains BstXI restriction sites which, following in vivo processing, are intact only in these artifactual clones.

Inclusion in the protocol of a BstXI digest serves to remove these products from subsequent cloning steps. Also, the single BamHI cloning site in the original vector has been replaced with a multiple cloning site, allowing greater flexibility in the types of genomic fragments which can be surveyed for exons. At least 90% of clones generated using this vector are likely to be parts of genes by Northern analysis, cDNA library screening, or by conservation among different species (24).

We are using this technique to identify genes in the region of 17q21 associated with familial breast and ovarian cancer. Germline mutations in a gene, BRCA1, are believed to be responsible for roughly half of all heritable, early onset breast cancer, and for most disease in families with strong histories of both breast and ovarian cancer (25,26). As a result of frequently observed loss of 17q21 alleles in breast and ovarian tumors, BRCA1 is referred to as a tumor suppressor gene (27). This gene is therefore believed to be expressed, and to play a role in regulating cellular growth, in normal breast and ovarian tissues. The critical interval believed to contain BRCA1 is currently estimated to be approximately 1-1.5 Mb in length, flanked by the polymorphic markers D17S857 and D17S78 (28,29). Cosmids from this region are being surveyed for candidate exons, and the clones generated are being used to isolate larger corresponding cDNAs. The cDNAs in turn are being used to search for mutations in DNAs from sporadic tumors, and in germline DNAs of affected members of breast- and breast/ovarian-cancer families. In this chapter we describe our results using exon amplification to identify candidate genes for BRCA1. Strategies are discussed for characterizing large numbers of exon clones for redundancy, for artifactual and repetitive sequences, and for regional localization. We will also discuss how exon amplification complements other techniques being applied to this region, and how an integrated approach has facilitated characterization of gene structure and development of a transcription map for this region.

METHODS

Generation of Exon Clones

Exon amplification was performed as described (4), using the modified splicing vector pSPL3 (24). In one experiment, DNAs were prepared from 15 ml cultures of 136 unordered cosmid clones from the 17q12-21 region. The DNAs were organized into thirteen pools of ten cosmids each, plus one pool of six. Approximately 10 μg aliquots of DNA from each pool were digested using restriction endonucleases BamHI plus BglII. Following phenol/chloroform extractions and ethanol precipitation, one-tenth of the resuspended DNAs were ligated at room temperature to 50 ng of BamHI-digested, phosphatased pSPL3-IV (another version of this vector, pSPL3-VI, carries the multiple cloning site in the reverse orientation). One-half of the ligation reaction was used to transform competent HB101 cells. After overnight growth on LB agar plates containing 50 μg/ml ampicillin, colonies were scraped into 10 ml of LB broth, harvested by centrifugation, and total plasmid DNAs were isolated by a standard alkaline lysis protocol and resuspended finally in 100 μl of TE buffer. One-tenth of each of the DNA pools was analyzed by PvuII digestion/agarose gel electrophoresis to visualize the complexity of genomic sequences cloned in the splicing vector. A fifteen microliter aliquot of each DNA pool was then transfected into COS-7 cells using a GenePulser electroporator (BIORAD; 1.2 kV, 25 mF). The cells were cultured two days, allowing transcription and processing of RNAs derived from the transfected pSPL3 constructs. Cytoplasmic RNA was isolated by lysing cells with Triton X-100, centrifugation to remove the nuclei, phenol/chloroform extraction of the supernatant, and recovery of the RNA by ethanol precipitation.

One-tenth of the resuspended RNA from each pool then served as template for generating single strand cDNA using reverse transcriptase (GibcoBRL), primed by the vector-specific oligonucleotide SA2 (ATCTCAGTGGTATTTGTGAGC). The first strand cDNAs were made double stranded and PCR-amplified with vector primers SA2 and SD6 (TCTGAGTCACCTGGACAACC), using an annealing temperature of 60°C. Amplification was interrupted after six cycles and the 1°PCR products were digested overnight with BstXI (New England Biolabs) to eliminate splicing artifacts. PCR amplification then proceeded for another 25 cycles using nested vector primers SA4 (CACCTGAGGAGTGAATTGGTCG) and SD2 (GTGAACTGCACTGTGACAAGC). The complexity of the 2°PCR products was again assayed by agarose gel electrophoresis. These products were then digested with SalI plus BglII (sites which immediately flank the vector splice sites), and the efficiency of the digests was evaluated by gel electrophoresis. The splice products were then ligated into SalI- plus BamHI-digested pBluescriptIIKS$^+$ (pBSIIKS$^+$), transformed into competent DH5α cells, and grown on LB/ampicillin plates containing X-Gal to screen for clones with inserts. For UDG-cloning, the 2°PCR primers SA4 and SD2 were also synthesized to include at their 5' ends the sequences (CUA)$_4$ or (CAU)$_4$, respectively. The resulting PCR products were annealed for 30 min to the vector pAMP1 (LifeTechnologies, GibcoBRL) in the presence of uracil DNA glycosylase, and transformed into DH5α as described above. White colonies generated by either approach were toothpick-inoculated into 96-well microtitre dishes containing LB/ampicillin broth, and viable stocks of the clones were maintained in 20% glycerol at -80°C.

Characterization of Exon Clones

To determine the sizes of individual exons, approximately 1-2 μl of cells from fresh cultures were included in 25 μl PCR reactions containing primers which anneal immediately adjacent to the exon boundaries: SD5, CCCTCGAGGTCGACCCAGC; SA5, CTAGAACTAGTGGATCTCCAGG. For UDG-cloned exons, inserts were amplified using the primers: AMP1t7, ACGCTCTAGAGTCGACCCAGC; AMP1sp6, CCCCTCGGGAGATCTCCAG. Reactions included 1x reaction buffer (10 mM Tris, pH 8.3; 1.5 mM MgCl$_2$; 50 mM KCl), 200 μM each dNTP, 2 ng/μl each primer, and 1 unit Taq DNA polymerase (Boehringer Mannheim). After denaturation at 94°C, amplification proceeded for 35 cycles of: 94°C (1 min), 60°C (1 min), 72°C (1 min), followed by a 5 min 72°C extension. After addition of loading dye, 4 μl aliquots of each reaction were analyzed by electrophoresis in 2.5% agarose gels and compared to size standards. Each primer pair contributes 43 bp to the observed size of each insert. The remainder of each amplified insert was saved for use as hybridization probes (below). When needed, inserts were gel-purified on 1.5-2% low melting point agarose gels, and the excised bands were melted at 65°C and diluted in two volumes of water. Five to ten microliter aliquots were used in random hexamer primed labelling reactions (30) including [α-^{32}P]dCTP (Amersham).

To test for exon clone redundancy, and to map exons within our cosmid contigs, purified inserts were used as hybridization probes to screen gridded arrays of exon or cosmid DNAs, respectively, on nylon filters (Hybond-N, Amersham). Hybridization filters were prepared using a 96-pin stamping device which replicates clones from four microtitre dishes (384 clones) onto a single 150 mm diameter nylon circle. The colonies were grown overnight on the nylon membranes overlaying LB/ampicillin agar plates. Colonies were lysed *in situ* on filter papers saturated with denaturing solution (0.5 N NaOH/1.5 M NaCl), the denatured DNAs were neutralized in 1 M Tris(pH 8)/1.5M NaCl, and the DNAs were fixed to the nylon with UV using a Stratalinker (Stratagene). Following prehybridization for several hours, hybridization of labelled inserts to these grids was allowed to proceed overnight at 65°C in 1 M NaCl/1% SDS/1xDenhardts/10% dextran sulfate/100 μg/ml denatured salmon sperm DNA, using denatured probe at 10^6 cpm/ml. Filters were washed at 65°C for 0.5-1 h in 2XSSC/0.1% SDS and exposed to Kodak XAR-5 film. To identify clones containing either highly repetitive sequences or products derived from cryptic splicing within the pSPL3 vector, copies of each grid were also probed with human Cot-1 DNA (GibcoBRL) and an HIV tat intron probe, respectively. Exon clones were also tested for single copy sequences using the labelled inserts to probe genomic DNA Southern blots. Hybridization and wash conditions were as described above for the filter grids.

Sequencing of cloned exons was performed on an automated sequencer (Applied Biosystems) using fluorescent methodology. DNAs were sequenced using primers specific to the T3 and T7 promoters flanking the cloning sites in pBSIIKS$^+$, and with primers specific to the SP6 and T7 promoters in pAMP1. Since the products are directionally cloned, the "sense" strand corresponding to the COS-7 mRNA containing the spliced exon can be determined by sequencing from the T3 side in pBSIIKS$^+$ and from the T7 side in pAMP1.

Screening cDNA Libraries

Purified exon inserts were used as hybridization probes to screen lambda phage libraries of cDNA clones by standard methods (31). Two cDNA libraries derived from normal breast have been screened: an oligo dT-primed library in the vector λgt11 (32), and a random-primed library in λgt10 (Clontech). Approximately 50,000 pfu were plated in top agarose onto each of ten 150 mm agar plates (total 500,000 pfu). After 8-12 h growth at 37°C and a brief period at 4°C, duplicate nylon filter lifts were made from each plate. The phage DNAs were denatured, neutralized, and fixed to the nylon, and hybridization conditions were as described above for the colony grids. Primary positive clones were purified through two additional rounds of lower density plating and hybridization. cDNA inserts were PCR-amplified from third round-purified phage in SM buffer, using primers which flank the EcoRI cloning sites in λgt10 (GAGCAAGTTCAGCCTGGTTAAGTC; G G C T T A T G A G T A T T T C T T C C A G G G) and λ g t 1 1 (ACGACTCCTGGAGCCCGTCAGTATC; CACCAGACCAACTGGTAATGGTAGC). Another set of these primers has also been synthesized to include the CAU and CUA repeats at their 5' ends. The amplified inserts were subcloned into plasmid vectors using either the TA Cloning kit (pCRII, Invitrogen) or by UDG cloning with the CloneAmp kit (pAMP1, GibcoBRL). For either vector, sequencing was performed using primers specific for the SP6 and T7 promoter sequences flanking the cloning sites.

RESULTS

Cloning Exons from the BRCA1 Region

In a single experiment, exon amplification was used to generate exon clones from 136 unordered cosmids derived from the BRCA1 region of chromosome 17q21. These cosmids were identified using Alu-PCR products derived from YACs to probe a gridded chromosome 17-specific cosmid library (provided by L. Deaven, Los Alamos National Laboratory). The cosmids span a distance of approximately 1-2 megabase pair (Mb) of genomic DNA. The purified cosmid DNAs were organized into thirteen pools of ten each, plus one pool of six (pools A-N). The pooled DNAs were digested with BamHI and BglII, and shotgun cloned into the BamHI site of the plasmid splicing vector, pSPL3. More than 2000 candidate exon clones were generated in this experiment. Clones were individually picked at random to fill one 96-well microtitre dish for each of the fourteen cosmid pools (total 1344).

Since this first experiment, approximately 250 additional cosmids from this region have been isolated and are being assembled into contigs of overlapping clones. In a second experiment, exon amplification was performed using forty-five cosmids representing several small contigs spanning a total of 0.5-1 Mb. The cosmid DNAs were organized into five pools of 8-10 each (pools O-S). In this experiment we took advantage of the multiple cloning site in pSPL3 and surveyed for exons using two restriction digest strategies. As before, the DNAs in each pool were digested with BamHI/BglII and were subcloned into BamHI-digested pSPL3. The other strategy involved digesting the same DNAs with PstI and cloning into similarly digested vector. Combining different restriction approaches offers greater opportunities for isolating exons which, perhaps due to restriction fragment length, may not have been efficiently cloned into the splicing vector. Also, for this experiment the protocol was modified to permit cloning the final PCR-amplified splice products using uracil DNA glycosylase (UDG) and uracil-containing primers. For this purpose, the primers SA4 and SD2 were synthesized to include at their 5' ends the sequences $(CUA)_4$ or $(CAU)_4$, respectively. The resulting PCR products were annealed to the vector pAMP1 (GibcoBRL) and transformed. This approach also generated large numbers of candidate exon clones (1500-2000), from which 950 have been arrayed in ten microtitre dishes. Preliminary characterization suggests a similarly high complexity of inserts were generated by this approach.

Characterizaton of Exon Clones

Sizes of exons were determined by gel electrophoresis of cloned inserts PCR-amplified directly from cells in fresh cultures. The average size of 368 PCR-amplified inserts was approximately 200 base pairs. Most sizes ranged from 50 to 400 bp (Fig. 1), however the largest number of clones was in the 100-150 bp range. A few clones greater than 600 bp are likely to contain several inserts per clone (described below).

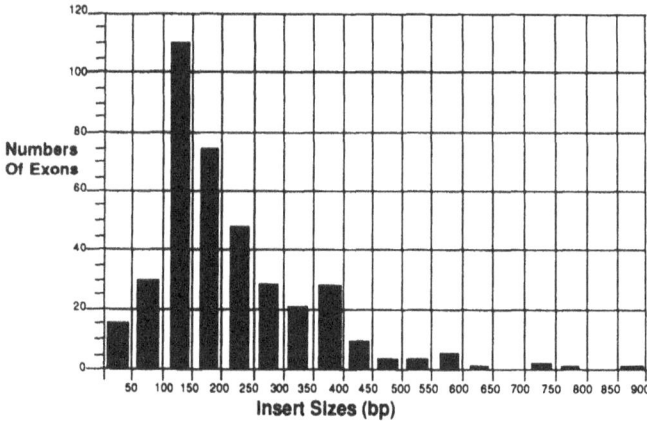

Figure 1. Sizes of cloned exons. Shown are the frequencies of observed exon sizes within 50-bp size intervals. Sizes were estimated by gel electrophoresis of PCR-amplified clone inserts and subtracting 43 bp for vector sequence included in each product.

Since our cosmids derived from a relatively narrow region and were highly redundant, we anticipated considerable redundancy among the exon clones. Therefore efforts were made to identify and eliminate clone redundancy, and to assemble a representative set of exons which would be used for screening cDNA libraries. This was accomplished using individual exons as hybridization probes to screen gridded DNAs of the exon library. Our initial results using exon probes derived from the "A" pool suggested an average redundancy of 5 to 10-fold. Figure 2A shows an example of this type of analysis, in which one probe (J.E2) detected itself plus eight other clones. We subsequently reduced our working set from fourteen to eight microtitre dishes of clones, excluding those already shown to be redundant with the earlier tested exons. To date, nearly 70 representative exon clones from the first experiment have been identified, and approximately 125 from the combined experiments. The average frequency of redundant clones was approximately 10-fold, however the largest number of representative exons were members of related groups of three or less (Fig. 2B). While some of the clones were found to be unique in the exon library, a few were represented as many as 20-50 times. These differences may reflect a bias for preferred splice sites in COS-7 cells, or may reflect differences in the depth of coverage in the region provided by the unordered cosmid clones.

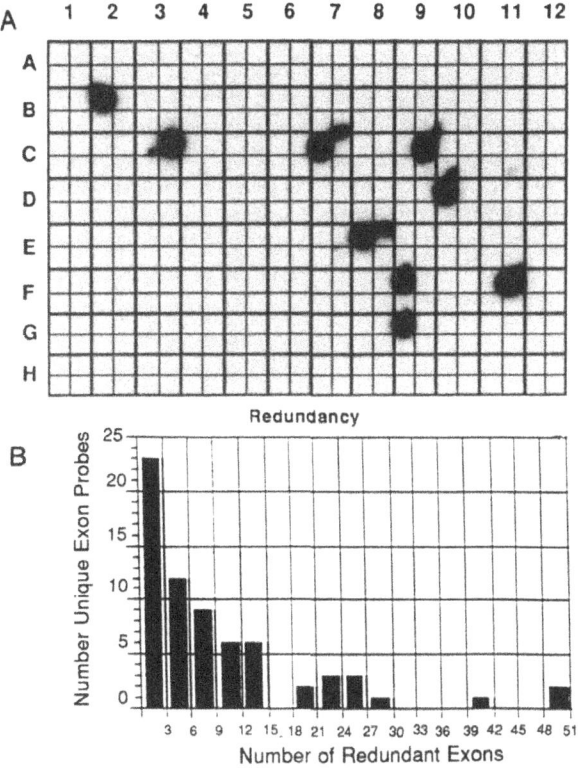

Figure 2. Characterization of exon clones for redundancy. (A) Panel shows hybridization results of probing a nylon filter containing 4x96 gridded exon clones with clone J.E2. This probe detected eight redundant clones. (B) Shown is the number of redundant clones, grouped by threes, identified by representative exons. The largest number of representative exons identified redundant clones totaling three or less.

Before being used as hybridization probes to screen cDNA libraries, representative exons were mapped back to confirm their region of origin and were tested for the presence of highly repetitive sequences. This latter possibility is an especially important concern when screening oligo dT-primed cDNA libraries, as Alu repeat elements are occasionally found within the untranslated regions of mRNAs. To address these concerns, exons were used to probe genomic Southern blots containing human DNA and DNAs from somatic cell hybrids with breakpoints flanking the BRCA1 region (Fig. 3A). We frequently observed conservation of the cloned sequences in rodent DNAs, adding support to the authenticity of the clone as a transcribed sequence. The exons were also used to probe gridded DNAs from our set of nearly 400 cosmids from this region (Fig. 3B). Grid hybridizations provided an independent test for sequence copy number and enabled us to localize exons

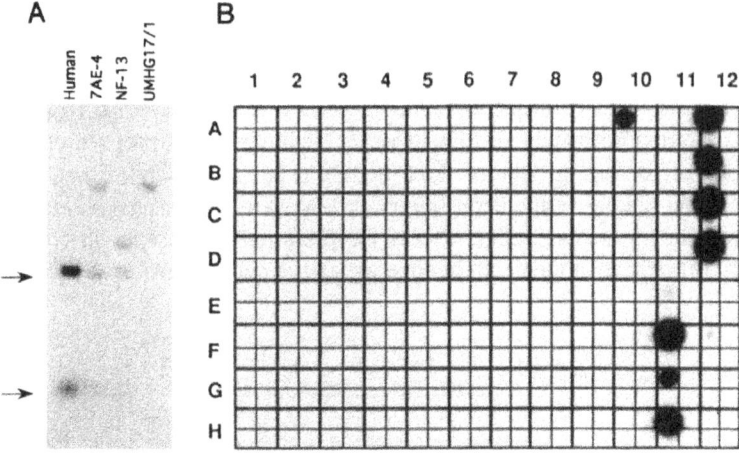

Figure 3. Localization of exon clones. (A) Southern analysis of a somatic cell hybrid mapping panel using the exon probe C.B12. Lanes: 1) human genomic DNA; 2)7AE-4, a human/rat hybrid retaining 17 as its only human chromosome (33); 3) NF-13, a human/mouse hybrid retaining 17q11.2-qter (34); 4) UMHG17/1, a human/hamster hybrid retaining 17q22-qter (35). Arrows identify human-specific fragments. (B) Hybridization results of screening a nylon filter containing 4x96 gridded cosmid DNAs with the exon probe J.E2. Cosmids were organized in the grids approximately by location, explaining the non-random pattern of detected clones.

with high resolution within existing contigs. They have also provided valuable information for assembling contigs in other regions using content mapping approaches. Localization by Southern analysis of hybrid mapping panels is sometimes difficult for several reasons. It occasionally has been difficult to discern restriction length variants across species, prohibiting unambiguous localization of the clone. Also, smaller inserts sometimes prove unsuitable as hybridization probes for detecting single copy sequences in genomic DNA, although they are generally useful for screening gridded cosmid DNAs. Recently we have found it convenient to substitute a small blot containing one lane each of human genomic DNA, pooled DNA from our set of regionally localized cosmids, and size markers, for the hybrid mapping panel. If these blots demonstrate the absence of repeated sequences, and if hybridization to the cosmid grids is consistent with single copy localization within our contigs, the exon clone is used as a probe for screening cDNA libraries. More than 90% of tested clones have been found to be both single copy and to derive from the correct chromosomal location.

Cloning Artifacts

A relatively common feature of exon amplification is the generation of chimeric clones, where several exons are co-ligated into a single plasmid. Since the PCR-amplified splice products are directionally cloned into pBluescript as SalI- and BglII-ended molecules, chimeric clones are usually observed as 3-way co-ligations to maintain compatibility between insert and vector ends. Whereas ligation of the BglII end of the exon into the vector BamHI site ordinarily destroys both sites, chimeric clones are easily detected by the presence of an intact BglII site (and a second SalI site). Forty-five of our representative exon clones were examined by both SalI digestion and sequence analysis, and six (13%) were found to be chimeric. While such clones can make refined localization difficult due to hybridization to non-overlapping groups of cosmids, it is relatively straightforward to subclone the component exons and proceed in the characterization of each independently. Alternatively, in some instances it is possible to partially sort out the cosmid hybridization patterns generated by chimeric clones when one or more of their component exons also exist as independent, non-chimeric clones. UDG-cloning may provide a means for eliminating chimeric clones due to the long (12 bp) non-complementary insert ends. Still, with either approach the possibility exists for chimerism of genomic fragments in the splicing vector. Although unlikely to be a major concern, this situation could potentially result in the joining of exons from different genes.

Since exon amplification is a functional assay, it is possible to clone DNAs which are not "true" exons, but which either contain or are flanked by sequences which behave as splice sites. During characterization of our clones, three (C.B6, D.B1, J.G5) appeared to contain highly repeated sequences when probed against both genomic DNA and gridded cosmids. Sequence analysis of two of these revealed the presence of Alu repeat sequences. Each clone contained only the 3' half of an Alu repeat in an inverse orientation relative to the inferred direction of transcription in the COS-7 cell. Also, each began at precisely the same nucleotide within a consensus Alu (36), suggesting that sequences on one strand of at least some Alu members can function as splice sites (18). Although this mechanism exists for cloning Alu repeats, we estimate that only a small fraction (3-4%) of our clones contain these repeats. This is based upon both the frequency of detection using individual exons to probe genomic DNA, and upon results of probing gridded exon DNAs with a cloned Alu repeat (ONCOR). Two other clones (J.G4, J.H1) generated faint hybridization smears with target genomic DNA, but hybridized to discrete cosmid clones, suggesting the inclusion of low copy repeat sequences.

We have also observed several examples of cloned sequences derived entirely from the pSPL3 intron. Sequence analysis of two clones revealed the presence in each of a 117 base pair sequence derived entirely from the HIV tat intron. This sequence is flanked in the vector by sequences which apparently act as cryptic donor and acceptor splice junctions. We estimate that these artifacts also represent only a small fraction of our exon library. When used as a probe, the tat clone K.G1 hybridized to 15/384, or 3.9%, of gridded exon DNAs. Isolation of the identical sequence by another group (18) suggests that this sequence may comprise a large proportion of the non-repetitive artifactual clones.

Using Cloned Exons to Identify Genes in the BRCA1 Region

Cloned exons represent an ideal resource for isolating larger cDNA clones. Since the proportion of artifactual clones is low, and since selection for internal exons reduces the likelihood of capturing repetitive sequences, most clones in the exon library are expected to be suitable as hybridization probes for screening cDNA libraries. To isolate candidate genes for BRCA1, cloned exons are being used to screen cDNA libraries derived from normal breast tissue. The cDNAs are in turn being used to search for mutations in constitutional DNAs of affected members of breast/ovarian cancer families and in DNAs of sporadic tumors. The exon probes are being used singularly, or in small pools to screen oligo dT- and random-primed breast cDNA libraries. To date more than 100 cDNA clones from this region, thought to be derived from 10-12 different genes, have been isolated.

In addition to the genes identified by cDNA library screening, numerous genes were identified using exon sequences to search nucleic acid and protein sequence databases for homologies using the programs BLASTN and BLASTX (37,38). Searches using exon sequences have identified four genes previously mapped to proximal 17q, including the gene for estradiol 17-β dehydrogenase (EDH17B) which is known to be in this region (28). Twelve other genes have been identified which share strong sequence homologies with exon sequences, but for which no localization has yet been assigned. The positions of these genes are being determined within physical maps of the region. Several of these were identified as probable human homologs of genes previously identified in other species. For example, four independent exons showed strong homology to a previously characterized Drosophila gene and are thus likely to be derived from a previously unknown human homolog of this gene. One of these exons (C.B12) identified human cDNAs we designate as MTO-88 and MTO-89. In turn, when these cDNAs were used to probe gridded DNAs of the exon library, three more exon clones were found to be derived from this gene. Alignment with the Drosophila gene has enabled us to determine the putative order of the exons within the human gene (Fig. 4). Also, by comparing single-exon clones with clones apparently containing multiple exons (H.A7/H.A11, and C.B3/ C.B12/ OBB.A2), it was possible to infer the locations of several exon/intron junctions. These results demonstrate that exon amplification can be useful not only in the identification of genes, but also in the description of gene structure and the characterization of evolutionary relationships among genes of different species.

The final class of exon clones are those which have neither identified homologous sequences in database searches, nor have identified cognate clones in breast cDNA libraries. Several have identified sequences in the databases with limited similarities, suggesting they may be distantly related genes. From the results of breast cDNA library screening and comparing exons with known genes, it is clear that a large fraction of the clones are the result of splicing at true exon/intron junctions. It is therefore likely that many of the remaining clones are similarly derived from genes expressed in other tissues.

In addition to exon amplification, the techniques of magnetic bead cDNA capture (5), direct cDNA screening using large insert genomic clones (1,7), and island rescue PCR (39) are being used to identify genes in this region. Our interests in combining different gene isolation strategies have been to survey the region more efficiently for all genes, and to extend the cloned coverage of each coincidentally identified gene. Identification of overlaps among cloned genes can also provide a measure of the degree to which a region has been exhaustively searched. To identify overlaps we have used individual exons and cDNAs as probes to screen nylon filters containing gridded transcript clones isolated by the various techniques. Overlaps have been observed between cloned exons and clones generated by each of the other techniques. Shown in Fig. 5, the cDNA MTO-80 (grid position H8) hybridized both to the exon which had originally detected that cDNA (position E10), and also to another cDNA isolated by magnetic bead capture (FJC-3A5, position C8). Itself used as a probe, FJC-3A5 in turn detected a set of cDNAs identified by a different exon clone (not shown). Integration of data from different techniques thus helped to assemble three groups of cDNAs previously thought to be unrelated into one gene. Another example is the gene shown in Fig. 4. Portions of this gene were also isolated by island rescue PCR and by direct cDNA screening using genomic probes. The cDNAs isolated by these techniques overlap those identified by cloned exons, extending the coverage of this gene by at least another 1 Kb. Finally, overlaps have also been observed by directly comparing the sequences of exons and other cDNA clones. These examples demonstrate how a combined approach involving different gene isolation techniques helps to obtain complete coverage of the transcribed portions of genes, and also to estimate the number of genes within a region.

Thus, a large fraction of our exon clones either detected homologous gene sequences in database searches, have directly isolated cDNA clones corresponding to known or previously unidentified genes, or were found to be portions of transcribed sequences isolated by other methods.

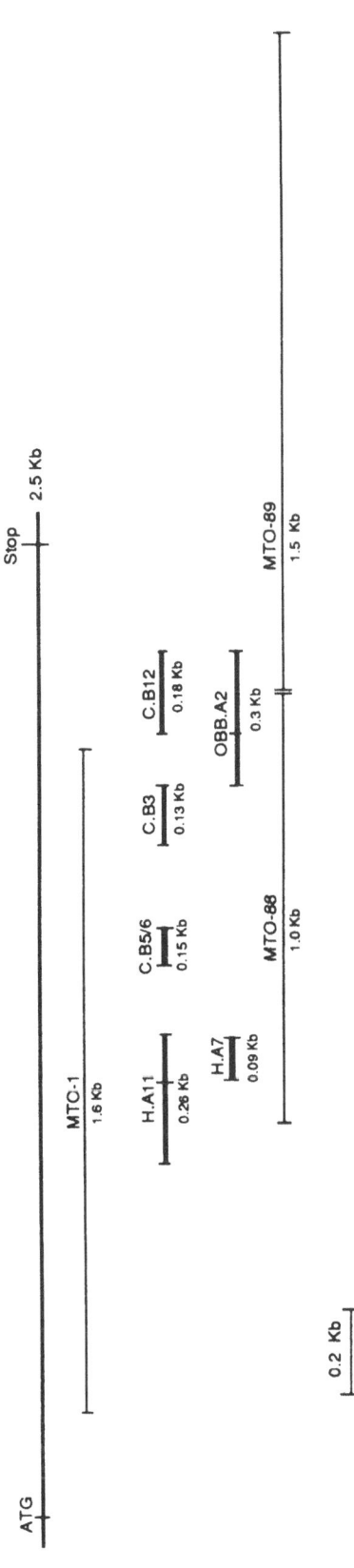

Figure 4. Identification of the putative human homolog of a previously identified Drosophila gene. Top line represents a Drosophila mRNA identified in database searches by several exon sequences. Alignment with respect to the Drosophila sequence of human exons and cDNAs is shown below. Cloned exons are shown as heavy bars. Human cDNAs MTO-88 and MTO-89 were isolated from a random-primed breast cDNA library using the exon C.B12 as probe. cDNA MTO-1 was isolated using the technique of island rescue PCR (39, see text).

193

Long-range Mapping

In addition to positional cloning strategies, exon amplification should also contribute toward efforts to build physical and gene maps of chromosomal regions and entire chromosomes. Since most cloned exons are free of repeat sequences, they can be included in hybridization based methods of contig assembly. Exons from the 17q21 region have been extremely useful in identifying overlaps among previously unordered cosmids, and for confirming the validity of contigs established by other approaches. Exons can also be useful for generating large numbers of sequence-tagged sites (STSs) from any chromosomal region. STSs are short tracts of single copy DNA sequence that can be easily recovered at any time by PCR as a landmark defining a position on the physical map (40). Since the

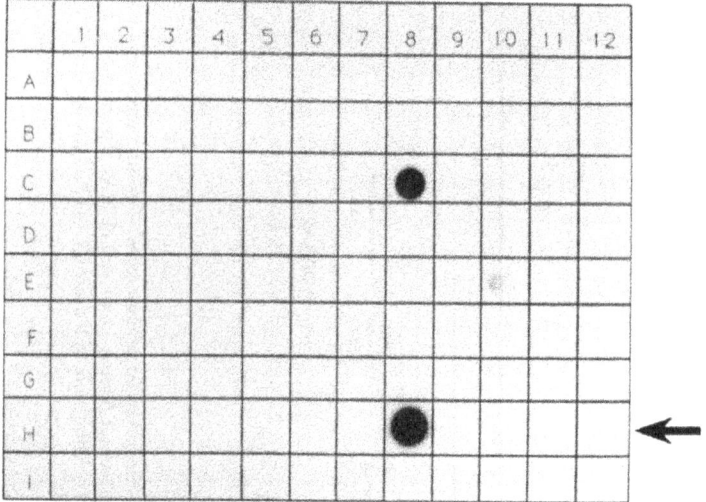

Figure 5. Identification of overlapping gene sequences cloned by different methods. Shown are hybridization results obtained from probing gridded DNAs of cloned gene sequences with the cDNA MTO-80 (identified by arrow). Grid positions: C8) cDNA FJC-3A5 isolated by magnetic bead capture; E10) exon A.F1 which was used to isolate cDNA MTO-80; H8) cDNA MTO-80.

cloned exons average 200 bp or less they are easily sequenced, typically in a single reaction. In preliminary experiments we have tested the feasibility of developing STSs from cloned exons. Figure 6 shows the results of using two exon-STSs to screen a portion of a chromosome 17 radiation-reduced hybrid panel (41). Both STSs amplified human-specific products which were easily distinguished from rodent controls. We have also included a number of exon-STSs in efforts to assemble YAC contigs using content mapping strategies (not shown). The ability to scale up exon amplification to survey several megabases of DNA at a time (24) means that efficient PCR-based methods such as these will be valuable for placing exon clones within long-range transcription maps of chromosomes.

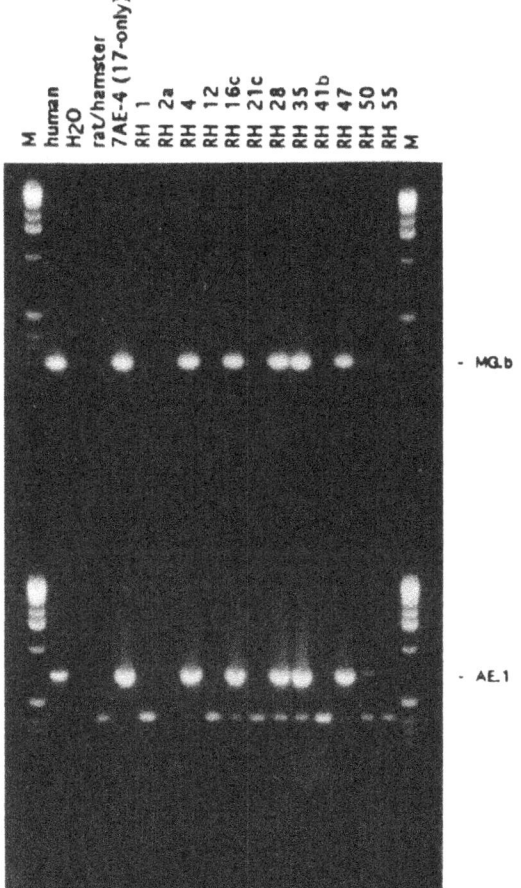

Figure 6. Localization of exons using radiation hybrid mapping. Two exon-STSs were used to PCR-amplify human sequences from a partial set of chromosome 17 radiation-reduced hybrids. Controls included human genomic DNA, water, genomic DNAs from rat and hamster (both of which are present in the hybrids), and DNA from the monochromosome-17 parental hybrid 7AE-4. (Names of the RH clones were arbitrarily assigned and do not correspond to those in Ref. 41.)

DISCUSSION

Exon amplification is being intensively applied as part of positional cloning strategies to identify genes in the BRCA1 region of human chromosome 17q21. Our studies to date suggest that a large proportion of the clones generated are "true" gene sequences. Approximately one-third of the exon clones tested have successfully isolated clones from breast cDNA libraries. These cDNAs are thought to be derived from as many as twelve genes. It is anticipated that a larger fraction of clones would detect cognate cDNAs if libraries derived from additional tissues were also screened. In addition to library screening, searching nucleic acid and protein databases with all exon sequences resulted in the identification of as many as sixteen genes. Several of these are represented in our collection of cDNAs. Finally, a number of clones have also been found to be parts of genes cloned by other techniques. Although some of the exons may ultimately prove to be portions of the same gene, we estimate this technique has identified as many as 20-30 genes within a region spanning 1-2 Mb.

Recent modifications to the original system have substantially reduced the frequencies of artifactual clones containing only vector sequences or cryptic splice products (24). These products were estimated to represent only ~4% of the clones in our exon library. A similar percentage of clones appeared to contain highly repetitive sequences; characterization of two of these demonstrated the presence of Alu repeat sequences. Otherwise, >90% of the clones appeared to be single copy and to map back to the correct chromosomal location. Although artifactual clones no longer comprise a large proportion of the clones generated, the technique has several potential limitations. First, the short average length occasionally makes use of these clones as hybridization probes for Northern and genomic Southern blots difficult. Second, since the technique utilizes only genomic DNAs, confirmation of exon clones as expressed sequences can sometimes be difficult when they are derived from genes expressed either at very low levels, or in limited cell types or developmental stages. Third, the requirement for two functional splice sites most likely precludes isolation of the first or last exons of genes, except in rare instances where these exons may be included in products arising from cryptic splice events. The technique has proved to be highly efficient for isolating internal exons; in several instances multiple exons from the same gene were isolated in independent clones. However, genes with fewer than three exons are likely to be missed by this technique. The use of techniques specifically targeting terminal exons may provide a means for identifying such genes (15,16).

While exon amplification has isolated portions of most genes known to be in this region, we have not identified exons from several genes identified by direct screening and/or magnetic bead capture. It is possible that the genes apparently missed by exon amplification either are comprised of less than three exons, that exons from these genes were not among the original transformants randomly picked to be arrayed in microtitre dishes, or simply that genomic clones containing these genes were not included among those surveyed for exons. Our experiences suggest that a combined approach involving several transcript identification techniques, utilizing both genomic and cDNA resources, provides the greatest opportunity to exhaustively search a region for genes.

REFERENCES

1. P. Elvin, G. Slynn, D. Black, A. Graham, R. Butler, J. Riley, R. Anand, and A.F. Markham, Isolation of cDNA clones using yeast artificial chromosome probes, *Nucl. Acids Res.* 18:3913 (1990).

2. S. Parimoo, S.R. Patanjali, H. Shukla, D.D. Chaplin, and S.M. Weissman, cDNA selection: efficient PCR approach for the selection of cDNAs encoded in large chromosomal DNA fragments, *Proc. Natl. Acad. Sci.* 88:9623 (1991).

3. M. Lovett, J. Kere, and L.M. Hinton, Direct selection: a method for the isolation of cDNAs encoded by large genomic regions, *Proc. Natl. Acad. Sci.* 88:9628 (1991).

4. A.J. Buckler, D.D. Chang, S.L. Graw, J.D. Brook, D.A. Haber, P.A. Sharp, and D.E. Housman Exon amplification: A strategy to isolate mammalian genes based on RNA splicing, *Proc. Natl. Acad. Sci.* 88:4005 (1991).

5. D.A. Tagle, M. Swaroop, M. Lovett, and F.S. Collins, Magetic bead capture of expressed sequences encoded within large genomic segments, Nature 361:751 (1993).

6. M.R. Wallace, D.A. Marchuk, L.B. Andersen, R. Letcher, H.M. Odeh, A.M. Saulino, J.W. Fountain, A. Brereton, J. Nicholson, A.L. Mitchell, B.H. Brownstein, and F.S. Collins, Type I neurofibromatosis gene: identification of a large transcript disrupted in three NF1 patients, *Science* 249:181 (1990).

7. J. Buxton, P. Shelbourne, J. Davies, C. Jones, T. Van Tongeren, C. Aslanidis, P. de Jong, G. Jansen, M. Anvret, B. Riley, R. Williamson, and K. Johnson, Detection of an unstable fragment of DNA specific to individuals with myotonic dystrophy, *Nature* 355:547 (1992).

8. The Huntington's Disease Collaborative Research Group, A novel gene containing a trinucleotide repeat that is expanded and unstable on Huntington's disease chromosomes, *Cell* 72:971 (1993).

9. Z. Sedlacek, B. Korn, D.S. Konecki, R. Siebenhaar, J.F. Coy, P. Kioschis, and A. Poustka,

Construction of a transcription map of a 300 Kb region around the human G6PD locus by direct cDNA selection, *Hum. Molec. Genet.* 2:1165 (1993).

10. J.M. Rommens, B. Lin, G.B. Hutchinson, S.E. Andrew, Y.P. Goldberg, M.L. Glaves, R. Graham, V. Lai, J. McArthur, J. Nasir, J. Theilmann, H. McDonald, M. Kalchman, L.A. Clarke, K. Schappert, and M.R. Hayden, A transcription map of the region containing the Huntington disease gene, *Hum. Molec. Genet.* 2:901 (1993).

11. W.-F. Fan, X. Wei, H. Shukla, S. Parimoo, H. Xu, P. Sankhavaram, Z. Li, and S.M. Weissman, Application of cDNA selection techniques to regions of the human MHC, *Genomics* 17:575 (1993).

12. G.M. Duyk, S. Kim, R.M. Myers, and D.R. Cox, Exon trapping: a genetic screen to identify candidate transcribed sequences in cloned mammalian genomic DNA, *Proc. Natl. Acad. Sci.* 87:8995 (1990).

13. D. Auch and M. Reth, Exon trap cloning: using PCR to rapidly detect and clone exons from genomic DNA fragments, *Nucl. Acids Res.* 18:6743 (1990).

14. M. Hamaguchi, H. Sakamoto, H. Tsuruta, H. Sasaki, M. Tetsuichiro, T. Sugimura, and M. Terada, Estabishment of a highly sensitive and specific exon-trapping system, *Proc. Natl. Acad. Sci.* 89:9779 (1992).

15. D.B. Krizman and S.M. Berget, 3'-terminal exon trapping: identification of genes from vertebrate DNA, *LifeTechnol. Focus* 15:106 (1993).

16. P.E. Nisson and P.C. Watkins, The 3' exon trapping system, *LifeTechnol. Focus* 15:108 (1993).

17. M.P. Duyao, S.A. Taylor, A.J. Buckler, C.M. Ambrose, C. Lin, N. Groot, D. Church, G. Barnes, J.J. Wasmuth, D.E. Housman, M.E. MacDonald, and J.F. Gusella, A gene from 4p16.3 with similarity to a superfamily of transporter proteins, *Hum. Molec. Genet.* 2:673 (1993).

18. M.A. North, P. Sanseau, A.J. Buckler, D. Church, A. Jackson, K. Patel, J. Trowsdale, and H. Lehrach, Efficiency and specificity of gene isolation by exon amplification, *Mammal. Genome* 4:466 (1993).

19. D.M. Church, L.T. Banks, A.C. Rogers, S.L. Graw, D.E. Housman, J.F. Gusella, and A.J. Buckler, Identification of human chromosome 9 specific genes using exon amplification, *Hum. Molec. Genet.* 2:1915 (1993).

20. J.B.J. Kwok, E. Gardner, J.P. Warner, B.A.J. Ponder, and L.M. Mulligan, Structural analysis of the human Ret proto-oncogene using exon trapping, *Oncogene* 8:2575 (1993).

21. A. Andreadis, P.E. Nisson, K.S. Kosik, and P.C. Watkins, The exon trapping assay partly discriminates against alternatively spliced exons, *Nucl. Acids Res.* 21:2217 (1993).

22. C. Vulpe, B. Levinson, S. Whitney, S. Packman, and J. Gitschier, Isolation of a candidate gene for Menkes disease and evidence that it encodes a copper-transporting ATPase, *Nature Genetics* 3:7 (1993).

23. J.A. Trofatter, M.M. MacCollin, J.L. Rutter, J.R. Murrell, M.P. Duyao, D.M. Parry, R. Eldridge, N. Kley, A.G. Menon, K. Pulaski, V. Haase, C. Ambrose, D. Munroe, C. Bove, J.L. Haines, R.L. Martuza, M.E. MacDonald, B.R. Seizinger, M.P. Short, A.J. Buckler, and J.F. Gusella, A novel moesin-, ezrin-, radixin-like gene is a candidate for the neurofibromatosis 2 tumor suppressor, *Cell* 72:791 (1993).

24. D.M. Church, C.J. Stotler, J.L. Rutter, J.R. Murrell, J.A. Trofatter, and A.J. Buckler, Isolation of genes from complex sources of mammalian genomic DNA using exon amplification, *Nature Genetics* 6:98 (1994).

25. J.M. Hall, M.K. Lee, J. Morrow, B. Newman, L. Anderson, B. Huey, and M.-C. King, Linkage of early-onset familial breast cancer to chromosome 17q21, *Science* 250:1684 (1990).

26. D.F. Easton, D.T. Bishop, D. Ford, G.P. Crockford, and the Breast Cancer Linkage Consortium, Genetic linkage analysis in familial breast and ovarian cancer: results from 214 families, *Am. J. Hum. Genet.* 52:678 (1993).

27. S.A. Smith, D.F. Easton, D.G.R. Evans, and B.A.J. Ponder, Allele losses in the region 17q12-21 in familial breast and ovarian cancer involve the wild-type chromosome, *Nature Genetics* 2:128 (1992).

28. D.P. Kelsell, D.M. Black, D.T. Bishop, and N.K. Spurr, Genetic analysis of the BRCA1 region in a large breast/ovarian family: refinement of the minimal region containing BRCA1, *Hum. Molec. Genet.* 2:1823 (1993).

29. J. Simard, J. Feunteun, G. Lenoir, P. Tonin, T. Normand, V. Luu The, A. Vivier, D. Lasko, K. Morgan, G. Rouleau, H. Lynch, F. Labrie, and S. Narod, Genetic mapping of the breast-ovarian cancer syndrome to a small interval on chromosome 17q12-21: exclusion of candidate genes EDH17B2 and RARA, Hum. Molec. Genet. 2:1993 (1993).

30. A.P. Feinberg and B. Vogelstein, A technique for radiolabeling DNA restriction fragments to high specific activity, *Anal. Biochem.* 132:6 (1983).

31. W.D. Benton and R.W. Davis, Screening λgt recombinant clones by hybridization to single plaques in situ, *Science* 196:180 (1977).

32. A. Swaroop and J. Xu, cDNA libraries from human tissues and cell lines, *Cytogenet. Cell Genet.* 64:292 (1993).

33. R.J. Leach, M.J. Thayer, A.J. Schafer, and R.E.K. Fournier, Physical mapping of human chromosome 17 using fragment-containing microcell hybrids, *Genomics* 5:167 (1989).

34. P.R. Fain, E. Solomon, and D.H. Ledbetter, Second International Workshop on Human Chromosome 17, *Cytogenet. Cell Genet.* 57:65 (1991).

35. W.L. Flejter, M. Watkins, K.J. Abel, S.C. Chandrasekharappa, B.L. Weber, F.S. Collins, and T.W. Glover, Isolation and characterization of somatic cell hybrids with breakpoints spanning 17q22-q24, *Cytogenet. Cell Genet.* 64:222 (1993).

36. B.F. Koop, M.M. Miyamoto, J.E. Embury, M. Goodman, J. Czelusniak, and J.L. Slightom, Nucleotide sequence and evolution of the orangutan epsilon globin gene region and surrounding Alu repeats, *J. Molec. Evol.* 24:94 (1986).

37. S.F. Altschul, W. Gish, W. Miller, E.W. Myers, and D.J. Lipman, Basic local alignment search tool, *J. Mol. Biol.* 215:403 (1990).

38. W. Gish, D.J. States, Identification of protein coding regions by database similarity search, *Nature Genetics* 3:266 (1993).

39. J.M. Valdes, D.A. Tagle, and F.S. Collins, Island rescue PCR: a novel method for isolating transcribed sequences from YACs and cosmids, *Proc. Natl. Acad. Sci.* (in press).

40. M. Olson, L. Hood, C. Cantor, and D. Botstein, A common language for physical mapping of the human genome, *Science* 245:1434 (1989).

41. K.J. Abel, M. Boehnke, M. Prahalad, P. Ho, W.L. Flejter, M. Watkins, J. VanderStoep, S.C. Chandrasekharappa, F.S. Collins, T.W. Glover, and B.L. Weber, A radiation hybrid map of the BRCA1 region of chromosome 17q12-21, *Genomics* 17:632 (1993).

ISOLATION OF CODING SEQUENCE FROM COSMIDS AND YACs BY EXON AMPLIFICATION

Michael North[1], Fernando Gibson[2], Stephen Brown[2], Beatrice Griffiths[3], Ellen Solomon[3] and Hans Lehrach[1]

[1]Genome Analysis Laboratory and [3]Somatic Cell Genetics Laboratory, Imperial   Cancer Research Fund, Lincoln's Inn Fields, London WC2A 3PX, United Kingdom
[2]Department of Biochemistry and Molecular Genetics, St. Mary's Hospital Medical School, Imperial College of Science, Technology and Medicine, London W2 1PG, United Kingdom

ABSTRACT

Exon amplification is an increasingly popular approach to the identification of transcribed sequences and complements other strategies to rapidly isolate coding sequence. The following chapter describes application of this system to amplify candidate exons from cosmids and YACs, including 8 cosmids which span a well characterised 185 kb region of the human major histocompatibility class II region on chromosome 6. We have examined the efficiency, specificity and reproducibility of the system in isolating exons from genes on cosmids and YACs and describe the nature and frequency of artefact amplifications in routine screening. We show that the exon amplification procedure can be used successfully with a wide variety of cosmids and YACs which have different numbers of genes and gene structures and describe an efficient approach to the isolation of full length cDNA sequences by the integration of exon amplification and cDNA enrichment technology.

INTRODUCTION

The identification of coding sequence from defined chromosomal regions is a central problem in mammalian molecular genetics. Traditional strategies include hybridisation of genomic fragments directly to cDNA libraries or Northern blots, isolation of conserved sequences between divergent species and the identification of CpG islands as potential markers of transcriptional units (1-4). These methods are labour intensive and can only be used for testing relatively small regions of the genome. Recently, several new approaches have been documented including the direct cloning of human transcripts from human-rodent somatic cell hybrids or the selective hybridisation of complex cDNA populations to genomic DNA (5-8). One of the promising new approaches, the exon amplification system described by Buckler *et al.* (9), allows exons to be isolated from genomic DNA following the selective removal of intervening sequences by the eukaryotic splicing mechanism. The method is technically straightforward and can be applied, in theory, to test genomic DNA in any form (cosmid and YAC clones, flow sorted chromosomes etc). An immediate advantage of the technique is that it can be used to isolate exons irrespective of the level of expression and tissue distribution of the cognate cDNA and can be used to efficiently screen large genomic regions for coding sequence.

Briefly, a mammalian expression vector (pSPL1) has been constructed allowing genomic DNA to be inserted into an intron flanked by the 5' and 3' splice sites of the HIV-1 tat gene. Recombinant clones are transfected into COS-7 cells and high levels of transcription are driven by the expression vector SV40 early promoter. During *in vivo* processing, splice sites of any exon contained within an inserted genomic fragment are paired with the tat splice sites so that intronic DNA is excised and the exon is retained in the mature RNA. Reverse transcription followed by PCR can then be used to amplify such 'trapped' exons. Other exon trapping/amplification methods following broadly similar principles have been described by Auch *et al.* (10) and Hamaguchi *et al.* (11).

The pSPL1 system is becoming an increasingly popular approach to identify coding sequence and we describe in this chapter a study of the efficiency and specificity of the method with examples of gene isolation from cosmids and YACs. A simple method for exon amplification from YAC DNA is particularly desirable given the technical difficulty of direct YAC hybridisation to cDNA libraries (suppression of low copy number repeats and low percentage of coding sequence in the YAC probe leading to a low signal to noise ratio). Additionally, all YAC/cDNA hybridisation approaches are dependent on the presence of the relevant gene sequence in the cDNA library tested and on the absence of significant hybridisation of pseudogenes on the YAC to expressed sequences in the cDNA library. Nevertheless, recent developments in the technology of cDNA/YAC hybridisation have greatly improved the efficiency of this approach and we describe below the integration of exon amplification with such technologies using, as an example, a region of human 17q21 containing a putative tumour suppressor gene (BRCA1) implicated in breast cancer.

MATERIALS AND METHODS

Cloning Cosmids into pSPL1 and Transfection

DNA from cosmids was prepared by alkaline lysis (12) and digested to completion with BamHI and BglII or partially (for a 1-6 kb size fraction) with Sau3AI. The digested cosmid DNA was ligated to the pSPL1 vector (which had been previously digested with BamHI and phosphatased). The ligation was transformed into *E. coli* strain DH5α and libraries of approximately 2000 primary recombinants were grown on LB agarose plates with ampicillin selection. DNA from each library was prepared by the alkaline lysis method and electroporated into COS-7 cells using a Bio-Rad Gene Pulser (1.2kV; 25μF) as described previously (9). The cells were grown for 48 h and cytoplasmic total RNA prepared. First-strand cDNA synthesis and primary PCR amplification were carried out as described by Buckler *et al.* (9).

Cloning YACs into pSPL1

YAC.F1 was separated from the host yeast chromosomes by pulsed field gel electrophoresis using a Pharmacia LKB Pulsaphor. Electrophoresis through a 1% low melting point gel (BRL) was performed in 0.5xTBE for 18 h at 200V, 120 mA with a switching time of 10 sec. Agarose gel slices (5 ml) containing 300-500 ng YAC.F1 DNA were excised under long wave U.V., equilibrated for 2 h with the appropriate restriction buffer and incubated with 600 units BamHI and 600 units BglII (New England Biolabs) for 5 h in a final volume of 9 ml. The digested DNA was ligated to the pSPL1 vector (which had previously been digested with BamHI and phosphatased). An aliquot of the ligation was transformed by electroporation into *E. coli* strain DH5α. A library of approximately 2000 recombinants, corresponding to a five-fold coverage of the YAC, was grown on LB agarose plates with ampicillin selection. DNA from the library was prepared by the alkaline lysis method and purified using a Qiagen column. Average insert sizes and non-recombinant background (3%) were determined by testing about 20 randomly picked colonies.

Primers, PCR Amplification and Direct Cloning of PCR Products

A secondary (nested) PCR reaction was carried out using an aliquot from the first PCR reaction (see above) and the primer pairs SA1 and SD1 described by Buckler *et al.* (9). Amplified products were then separated on 1.8% agarose gels and visualised by staining with ethidium bromide. The PCR products were purified from the gel as described by He *et al.* (13) and either reamplified or subcloned directly in a 'no-ligase' reaction following the generation of compatible single-stranded ends between the PCR product and the cloning vector by uracil DNA glycosylase (Cloneamp, BRL) as described by Nisson *et al.* (14).

Sequencing of Cloned PCR Products

Plasmid DNA was prepared by using Qiagen columns (Diagen), according to the manufacturer's recommendations. The sequencing reaction was performed by the dideoxynucleotide chain-termination method (15).

Southern Blot Analysis

DNA probes were purified as described by He *et al.* (13) and radiolabeled to a specific activity of 10^8-10^9 cpm/µg by random hexamer priming (16) using [a^{32}P] dCTP (Amersham). Hybridization to filters was carried out overnight in 50% formamide, 50 mM sodium phosphate buffer, pH 7.2, 1 mM EDTA, 10X Denhardt's, 4XSSC, 1% SDS, 10% dextran sulphate, 50 µg/ml sonicated salmon sperm DNA, and 100 µg/ml sonicated human placental DNA at 42°C. Cosmid and genomic filters were washed to 0.1XSSC, 0.1% SDS at 65°C and exposed to Kodak XAR 5 X-ray film at -80°C between 45 min and 72 h. For zooblot hybridisations, exon inserts were reamplified with primers SDZ1 and SAZ1, which immediately flank the candidate exon:
SAZ1: 5' GTCCCCTCGGGATTGGGAGGTGGGT 3'
SDZ1: 5' ACTCATAAGTTTCTCTATCAAAGCA 3'
PCR conditions were as for primers SA1 and SD1 and zooblot hybridisations were carried out as described above except that the hybridisation temperature was 30°C for 48 h and filters were washed at room temperature in 1XSSC/0.1%SDS.

RESULTS

Efficiency of Recovery of Known Genes

We tested eight cosmids spanning eight known genes in the class II region of the human major histocompatibility complex located on chromosome 6 (17), as illustrated schematically in Fig. 1.

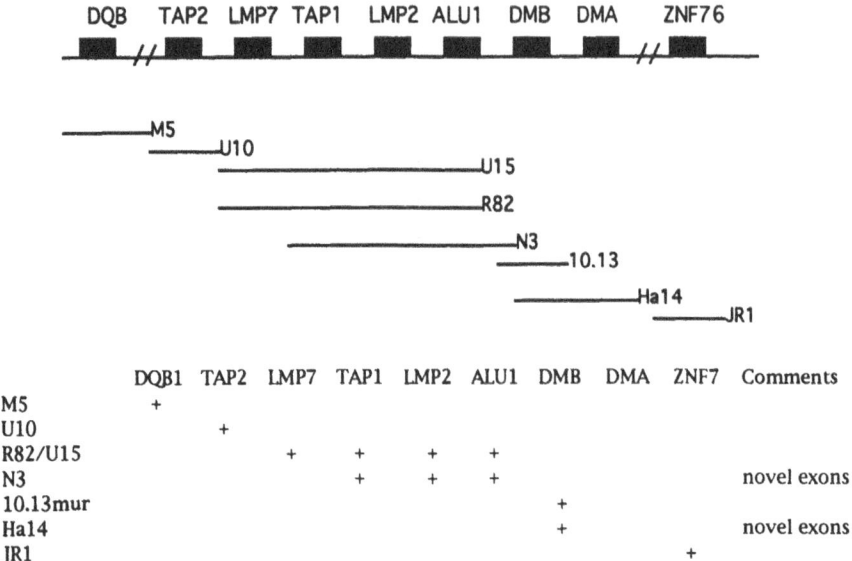

Figure 1. Schematic of genes (closed boxes) and cosmids from the class II region of the MHC used in the recovery assay. The table shows the known genes recovered (+) from each cosmid. All the cosmids are human except the mouse cosmid 10.13 (24).

The total region examined for genes amounted to about 185 kb. Amplified products from each cosmid were tested either by hybridisation to panels of cDNAs mapping to the region, by direct sequencing or by zoo blot hybridisation. Examples of exon amplification products from cosmids using the pSPL1 system are shown in Fig. 2.

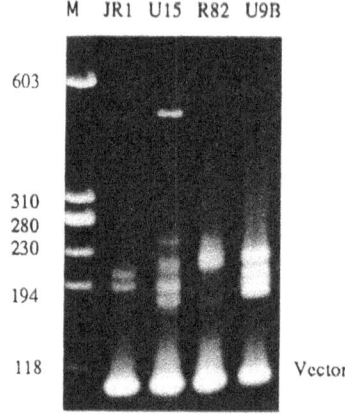

Figure 2. Examples of pSPL1 exon amplification from cosmids JR1, U15, R82 and U9B. Products of size greater than the vector band correspond to trapped exons.

With one exception we were able to identify at least one exon from each gene known to be present on a given cosmid. The exception was in the case of cosmid Hal4, which contains the genes DMB and DMA (18). Although two exons from DMB and three novel products were isolated, we obtained no visible exons (on a gel) from the DMA gene.

Four products were isolated from cosmids Ha14 and N3 which do not correspond to any known genes in this region. Flanking genomic sequences were good matches to the consensus splice junctions (Fig. 4). Products Ha14-E2 (51%G/C) and Ha14-E4 (53%G/C) were conserved on zooblots and both hybridised specifically to the mouse cosmid 4.24, which maps to an equivalent position to that of Ha14 in the mouse major histocompatibility complex (Fig. 3).

Novel Ha14 products

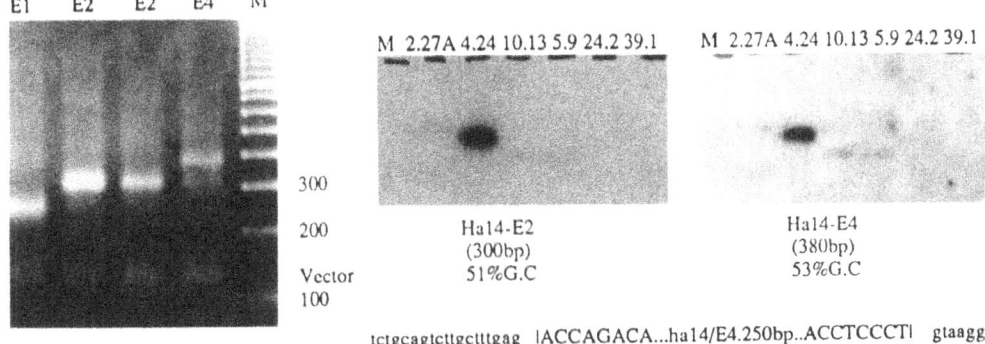

Ha14-E2
(300bp)
51%G.C

Ha14-E4
(380bp)
53%G.C

tctgcagtcttgctttgag |ACCAGACA...ha14/E4.250bp..ACCTCCCT| gtaagg
tgtagcttgtctttgttttag |CTAACCCA...ha14/E2.170bp..TGTCCATG| gtgatt

Figure 3. Three PCR products from Ha14 (in addition to two DMB exons) were visible on an agarose gel: E1, E2 and E4. They were gel purified and reamplified as shown above. They were specific to the cosmid and did not hybridise to the two known genes located on Ha14: DMA and DMB. Ha14-E2 and Ha14-E4 hybridised, as shown, specifically to the equivalent mouse MHC cosmid 4.24 and not to a variety of other mouse MHC cosmids present in the panel (32). Their flanking genomic sequences are shown and are a good match to the consensus 5' and 3' splice junctions.

These exons were not identified previously in conventional studies of the region (direct screening of cDNA libraries) because the genes encoding these exons are expressed at very low levels in B-cell or T-cell cDNA libraries.

Cloning and sequencing the three major exon amplification products from cosmid M5 revealed, in comparison to the published cDNA and genomic sequences, that exons 1, 2, 4 and 5 of the DQB gene had been isolated (19). Exon 5 is small (eight amino acids) and there seems to be some variability in whether it is spliced to exon 4 so that products were isolated containing only exon 4, or exon 4 plus exon 5. This is an example of 'exon skipping', a phenomenon observed with normal cellular genes (20) and which can clearly occur during splicing of pSPLI-derived transcripts in COS-7 cells (although the pSPL1 vector has been constructed to minimise these events (9)). DQB transcripts have been isolated with and without exon 4, a transmembrane domain (19). Exon amplification can therefore be used to aid characterisation of differentially spliced transcripts.

In the overall study of 32 cosmids (including those described above) we obtained genic products (defined by a positive hybridisation to a Southern blot of mouse and primate genomic DNA or by significant homology to known genes) from 13/32 of the cosmids, non genic (artefact) products from 6/32 of the cosmids (see below) and no amplification products of greater size than the vector-only PCR product from 15/32 of the cosmids: including four cosmids in which extensive conventional analysis does not detect any genes.

Fidelity of Splicing

The specific recognition of donor and acceptor splice junctions is vital to the widespread applicability of this system. We have determined the flanking genomic sequence for eight products isolated by exon amplification from MHC class II region described above. As shown in Fig. 4, the flanking sequences show good homology to the 5' or 3' splice consensus.

Fidelity of the splicing mechanism

```
                 3'splice junction                        5'splice junction
                         |                                        |
        INTRON                            EXON                 INTRON

tgtagcttgtctttgttttag     |CTAACCCA...ha14/E2...TGTCCATG|      gtgatt
cttctgggattttttgctcag     |CTGCTCGT...N3/E4.....ggatcc
attctgcagtcttgctttgag     |ACCAGACA...ha14/E4...ACCTCCCT|      gtaagg
gctttttttctcgttcttcag     |GTGAGAGG...N3/E2.....GTACTATG|      gtaagg
cactcaggttctgcttcttag     |GGGCTCAG...M5/E1.....TCAGAAAG|      gtgagg
ccatattctgtgtctctgcag     |ACCACCAT...U15/E1....TCTGCAGG|      gtgagt
tttgtattttttatttttag      |AAATACGG...N3/E1.....CCTCCCAA|      gtaggt
ctccaccttgtcctcacccag     |GCTGTACT...U15/E2....ATAAGAAG|      gtgggtg

                          5' splice consensus      CAG|      gtaagt
                                                              g
```

Figure 4. Sequences of the flanking genomic regions of eight MHC products isolated by exon amplification. The flanking sequences show characteristic structures for splice junctions (eg. a 3' splice junction is characterised by a polypyrimidine stretch ending with the conserved AG dinucleotide immediately upstream of the start of the exon). N3/E4 is interrupted by a BamHI site and is therefore a chimaeric product with the structure shown in Fig. 6. Product U15/E2 is an exon from the MHC gene LMP7 is partially incorrectly spliced as described in the text, the variable cag is underlined. None of the products were artefacts arising from cryptic splicing (25).

In only one occasion the splicing event was (partially) aberrant. MHC product U15-E2 appeared to be a mixed PCR population containing both the correctly spliced product and a product derived from a splicing event three bps upstream so that the codon cag was included in the amplified product. It seems likely that the latter product was the result of an aberrant splicing event since the 3' splice junction flanking this sequence is a poor match to the consensus. Such a product is, however, spliced correctly at the 5' junction and corresponds to exon 4 of LMP7.

Artefacts

During the study described above we identified a small number of PCR products arising from the amplification of noncoding sequences which do, however, contain regions with high homology to acceptor and donor splice junctions. Examples of artefacts of this nature are illustrated in Fig. 5.

N3/E1 80.8% identity to Alu repetitive element (151bp overlap)

```
         40        50        60        70        80        90
N3/E1  ggtcctacctacTTGGGAGGCCAAGGCAGGAAGATTGCTTGAGCCCAGGAGTTTGAGCTT
       |  ||| |  |||||||||||||  ||||||||  || ||  ||||||||||| |  ||| ||
Alu    GATCCCAGCTACTTGGGAGGCTGAGGCAGGAGGATCGCCTGAGCCCAGGAGGTGGAGGTT
         1820      1830      1840      1850      1860      1870

         100       110       120       130       140       150
N3/E1  ACTGTGAGCTGTGATCACACCACTGCACTCCAGCCTGGGTGACAAAGGAAGACCGTATTT
       | |||||||  |||||    |||||||||||||||||||||  ||| | |||||||| ||||||
Alu    GCAGTGAGCCATGATCGAGCCACTGCACTCCAGCCTGGGCAACAGATGAAGACCCTATTT
         1880      1890      1900      1910      1920      1930

N3/E1  ctaaaaaataaaaaatacaaatacaactacaaac
       |  ||| | ||  |  | ||| || || |  |
Alu    CAGAAATACAACTATAAAAAAAATAAATAAATCCTCCAGTCTGGATCGTTTGACGGGACT
         1940      1950      1960      1970      1980      1990
```

GO864/E1 Hiv tat only product

```
gtgttagTTTAAAGTGCACTGATTTGAAGAATGATACTAATACCAATAGTAGTAGCGGGA
GAATGATAATGGAGGAAGGAGAGATAAAAAACTGCTCTTTTAATATCAGCACAAGCATAA
GAGgtaagg
```

Figure 5. Examples of an artefact amplification product. N3/E1 is an Alu repeat cloned from MHC cosmids U15 and N3. Lowercase nucleotides indicate flanking sequences which can act as splice junctions. GO864/E1 (see Fig. 1) is a product derived solely from the HIV tat intron (again, lowercase nucleotides are flanking sequences which act as splice junctions). It is distinct from the cryptic splicing event shown in Fig. 6.

The Alu repeat was amplified independently in separate experiments using two overlapping MHC cosmids (U15 and N3). The product was sequenced and analysis of the flanking genomic region showed that the Alu repeat contains regions which can act as 5' and 3' splice junctions so that the intervening 108 bp can be amplified. This particular Alu structure is presumably fairly rare in the genome since the vast majority of Alu repeats are not amplified by this system (30/32 cosmids tested negative). Fig. 5 also illustrates a 241 bp amplification product from chromosome 22 cosmid G0864. This product is derived purely from the pSPL1 tat intron following splicing directed by cryptic donor and acceptor junctions. Products derived purely from the pSPL1 tat intron were also obtained from 4/32 cosmids. In our overall study, 15/32 cosmids gave no products of greater size than the wild type vector splice (including four cosmids where extensive conventional analysis does not detect any genes) while 6/32 gave an artefact product (a repetitive element or tat intron sequence). Hence the system appears, in general, to provide a reasonably stringent test for the presence of exon sequences and artefactual amplifications such as those detailed above are relatively rare (19%). Importantly, artefacts seem to fall into a small number of categories and can be eliminated at an early stage of analysis. The PCR products can be conveniently tested by hybridisation to a BamHI/BglII digest of the original cosmid and to a zooblot. Amplification of repeat elements are immediately apparent following hybridisation to human genomic DNA while products from the tat intron do not hybridise to the cosmid digest. In spite of their small size, most single-copy products we have tested can be shown to be conserved (at least in primates) by this method, although some products do give faint bands and require several days autoradiographic exposure. Products shown to be single copy and conserved in such an experiment can be rapidly cloned and sequenced as described previously.

Chimaeric Exons

A relatively common event in the pSPL1 system is the isolation of PCR products containing both unique sequence and material derived from the tat intron. This can occur when an exon is interrupted by a BamHI or BglII restriction site so that the exon is only amplified following compensation for the loss of the normal 5' splice junction by the activation of a cryptic 5' splice junction in the tat intron 66 bp downstream of the BamHI cloning site, as illustrated in Fig. 6. In many cases the specificity of the system for amplification of exons does not appear to be significantly compromised by the loss of a splice junction since the unique sequence either shows homology to a known gene or is conserved on zooblots. Two examples of chimaeric products with homology to known genes are shown in Fig. 6 and a successful zooblot hybridisation is shown in Fig. 8.

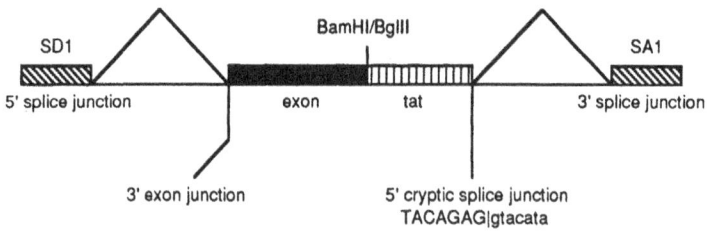

Hiv tat
GGATCCCCTGTGTGGAAGGAAGCAACCACTCTATTTTGTGCATCAGATGCTAAAGCATATGATACAGAG
(corresponding to residues 6343-6413 in the HIV envelope polyprotein)

```
                        *****D**********S*D*******G*U*******I*P***
glucuronidase
CO650/E1                SGPTUHMPUPSSFNDIGQGWRLRHFUSWLWYEREUTLLERW+HIV tat

                        **_ _* ***_*
102BY/17.25             SLADSHRKRKLCRCCDEFHPYCUEDH+HIV tat
Drosophila trithorax    CCEPYHQYCUQD
```

Figure 6. Cryptic splicing leading to a chimaeric exon: A genomic fragment containing an exon interrupted by a BamHI or BglII site is cloned into the pSPL1 vector. The absence of the exon 5' splice junction is compensated by the HIV tat 5' cryptic splice junction 66 bp downstream of the BamHI cloning site. Also shown are two chimaeric amplification products with homology to known genes (CO650/E1 and 102BY/17.25). Studies with known MHC genes indicate that very few mutations are introduced by the PCR steps involved in the amplification strategy so that, for example, CO650/E1 is distinct from, but homologous to the published glucuronidase sequence.

YAC EXON AMPLIFICATION

YAC.F1 is a mouse 175 kb YAC derived from a partial EcoRI C57BL/6 female YAC library with average insert size 250 kb and constructed in pYAC4. A subcloned BamHI/BglII double digest library of YAC.F1 constructed in pSPL1 was transformed into COS-7 cells. Total cytoplasmic RNA was isolated 48 h after transfection and subjected to RNA-based PCR as described above. Secondary amplified products were fractionated on agarose gels and subcloned. Over 80 clones were examined by restriction digestion and a minimal set of 9 unrelated products distinguished. Clones from each of these groups were sequenced. As summarised in Fig. 7, two of the clones (ET1 and ET26) displayed a high degree of homology to known repetitive elements (90% identity to the mouse B1 repetitive element and 80% identity to the HSAG-1 middle repetitive element, respectively). One product (ET1), though not repetitive, did not contain an open reading frame in the correct orientation. The amplified products were, on average, 50% GC rich and had an average size of 90 bp.

An interesting feature of the products is that 6 of the 9 were chimaeric due to the activation of the cryptic tat splice junction (as described above). This significantly higher

Product	Size	%GC	ORF	Cryptic	Repeat	Zooblot
ET1	106	54	No	No	Yes	(-)
ET15	71	51	No	Yes		(-)
ET17	117	62	Yes	Yes		(+)
ET26	53	57	Yes	Yes	Yes	(-)
ET29	85	53	Yes	No		(+)
ET35	44	55	Yes	Yes		(-)
ET42	70	51	Yes	Yes		(+)
ET44	84	60	Yes	Yes		(-)
ET58	186	54	Yes	No		(+)

Figure 7. Summary of the nine major exon amplification products from YAC.F1 (ET1-ET58).

level of chimaerism to that observed in trapping experiments with MHC cosmids may be due to an abnormal distribution of BamHI and BglII sites in the YAC so that many exons are interrupted by a BamHI or BglII site. However, one of the highly conserved exons (ET17) arose from just such an event suggesting that the specificity of the system may not have been significantly compromised.

In order to assess the potential recovery of coding sequences, the six amplified products that contained potential open reading frames and no homology to known repetitive sequences were hybridised to zooblots (ET17, 29, 35, 42, 44, 58). Three clones (ET29, 35, 44) failed to hybridise reproducibly and cleanly to zooblots presumably due to their small size. One clone (ET42) hybridised to mouse and rat DNA but failed to detect sequences in other species. Two clones (ET17, 58) clearly detected sequences conserved across a wide variety of species (Fig. 8). Both ET17 and ET58 hybridised specifically back to the tested YAC and mapped to different regions of the YAC (F.G, unpublished).

A modification of this procedure was evaluated in which genomic DNA was prepared from a AB1380 culture containing a mouse YAC and subjected to exon amplification. While this method removes the necessity of gel purifying the YAC, amplification products were predominantly (>90%) derived from the yeast genome. Sequencing a set of such clones identified several yeast ribosomal genes and previously characterised yeast genes such as the ferrochelatase gene (F.G, unpublished).

INTEGRATION WITH cDNA ENRICHMENT STRATEGIES

Recent advances in YAC to cDNA hybridisation technology (variously referred to as cDNA enrichment or hybrid selection) have increased the power of this approach considerably and are described elsewhere in this volume. Any cDNA hybridisation strategy should take in consideration that only some 10-20% of all mRNAs may be expressed in any differentiated cell type (about 10,000 genes per cell type) and therefore use a complex mixture of RNA sources and/or normalised cDNA libraries to increase the likelihood of transcript identification. Both exon amplification and cDNA enrichment can be used separately but a particularly efficient way of screening large genomic regions for genes is

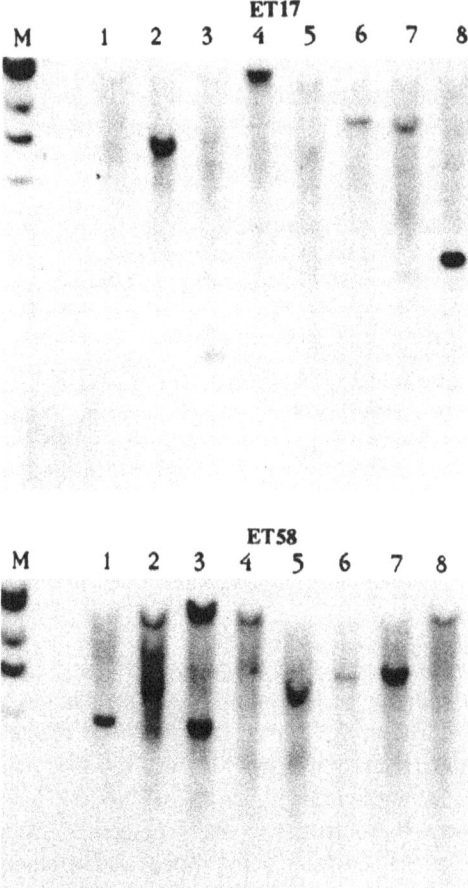

Figure 8. Examples of zooblot hybridisations of YAC.F1 exon amplification products ET17 and ET58 (see methods for hybridisation conditions). lane 1:human; 2 baboon; 3 cow; 4 pig; 5 sheep; 6 mouse; 7 rat; 8 gerbil.

to use these two techniques in parallel. For example, a set of cosmids or a YAC spanning a genomic region is used in the cDNA enrichment protocol to generate a minilibrary (approximately 1 recombinant/kb input genomic DNA) of cDNAs which are picked and stored in microtitre plates and spotted onto hybridisation membranes. Clones containing repetitive elements, ribosomal sequences, mitochondrial sequences etc. can be identified by hybridisation (eg. Cot-1 DNA to identify medium and high copy number repeats). The same set of cosmids or YACs are exontrapped (either individually or in pools) following the methods described above and individual exon amplification products cloned. The set of candidate exons are tested by hybridisation to a Southern blot of a BamHI/BglII digest of the cosmid set (to identify tat-only artefacts), including a track of genomic DNA (to identify repetitive elements). A minilibrary membrane is included in this hybridisation and positively hybridising cDNAs with each (non-artefact) exon amplification product are recorded.

This strategy can be used to identify a minimal set of cDNAs mapping to the region since, for example, different exons from the same gene detect overlapping sets of cDNAs in the minilibrary. cDNA walking to isolate a full length transcript can be performed extremely rapidly within the enriched minilibrary. Hybridisation of even very small exons to such library filters is very straightforward and the randomly-primed cDNAs detected typically fall in the 500-1000 bp size range. The minilibrary array can be additionally hybridised with protein motif oligonucleotides, trinucleotide repeats, single-copy genomic fragments etc. All the information from such additional screens can be easily integrated and a minimal set of cDNAs chosen for sequencing. An important advantage with this approach is that it eliminates two common causes of difficulties with these techniques: the identification of a full-length cDNA starting from a single exon amplification product and the complexity of the cloned product from cDNA enrichment. Additionally, artefacts such as processed pseudogenes (potentially isolated by cDNA enrichment) are generally not isolated by exon amplification.

An example of this methodology is shown in Fig. 9. Here a pool of 10 cosmids spanning approximately 300 kb of genomic DNA within the 17q21 region containing the BRCA1 tumour suppressor gene was used in a cDNA enrichment experiment following the method of Korn et al. (8), with fetal brain and adult breast as sources of RNA (B.K, pers. comm.). cDNAs specifically hybridising to a particular genomic DNA sequence are selected by a biotin-streptavidin interaction (using streptavidin-coated magnetic beads) and the non-specific hybrids are dissociated by stringent washing. The cDNAs are eluted, amplified, cloned, and analyzed as "region-specific sublibraries". The technique typically results in an enrichment of the selected cDNAs by factors of 10^3 to 10^6.

cDNA clones were manually picked into two 384 well microtitre plates and hybridisation membranes prepared. Repeats were identified by hybridisation of Cot-1 DNA (Fig. 9A and F) and a gene known to be present in the 300 kb region, Pancreatic Polypeptide Y (PPY), identified three cDNA clones within the minilibrary array: acting as a positive control (Fig. 9B). The 10 cosmids spanning this region were each individually subjected to exon amplification in 10 parallel reactions and a set of discrete amplified exons subcloned and sequenced. Individual products were hybridised to the minilibrary arrays and all products presently tested (6/6) have identified one or more minilibrary cDNAs (cf. Figs. 9C, D and E). Hybridising each cDNA back to the minilibrary array allows rapid walking (as illustrated schematically in Fig. 9) and allows the complexity of the minilibrary to be reduced.

A targeted entry approach of this nature allows the identification of transcripts of low abundance in the minilibrary. For example, identification of PPY transcripts (3/768) by sequential examination of each member of the minilibrary would be extremely inefficient. The enrichment factor in the procedure is substantial: for example, the exontrap product used in panel D (Fig. 9) is derived from a gene which is expressed at approximately $1/10^6$ in fetal brain cDNA and is represented at approximately 1% (6/768) in the minilibrary.

DISCUSSION

We have shown that the exon amplification procedure can be used successfully with a wide variety of cosmids and YACs which have different numbers of genes and gene structures. Novel exons and exons from well characterised genes have been isolated. A central issue with such approaches is the degree to which the splicing process is specific for genic sequences. We address this question by examining the products obtained with 32 cosmids, including 8 cosmids which span a well characterised 185 kb region of the human major histocompatibility class II region. We have tested the efficiency of the system in isolating exons from genes known to be present in this region and have determined the nature and frequency of artefact amplifications in routine cosmid screening. Out of 32 cosmids tested, 6/32 gave non genic amplification products, 4/32 gave exons chimaeric with the tat intron and 15/32 gave no products other than the vector-only splice. Such

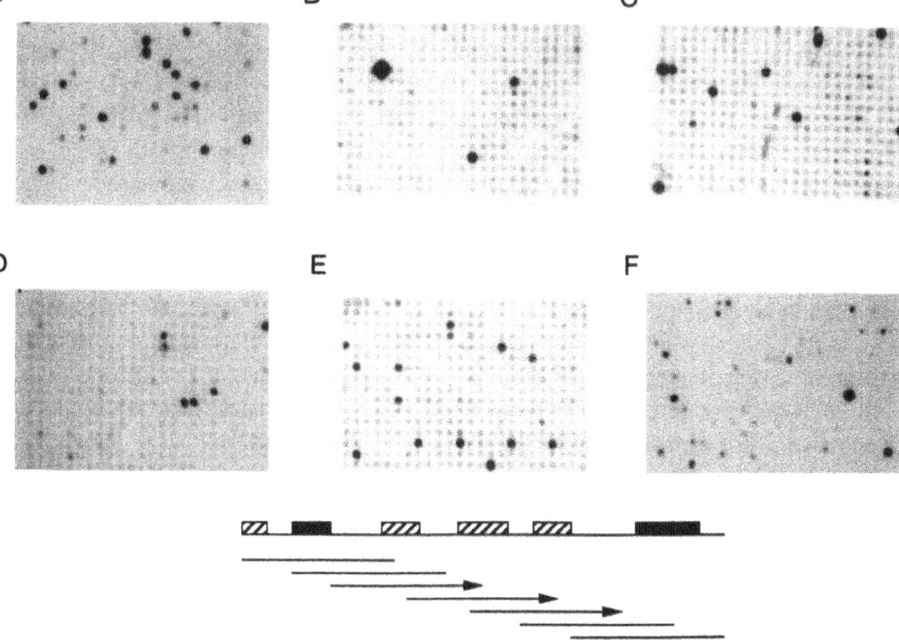

Figure 9. Examples of hybridisations of exon amplification products to gridded arrays of region-specific cDNA minilibraries (prepared by enrichment methods). A and F are Cot-1 hybridisations; B is the hybridisation of PPY, a gene known to be present in the region; C, D and E are exon hybridisations. A schematic is shown to illustrate cDNA walking within the minilibrary between cDNAs detected by the hybridisation of different exons from the same gene.

considerations are also important in determining the maximum complexity of genomic DNA which can be tested for genes in this manner. We have no evidence that artefact products are preferentially amplified or cloned in comparison to real exons (in terms of band intensities of the secondary PCR products) and therefore levels of artefact products should remain at similar levels to those described above irrespective of the complexity of the input genomic DNA. Artefact levels in other exon amplification/trapping systems have not been documented in detail although, in a first test of the retroviral shuttle vector system produced by Duyk *et al.* (21) a 50% artefact level was observed while Hamaguchi *et al.* (11) were able to confirm 4 out of 7 PCR products isolated in their studies. Our experiments do not support the 75% false positive frequency for this system described by Hamaguchi *et al.* (11).

For most purposes the system can be regarded as successful if it results in the amplification of a minimum of one exon from each gene present in a complex input genomic sample and if these products are of approximately equal abundance. Losses are inevitable and can arise for a variety of reasons. As described previously, the 5'-most and 3'-most exons of a gene can theoretically not be amplified although exceptions can arise through the activation of compensating cryptic splice junctions in the tat intron (cf. M5/E2). Also, exon skipping has been observed (DQB exon amplification products M5/E1 and M5/E3) and it seems possible that different exons may splice to the flanking pSPL1 sequences with different efficiencies leading to a preferential amplification of particular exon sequences. Exons may also be lost if an appropriate BamHI/BglII genomic fragment is too large to be efficiently subcloned into the pSPL1 vector (insert sizes of greater than 5 kb are rarely observed). This particular bias can be reduced if genomic DNA is partially digested with Sau3AI and a size fraction between 1 and 6 kb subcloned into pSPL1. Even without varying the method of digestion of the input genomic DNA, we were able to clone at least one exon from 88% of all known genes tested. If the system was applied to more complex input sources such as flow sorted chromosomes or whole genomes a similar success rate would generate very large numbers of new gene markers. Markers derived by this method are unlikely to be obtained in equal numbers from all genes in the region. For example, considering MHC cosmid U15, we reproducibly amplified three exons from the LMP2 gene but only one exon from TAP1. Exon amplification from defined genomic regions (eg. flow sorted human chromosomes) will be particularly powerful if the transcriptional information generated can be integrated with detailed physical and genetic maps. The reference library database is a simple and powerful system to integrate data from different types of experiments in this way (22,23).

ACKNOWLEDGEMENTS

We thank Professor B. Young and Dr M. Bower for the sequence analysis of products 108BY-E1 and 102BY-E1 and Drs. B. Korn and A-M. Poustka for preparing enriched minilibrary arrays.

REFERENCES

1. P. Elvin, G. Slynn, D. Black, A. Graham, R. Butler, J. Riley, R. Anand, and A.F. Markham, Isolation of cDNA clones using yeast artificial chromosome probes, *Nucl. Acids Res.* 18:3913 (1990).
2. B.G. Herrmann, S. Labeit, A. Poustka, T.R. King, and H. Lehrach, Cloning of T gene required in mesoderm formation in the mouse, *Nature* 343:617 (1990).
3. A.P. Monaco, R.L. Neve, C. Colletti-Feener, C.J. Bertelson, D.M. Kurnit, and L.M. Kunkel, Isolation of candidate cDNAs for portions of the Duchenne muscular dystrophy gene, *Nature* 323:646 (1986).
4. A. Bird, CpG-rich islands and the function of DNA methylation, *Nature* 321:209 (1986).
5. L. Corbo, J.A. Maley, D.L. Nelson, and C. Caskey, Direct cloning of human transcripts with HnRNA from hybrid cell lines, *Science* 249:652 (1990).
6. P. Liu, R. Legerski, and M.J. Siciliano, Isolation of human transcribed sequences from human-rodent somatic cell hybrids, *Science* 246:813 (1989).
7. K. Abe, Rapid isolation of desired sequences from lone linker PCR amplified cDNA mixtures: application to identification and recovery of expressed sequences in cloned genomic DNA, *Mammal. Genome* 2:252 (1992).
8. B. Korn, Z. Sedlacek, A. Manca, P. Kioschis, D. Konecki, H. Lehrach, and A. Poustka, A strategy for the selection of transcribed sequences in the Xq28 region, *Hum. Molec. Genet.* 1:235 (1992).
9. A.J. Buckler, D.D. Chang, S.L. Graw, D.J. Brook, D.A. Haber, P.A. Sharp, and D.E. Housman, Exon amplification: A strategy to isolate mammalian genes based on RNA splicing, *Proc. Natl. Acad. Sci. USA* 88:4005 (1991).
10. D. Auch and M. Reth, Exon trap cloning: using PCR to rapidly detect and clone exons from genomic DNA fragments, *Nucl. Acids Res.* 18:6743-6 (1990).
11. M. Hamaguchi, H. Sakamoto, H. Tsuruta, H. Sasaki, T. Muto, T. Sugimura, and M. Terada, Establishment of a highly sensitive and specific exon-trapping system, *Proc. Natl. Acad. Sci. USA* 89:9779 (1992).
12. T. Maniatis, E.F. Fritsch, and J. Sambrook, "Molecular Cloning: A Laboratory Manual," Cold Spring Harbour University Press, Cold Spring Harbour (1982).
13. M. He, H. Liu, Y. Wang, and B. Austen, Optimised centrifugation for rapid elution of DNA from agarose gels, *Gene Analysis and Techniques* 9:31 (1992).
14. P.E. Nisson, A. Rashtchian, and P.C. Watkins, PCR methods and applications, 1:120 (1991).
15. F. Sanger, S. Nicklen, and A.R. Coulson, DNA sequencing with chain-terminating inhibitors, *Proc. Natl. Acad. Sci. USA* 74:5463 (1977).
16. A.P. Feinberg and B. Vogelstein, A technique for radiolabelling DNA restriction endonuclease fragments to high specific activity, *Anal. Biochem.* 132:6 (1984).

17. J. Trowsdale, J. Ragoussis, and R.D. Campbell, Map of the human MHC, *Immunology Today* 12:443 (1991).

18. A.P. Kelly, J.J. Monaco, S. Cho. and J. Trowsdale, A new human HLA class II related locus, DM, *Nature* 353:571 (1991).

19. P. Briata, S.F. Radka, S. Sartoris, and J.S. Lee, Alternative splicing of HLA-DQB transcripts and secretion of HLA-DQB-chain proteins: Allelic polynorphism in splicing and polyadenylation sites, *Proc. Natl. Acad. Sci. USA* 86:1003 (1989).

20. A. Andreadis, M.E. Gallego, and B. Nadal-Ginard, Generation of protein isoform diversity by alternative splicing: mechanistic and biological implications, *Annu. Rev. Cell Biol.* 3:207 (1987).

21. G.M. Duyk, S. Kim, R.M. Myers, and D.R. Cox, Exon trapping: a genetic screen to identify candidate transcribed sequences in cloned mammalian genomic DNA, *Proc. Natl. Acad. Sci. USA* 87:8995 (1990).

22. H. Lehrach, R. Drmanac, J.D. Hoheisel, Z. Larin, G. Lennon, A.P. Monaco, D. Nizetic, G. Zehetner, and A. Poustka, Hybridisation fingerprinting in genome mapping and sequencing, *in* "Genome Analysis Volume I: Genetic and Physical Mapping," K.E. Davies, and S. Tilghman, eds., Cold Spring Harbour University Press (1990).

23. D. Nizetic, G. Zehetner, A.P. Monaco, L. Gellen, B.D. Young, and H. Lehrach, Construction, arraying and high density screening of large insert libraries of human chromosomes X and 21: their potential use as reference libraries, *Proc. Natl. Acad. Sci. USA* 88:3233 (1991).

24. M. Steinmetz, D. Stephan, and A. Fischer-Lindhal, Gene organisation and recombinational hotspots in the murine major histocompatibility complex, *Cell* 44: 895 (1986).

25. H. Hornig, M. Aebi, and C. Weissman, Effect of mutations at the lariat branch acceptor site on beta-globin pre-mRNA splicing in vitro, *Nature* 324:589 (1986).

INTEGRATED TRANSCRIPTIONAL MAPS OF LARGE DNA REGIONS: TOWARDS A TRANSCRIPTIONAL MAP OF HUMAN CHROMOSOME 21

Marie-Laure Yaspo[1], Philippe Sanséau[2], Dean Nizetic[1], Bernhard Korn[3], Annemarie Poustka[3] and Hans Lehrach[1]

[1]Genome Analysis Laboratory and [2]Human Immunogenetics Laboratory
Imperial Cancer Research Fund, 44 Lincoln's Inn Fields, P.O. Box 123
London WC2A 3PX, United Kingdom
[3]Deutsches Krebsforschungszentrum, Im Neuenheimer Feld 280, 6900
Heidelberg, Germany

ABSTRACT

We present here a relational mapping strategy to establish integrated physical and transcriptional maps of large DNA regions. In this study, human chromosome 21 is used as a model because a physical map spanning 40 megabases on the 21q arm has already been achieved. The first step of this global analysis is to extract most of the genic content encoded by this chromosome. For this purpose, exon-amplification and cDNA selection methods are applied in conjunction to derive expressed sequence libraries from chromosome 21-specific cosmids. These resources will constitute one of the tools necessary for the relational analysis, allowing cross-screenings between genomic and coding sequence libraries. Eventually, the new genes will represent a resource to hunt for potential candidate genes implicated in the determinism of severe diseases associated to chromosome 21.

INTRODUCTION

Coding sequences represent only a small fraction (3-5%) of the human genome, and identifying the 100,000 estimated genes (1) is still a formidable task. If the main aspect of gene hunting is related to the positional cloning of the more than 3,000 human genetic diseases described (2), a systematic large scale analysis is expected to provide new clues towards the global knowledge of the genetic message. More than 6,000 random cDNAs have already been partially sequenced (3-7), but few of the generated Expressed Sequence Tags (EST) have been mapped or assign to a particular chromosome (8,9). In this context, building integrated physical and transcriptional maps represents another level of information in regard to the functional organisation of the genome.

The backbone for such an analysis is provided by genomic libraries constructed in Yeast Artificial Chromosomes (YACs), P1 or cosmid clones (10,11) used to construct physical maps of large genomic regions or chromosomes (12-15); these collections of

Identification of Transcribed Sequences, Edited by
U. Hochgeschwender and K. Gardiner, Plenum Press, New York, 1994

clones constitute the basic framework for gene identification techniques. Indirect methods such as identification of cross-species conserved sequences (16,17), CpG islands (18,19), or potential splice sites (20) are non-exhaustive and lead to major drawbacks for a large scale analysis. The recent development of cDNA selection (21-24) and exon-trapping approaches (25-29), allowing the direct isolation of coding stretches from genomic clones, could greatly advance the construction of transcriptional maps. cDNA selection is a hybridization-based method leading to a specific enrichment for cDNAs encoded by the targeted genomic region, while exon-trapping uses a shuttle vector to retrieve spliced exons irrespective of their expression pattern. When used simultaneously, these two techniques may be expected to identify most of the genes contained in the region analysed.

Our aim is to derive expressed sequence libraries specific for human chromosome 21 by the aforementioned strategies. With a size of only 50 Mb, this chromosome is a model of choice for such a study because it is the paradigm of an unprecedented mapping effort (30). A quasi continuum of YACs (30,31) and cosmid clones (D. Nizetic *et al.*; in press) spanning the 21q arm provide an immediate genomic source for gene hunting and mapping. In this paper, we will discuss a global approach to integrate transcriptional and physical data. This can be achieved by gridding the expressed sequence libraries and compiling data obtained by cumulative cross-screenings between these new resources, genomic and cDNA libraries (32). Ultimately, the new coding sequences will provide a resource to hunt for candidate genes associated with chromosome 21-linked pathologies, such as Down Syndrome (33,34).

MATERIAL AND METHODS

cDNA Library Construction

PolyA+ messenger RNAs were prepared from various tissues originating from a 21 week human foetus using the Fastrack kit (Invitrogen). cDNAs were prepared by poly dT priming and size-selected before being directionally cloned into NotI-SalI digested pSPORT vector (Superscript cloning kit; Gibco-BRL). The average insert size was approximately 2 kb.

Colony Picking

A robotic device with a 96 pin spring loaded picking manifold was used to pick clones from 22x22 Nunc agar plates into 384 wells microtiter dishes.

Cosmid Libraries

Two ordered chromosome 21-specific libraries were used: the ICRF library (ED8767 host) for the regional analysis and the Lawrence Livermore National Laboratory library constructed by Pieter deJong (DH5α host), referred as LL21NC02, for the random analysis. These libraries each represent a 5 fold coverage of chromosome 21. The ICRF reference library (35) is ~60% pure and has been extensively screened; all the data are stored in the Reference Library DataBase (RLDB; G. Zehetner and H. Lehrach, *Nature* 364:489 (1994)) which provides access to the loci information for a large number of clones. The LL21NC02 (abbreviated as LL21) library is ~87 % pure (Peter deJong; personal communication) allowing the preparation of cosmid pools directly from complete microtiter dishes. The ribosomal clones have been removed for our analysis.

Preparation of High Density Cosmid and Plasmid Membranes

A robotic device equipped with a 96 pin or 384 pin transferring device was used to spot clones from microtiter dishes containing saturated cultures of cosmids or plasmids. The preparation of high density cosmid or plasmid membranes was carried out as described previously (35). Colonies were grown overnight on Hybond N+ membranes (Amersham) at 37°C on 2YT agar with 30 μg/ml kanamycin for cosmids or with 50 μg/ml ampicillin for plasmids. The colonies on membranes are then subjected to colony lysis (36). In this study 9,000 ICRF cosmids and 960 LL21 cosmids were spotted onto filters for screening purposes.

Hybridizations

Probes were labelled by random hexamer priming (37) with αP^{32} dATP and competed with human placental DNA (Sigma) in a volume of 500 ml in 4XSSC final and 1 mg of competitor DNA for 90 min at 68°C. Hybridizations were carried out overnight in 5XSSC, 1XDenhardt's, 0.005M Sodium Phosphate, 10% Dextran Sulfate and 50% formamide. Filters were washed to 0.1XSSC at 65°C and exposed to Kodak X-OMAT AR films at -80°C.

Hybridization of Gridded Filters

In order to light up the background for an exact positioning of the positives clones, 50 ng of vector DNA (Lawrist 4 for cosmid filters and pSPORT 1 for cDNA filters) was labelled with 50 μCi of S35 dATP by random priming and added in the hybridization solution together with the P32 specific probe.

Isolation of Cosmid DNA

Cosmids were grown overnight in 2YT medium with 30 μg/ml kanamycin. Individual cosmids were prepared by standard alkaline lysis miniprep (Maniatis). Cosmid pools were prepared either with Qiagen tips (Diagen) or by CsCl gradient centrifugation.

Gel Analysis and Blotting

Cosmids were digested for 2 h with 10 units of the appropriate enzyme per μg of DNA. The digested DNA was run on a 0.8% agarose gel in 1xTAE (Tris-Acetate EDTA) buffer. After migration, the DNA was transferred onto Hybond N+ membranes (Amersham) according to the manufacturer recommendations.

cDNA Selection

Pools of 10 to 87 cosmids were labelled by nick translation (1 μg of each pool) with biotine dUTP according to a protocol already described (23), using DNA polymerase I and DNAse I, to obtain a size of the DNA comprised between 0.4 and 4 kb. The cosmids were depleted of repetitive sequences by prehybridization with 50 μg of CotI DNA/pool (Gibco-BRL) for 2 h at 42°C. Biotinylated cosmids were bound to streptavidine coated magnetic beads (Dynabeads-Dynal) for 1 h. Cosmids were then hybridized overnight at 42°C with 5 μg of a mixture containing brain, liver and muscle amplified cDNAs. After washing to 0.1XSSC at 65°C, the cDNAs were eluted, size selected, and subjected to 23 cycles of amplification using primers and conditions described previously (23). The PCR amplified cDNAs were then used as a selected probe for a second round of selection-amplification. The final eluted cDNA products were cloned and used to transform DH5α bacteria.

Exon-amplification

The exon-amplification procedure developed by Buckler *et al.* (27) was used with the modified vector pSPL3 improving the overall efficiency of the procedure (A.Buckler, personal communication). For subcloning, pools of 5 to 6 cosmids were digested with BamHI+BglII for 1 h and cloned into BamHI-digested pSPL3 vector. The ligation product was used to transform XL1-blue bacteria. Ten individual clones were picked out of each transformation to ensure a random distribution of the cloned DNA fragments in the sublibrary. All colonies were then scraped from the agar plate and the DNA was prepared by standard alkaline lysis. 20 μg of DNA from each transformation assay were used to transfect COS-7 cells by electroporation using a Biorad apparatus (5×10^6 cells/transfection). 48 h after the transfection, the RNAs were prepared with RNAzol (Bioprobe systems). Reverse first strand synthesis was carried out from 2 μg RNA for 30 min at 42°C, using Superscript reverse transcriptase (Gibco-BRL) and 20 μM SA2 primer. Half of the material was then amplified by PCR (94°C 1 min/60°C 1 min/72°C 3 min for 35 cycles) using primers SA2 and SD6; usually gel analysis showed a smear containing prominent bands above the vector amplification product. This primary product was digested with BstXI for 4 h to eliminate bands arising from vector only and from the activation of the cryptic splice site, and a second PCR using SA4 and SD2 primers (for 25 cycles) was performed with 1/10 of the digested first PCR. Eventually, a third amplification of 30 cycles was carried out using modified SA5 and SD5 primers allowing the cloning of the PCR products into the pAMP1 vector (Gibco-BRL).The primer sequences were communicated by Alan Buckler.

RESULTS

Overall strategy

We aim to apply both cDNA selection and exon-amplification strategies to chromosome 21 cosmid pools. Although a YAC contig is virtually completed for chromosome 21 (30,31), YAC clones present a major inconvenience for this type of analysis. Besides the tedious isolation of YAC DNA by pulsed field gel electrophoresis (PFGE), the notable chimaerism and instability observed can dramatically confuse the transcript mapping. Moreover, we experienced that exon-amplification is difficult to apply to YACs. We previously showed that small YACs (150 kb) could be effectively trapped (38), but exons were recovered together with non-specific products likely to be derived from the yeast host. As an alternative, we are using regionally assigned cosmid pools (ICRF) referred to as "cosmid pockets"; we are also using pools of random cosmids (LL21) to achieve a greater genomic representation of chromosome 21.

The complex products generated by cDNA selection and exon-amplification will be used as probes against gridded libraries and cloned for storage in microtiter dishes.This strategy allows the construction of ordered exonic sequence libraries and to obtain simultaneously cumulative data by hybridization.

Definition of "Cosmid Pockets" on Chromosome 21

Individual chromosome 21 YACs (from ICRF, MRC, CEPH and JYSE; 39-42) were purified by PFGE and used as probes against the ICRF cosmid library (43); screening chromosome specific cosmids with the YACs overcomes the problem of chimaerism. The scored positive clones identified groups of cosmids corresponding to overlapping or non-overlapping intervals in the YAC map; these groups are called "pockets", irrespective of the fact that they form a contig or not (Nizetic et al.; in press). Pockets were validated with several methods, including hybridizations with single copy markers (RLDB), fingerprinting and fluorescent in situ hybridization (FISH).

In the present study, a pocket of 11 cosmids mapping to YAC ICRF y900F1026 (1.2 Mb) encompassing the D21S60-ERG locus in 21q22.2 (45) was used. This pocket has been unambiguously mapped by FISH analysis (data not shown).

Exon Amplification

The ERG pocket used for exon-amplification was divided in two pools of 5 and 6 cosmids, a and b respectively. The main motive for dividing the pool was to minimised losses in the exon recovery. Indeed, this system was originally demonstrated with single cosmids (27), and pooling several cosmids leads to an inevitable decrease in the efficiency of the procedure. Our experience shows that the number of exons recovered from single cosmids is almost always diminished when these cosmids are processed in a pool (M.L.Y; unpublished data). Some hypotheses that might explain this observation are discussed below. We estimated that an average of 10 cosmids per pool was a maximum limit for the retention of at least one exon from each cosmid carrying a gene.

Cosmid pools a and b were subcloned and processed as described in Methods. The third PCR amplification showed a distinctive smear containing several intense products above the vector band, demonstrating the presence of potential exons in each pool (Fig. 1). It is nevertheless unclear if the common lower band arise from the vector alone in the case of an incomplete BstXI digestion of the primary PCR, or from primer dimerisation. Indeed, those products have a similar size since the SA5 and SD5 primers are directly flanking the cloning site. Figure 1 shows that we have isolated at least 9 potential exons from the ERG pocket. However, at this stage of the analysis it is difficult to determine the number of exons in the PCR smear unless the products are cloned and analyzed in detail. Intense bands might represent exons preferentially amplified against other exons indeed present in the PCR product. The smears a and b have been cloned and retained as part of the merged chromosome 21 expressed sequence resource.

In order to identify the genomic fragments carrying these exonic sequences, complex products a and b were use to probe a Southern blot containing DNA from the original cosmids. As an example, Fig. 2 illustrates the results obtained with pool b; a similar hybridization pattern was partially shared between these cosmids reflecting the genomic overlap in the pocket. The specificity of the amplified exons was demonstrated by the absence of signal in the control cosmids.

As the final aim is to pool out the corresponding cDNAs, we used the complex exonic probes a and b to screen gridded cDNA libraries. A human foetal thymus cDNA library containing about 40,000 clones was screened and Fig. 3 shows a typical picture of 9000 clones spotted in duplicate in one filter. We identified 10 positive clones in the entire library, which are now under current analysis.

However, results of screening cDNA libraries are directly dependent upon   the

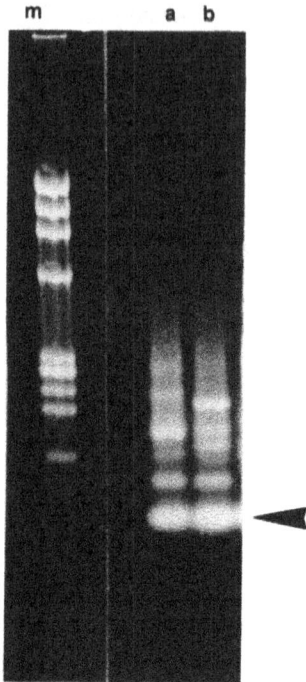

Figure 1. PCR amplification of exons using SA5 and SD5 primers; lines a and b corresponds to pools of 5 and 6 cosmids, respectively. PCR products are sized relative to the marker (m) ϕX174- HaeIII. The arrow indicates the 70 bp vector band.

Figure 2. Southern blot from EcoRI-digested cosmid DNA probed with the PCR product b. Lanes 1 to 6 correspond to the original cosmids of pool b. Lanes 7 to 9 correspond to chr. 21 cosmids mapping in the telomeric region and used as negative control.

Figure 3. Gridded thymus cDNA library spotted in duplicate. Filter was probed with both PCR products a and b. Arrows point out the scored positives.

expression pattern of the corresponding genes, and this leads *ipso facto* to a loss of information at this level. Exons should be retained as an independent resource of potentially expressed sequences.

In order to refine the mapping of the transcribed elements, the cosmid fragments identified in Fig. 2 could also be hybridized back to the cDNA libraries, but this approach is difficult to apply in the context of a large scale analysis.

The results presented here illustrate integration of genomic and transcribed sequence information. The basic strategy is to derive exon libraries from the regional pockets and to identify a subset of corresponding cDNAs.

cDNA SELECTION

The cosmid pools used for cDNA selection were either derived from the ERG pocket or randomly located along chromosome 21q. The composition of the pools is summarized in Table 1.

Table 1. Description of the cosmid pools

	Original library	Number of clones in each pool
Pool I	LL21	87
Pool II	LL21	28
Pocket ERG	ICRF	11

Each pool has been processed through the method described, using a mixture of human foetal brain, human foetal liver and human adult muscle cDNAs. The amplified selected cDNAs consisted of a distinctive complex product centered around 600 bp, as illustrated in Fig. 4. These PCR smears have been cloned and arrayed onto small filters for further screenings with various probes such as cosmid fragments and exons. The proportion of repetitive elements in the selected cDNAs was estimated to be about 10 % after screening these filters with total human DNA (data not shown).

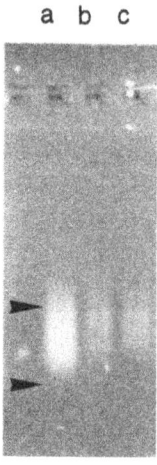

Figure 4. Analysis of the amplified selected cDNA of pools I (a), II (b) and ERG (c). Arrows indicates the 100 pb and 800 bp positions.

In order to achieve an integrated mapping and a correspondence between the ICRF and LL21 libraries, the complex PCR smears shown in Fig. 4 were used to probe the gridded cosmid filters. An important advantage of this step is to assess immediately which cosmids from the original pool effectively participated in the cDNA selection ; only the scored cosmids will be further analyzed. Figure 5 illustrates the hybridization patterns obtained with the complex probe derived from pool II; 22 positives were found in the ICRF library while 21 were detected in the LL21.

We are presenting here a more detailed study of the pool II integrating the mapping information provided by the RLDB. The positive cosmids (Fig. 5) have been pooled out and the corresponding EcoRI digestion patterns are shown in Figs. 6a and 6b. We observed that several cosmids were overlapping, demonstrating that the number of selected loci is in fact lower than the number of scored positives. Fifteen of the LL 21 cosmids define 6 small contigs and the 6 remaining clones showed an individual distinctive pattern, allowing us to propose that 12 loci in total were selected from the LL 21 library (Fig. 6a).

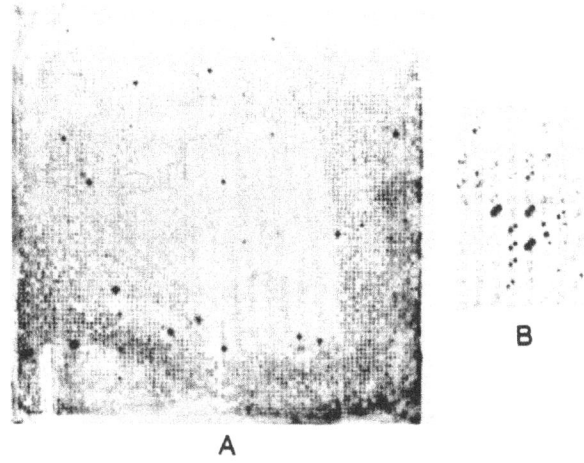

Figure 5. Comparative hybridization of the ICRF library (A) consisting of 9,000 clones spotted in single density and of the LL 21 library (B) consisting of 960 clones spotted in duplicate. The probe used is derived from pool II.

20 out of 22 cosmids scored in the ICRF library were referenced in the RLDB: A and F were not found; B was co-recognized with a total cortex human cDNA probe; C contains a zinc finger motif; D and E are mapping close to D21S16 but corresponds to a repetitive locus; G corresponds to a repetitive element in 21pter-q11; 10 clones are clustered in a pocket defined by YAC ICRF y904DO1225 in q11.2 (Cluster 1) and 5 clones belongs to a pocket defined by YAC ICRF yHO3749 in 21 q22.2-q22.3 (Cluster 2). These results suggests that the cDNA enriched probe derived from pool II recognized a maximum of 9 loci in the ICRF library.

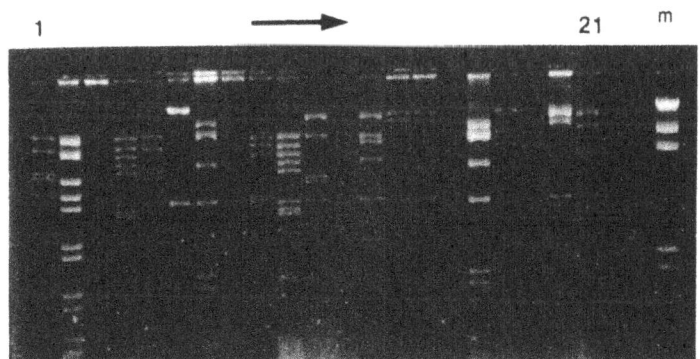

Figure 6a. EcoRI-digested cosmids from LL-21 numbered arbitrary from 1 to 21 (Fig. 6a) and from ICRF (Fig. 6b); size markers are λ-HindIII (m). Legends for Fig. 6b are in text. Some cosmids from cluster 1 are overlapping, while cosmids from cluster 2 are almost all identical.

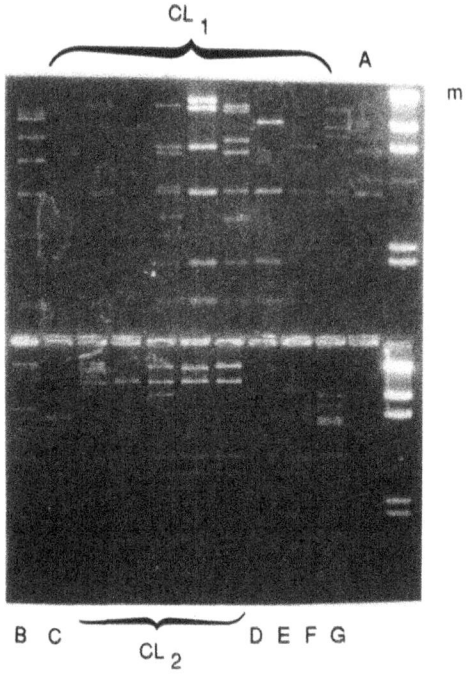

Figure 6b.

In order to identify the genomic fragments that participated in the cDNA selection process, the gels described in Fig. 6 were blotted and the DNA was hybridized successively with the complex selected cDNA probe and total human DNA. The results obtained are presented in Figs. 7a and 7b .

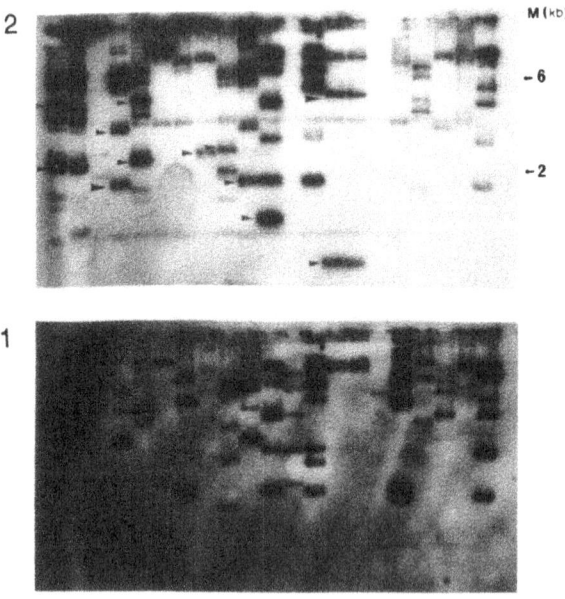

Figure 7a. Southern blots of the gels presented in Fig. 6 probed successively with total human DNA (1) and the complex cDNA product (2). Arrows are pointing out the potential genic subfragments. 7a: LL21; 7b: ICRF.

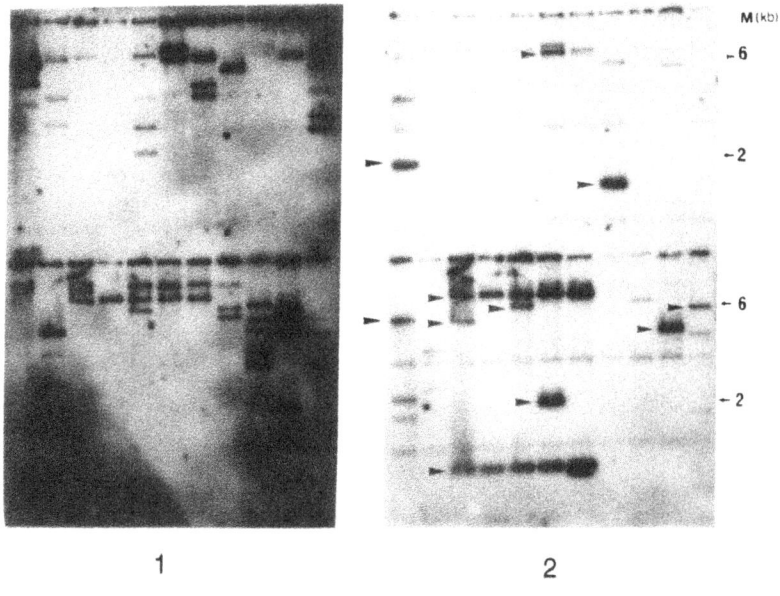

Figure 7b.

Hybridizations with total human DNA and complex cDNAs shows a substantially different pattern that validates the significance of using the selected cDNAs as a probe. This latter probe revealed discrete bands of strong intensity which are likely to contain part of the corresponding coding sequences. Some of the detected bands are common to several cosmids belonging to the same contig; this is particularly true in the case of cluster 2. In contrast, cosmids D, E and F assessed to repetitive loci (RLDB) do not show any significant hybridization signal with the cDNAs. Similarly, very few bands are detected in cluster 1; this observation is consistent with the paucity in genes in region q11.2 (45, 46). It is likely that these cosmids, as well as some "negatives" from the LL21 did contain particular repetitive sequences. Interestingly, the various signal intensities observed on the cosmid grid of Fig. 5 do not correlate with the presence or absence of genes in these cosmids and are likely to reflect the relative cDNA abundance in the complex PCR product.

In summary, these data suggest that a maximum of 12 loci were selected from the original pool of 28 cosmids, that approximates an overall efficiency of 42%.

As another example, hybridization of the LL21 grid with the selected cDNAs obtained from pool I (87 clones) is presented in Fig. 8.

We observed 34 positives (39 %) but since a certain degree of overlapping might similarly be expected among some of the scored cosmids, the effective number of loci carrying cDNAs is probably lower. The recovery appeared to be reduced when the size of the pools was scaled up. This observation suggests that a greater target complexity requires improvement in the procedure; the competition step might not be optimal and inefficiently blocked repetitive sequences could introduce a bias during the cDNA hybridization and subsequently in the PCR steps, affecting the representation of the specific cDNAs.

The cosmid ERG pocket was also used for the selection procedure. To verify the experiment, a blot containing digested DNA from YACs B19C12 (150 kb) and 259H11 (700 kb) encompassing the ERG locus (47) was hybridized successively with the selected cDNA probe and ERG cDNA (Fig. 9). A completely identical pattern was observed with both probes suggesting that ERG was the only gene being selected in the 700 kb region represented in these YACs. Since the cosmids chosen in the pocket did overlap, it is

223

unlikely that other genes mapping outside 259H11 could have been selected. This experiment illustrates the usefulness of PFGE from YACs to rapidly assess the genomic distribution of the selected cDNAs.

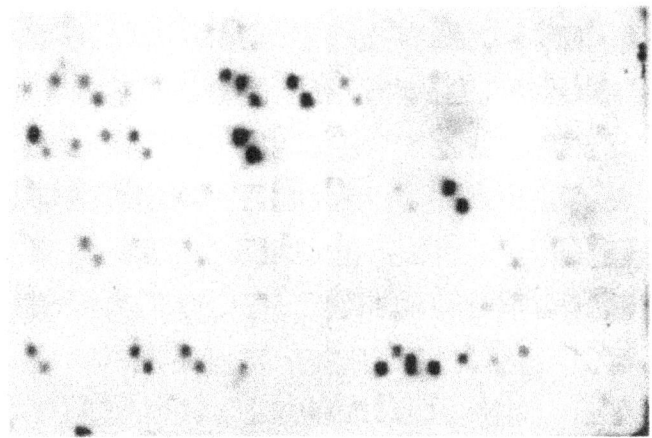

Figure 8. Hybridization of the LL21 grid with pool I selected cDNA probe.

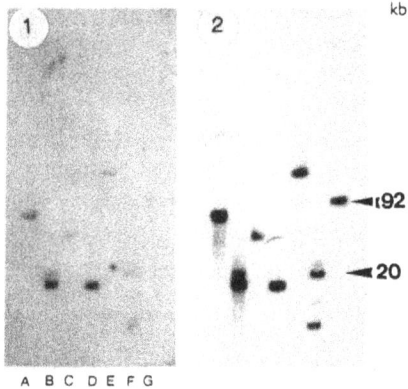

Figure 9. Sequential hybridization of a PFGE blot with 1: cDNA selected probe and 2: ERG cDNA. A to E correspond to YAC B19C12 digested with MluI, NruI, SacII, Sal I and SfiI respectively. F and G correspond to YAC 259H11 digested with NruI (F) and SfiI (G). YAC DNA preparation and PFGE analysis were performed as described in (47).

In order to compare exon-amplification and cDNA selection for the ERG pocket, the cosmid blot depicted in Fig. 2 was hybridized with the probes obtained from these two procedures. Interestingly, a different pattern was obtained: four bands were found in common whereas three to four bands were specific for each probe (data not shown). A more detailed analysis is required to determine the nature of this pattern and evaluate the respective complementarity of exon-amplification and cDNA selection.

224

DISCUSSION

A relational mapping strategy combining information from different libraries is an option in establishing integrated transcriptional maps. As an essential step, constructing chromosome specific expressed sequence resources represents the more direct tool for gene identification and mapping. Ideally, this could be achieved by deriving cDNA libraries from somatic hybrid cell lines containing single human chromosomes (48,49) but this approach is practically not satisfactory: rodent contamination is unavoidable and the expression of the human component is uncertain. The recent developments of a set of efficient protocols to retrieve coding elements from genomic clones oriented our basic strategy. We aim to define a subset of "gene containing" chromosome 21 cosmids and the corresponding expressed sequences using exon-amplification and cDNA selection techniques. However, the central problem in the context of a global approach is to isolate exhaustively genes from complex sources. Indeed, our results suggest that the size of the genomic target constitutes a striking limitation for these procedures, particularly for exon-amplification. In fact, the exon recovery is not optimal for pools of 10 cosmids representing a maximum complexity of 350 kb; the occurrence of this phenomenon is not explained so far. Since cDNAs libraries can be amplified by PCR with a good final representation, the exons should theoretically follow the same principle. It is thus likely that one or several steps of the procedure undergo real limitations: for instance, the isolated RNAs may not represent the original DNA complexity if the uptake of DNA is not optimal during the COS-7 transfection step. It is also possible that the clustering of the exons in a range of 100-200 bp might introduce a bias during the PCR. cDNA selection is more adapted to complex DNA sources, since pools of as many as 100 cosmids can be used, but a loss of efficiency was nevertheless observed in correlation with the complexity of the pool.

In turn, these techniques undoubtedly need to be improved for long range genome mapping in a "one go" experiment, but are extremely powerful for a cumulative analysis with smaller cosmid pools. The regional cosmid pockets will constitute a good target, especially for extracting candidate genes in the regions associated to Down syndrome and progressive myoclonus epilepsia where the phenotypic maps are more and more refined (30, 50). Eventually, the merged libraries obtained will provide a chromosome 21 coding sequences resource. Indeed forty genes are cloned so far (30) but it is expected that about 500 genes are lying along 21q (51). Nevertheless, the distribution of genes is not uniform along the chromosome and this number might be overestimated since coding sequences appeared to be mainly clustered in the telomeric fraction (46).

In the anticipation of future developments, isolation of genes along 21 might give clues for understanding chromosomal sequence organisation and the relations between the spatial organisation and the biological functions of genes. It has been observed that R bands are G+C rich, explaining the very high gene density in 21q22.3; in contrast, genes located in the light A+T rich component of the genome are rarely associated with CpG islands (45, 52). It has been suggested that a significant number of chromosome 21 genes may not be associated to CpG islands (47). In conclusion, it will be of particular interest to investigate the proportion of genes presenting these characteristics and to refine the compositional mapping of human chromosome 21.

ACKNOWLEDGEMENTS

We thank Dr. Alan Buckler for kindly providing the pSPL3 vector and primer sequences prior publication. This work was supported by EEC grant GENO 913013 (M.L.Y).

REFERENCES

1. W.F. Bodmer, Cold Spring Harbor Symp., *Quant. Biol.*, New York 51:1 (1989).
2. V.A. McKusick, "Mendelian Inheritance In Man," The Johns Hopkins University Press, Baltimore, 10th edn. (1992).
3. M.D. Adams, M. Dubnick, A.R. Kerlavage, R. Moreno, J.M. Kelley, T.R. Utterback, J.W. Nagle, C. Fields, and J.C. Venter, Sequence identification of 2.375 human brain genes, *Nature* 355:632 (1992).
4. M.D. Adams, A.R. Kerlavage, C. Field, and J.C. Venter, 3,400 new expressed sequence tags identify diversity of transcripts in human brain, *Nature Genetics* 4:256 (1993).
5. M.D. Adams, B.M. Soares, A.R. Kerlavage, C. Fields, and J.C. Venter, Rapid cDNA sequencing (expressed sequence tags) from a directionally cloned human infant brain cDNA library, *Nature Genetics* 4:373 (1993).
6. J. Takeda, H. Yano, S. Eng, Y, Zeng, and G.I. Bell, A molecular inventory of human pancreatic islets: sequence analysis of 1000 clones, *Hum. Mol. Genet.* 2:1793 (1993).
7. A.S. Kahn, A.S. Wilcox, M.H. Polymeropoulos, J.A. Hopkins, T.J. Stevens, M. Robinson, A.K. Orpana, and J.M. Sikela, Single pass sequencing and physical and genetic mapping of human brain cDNAs, *Nature Genetics* 2:180 (1992).
8. M.H. Polymeropoulos, H. Xiao, A. Glodek, M. Gorski, M.D. Adams, R.F. Moreno, M.G. Fitzgerald, J.C. Venter, and C.R. Merril, Chromosomal assignment of 46 brain cDNAs, *Genomics* 12:492; (1992).
9. M.H. Polymeropoulos, H. Xiao, J.M. Sikela, M.D. Adams, J.C. Venter, and C.R. Merril, Chromosomal distribution of 320 genes from a brain cDNA library, *Nature Genetics* 4:381 (1993).
10. H. Lehrach, R. Drmanac, J. Hoheisel, Z. Larin, G. Lennon, A.P. Monaco, D. Nizetic, G. Zehetner, and A. Poutska, Hybridization fingerprinting in genome mapping and sequencing. *Genome Analysis*, K.E. Davies and S.M. Tilghman, eds. (Cold Spring Harbor, New York: Cold Spring Harbor Laboratory Press), 1:39 (1990).
11. N. Sternberg, J. Ruether, and K. DeRiel, Generation of a 50,000-member human DNA library with an average DNA insert size of 75-100 kb in bacteriophage P1 cloning vector, *New Biol.* 2:151 (1990).
12. J.D. Hoeishel, E. Maier, R. Mott, L. McCarthy, A.V. Grigoriev, L.C. Schalkwyk, D. Nizetic, F. Francis, and H. Lehrach, High resolution cosmid and P1 maps spanning the 14-Mb genome of the fission yeast Schizosaccharomyces pombe, *Cell* 73:109 (1993).
13. A. Coulson, J. Sulston, S. Brenner, and J. Karn, Toward a physical map of the genome of the nematode Caenorhabditis elegans, *Proc. Natl. Acad. Sci. USA* 83:7821 (1993).
14. S. Baxendale, M.E. Mac Donald, R. Mott, F. Francis, C. Lin, S.F. Kirby, M. James, G. Zehetner, H. Hummerich, J. Valdes, F.S. Collins, L.J. Deaven, J.F. Gusella, H. Lehrach, and G.P. Bates, A cosmid contig and high resolution restriction map of the 2 megabase region containing the Huntington's disease gene, *Nature Genetics* 4:181 (1993).
15. D. Cohen, I. Chumakov, and J. Weissenbach, A first-generation physical map of the human genome, *Nature* 366:698 (1993).
16. A.P. Monaco, R.L. Neve, C. Colletti-Feener, C.J. Bertelson, D.M. Kurnit, and L.M. Kunkel, Isolation of candidate cDNAs for portions of the Duchenne Muscular Dystrophy gene, *Nature* 323:646 (1986).
17. D.C. Page, R. Mosher, E.M. Simpson, E.M.C. Fisher, G. Mardon, J. Pollack, B. McGillivray, A. De La Chapelle, and L.G. Brown, The sex determining region of the human Y chromosome encodes a zinc-finger protein, *Cell* 51:1091 (1987).
18. A.P. Bird, CpG-rich islands and the function of DNA methylation, *Nature* 321:209 (1986).
19. A.P. Bird, CpG islands as gene markers in the vertebrate nucleus, *Trends Genet.* 3:342 (1987).
20. G. Melmer, and M. Buchwald, Identification of genes using oligonucleotides corresponding to splice site consensus sequences, *Hum. Molec. Genet.* 1:433 (1992).
21. S. Parimoo, S.R. Patanjali, H. Shukla, D.D. Chaplin, and S.M. Weissman, cDNA selection: Efficient PCR approach for the selection of cDNAs encoded in large chromosomal DNA fragments, *Proc. Natl. Acad. Sci. USA* 88 9623 (1991).
22. M. Lovett, J. Kere, and L. M. Hinton, Direct selection: A method for the isolation of cDNAs encoded by large genomic regions, *Proc. Natl. Acad. Sci. USA* 88:9628 (1991).
23. B. Korn, Z. Sedlacek, A. Manca, P. Kioschis, D. Konecki, H. Lehrach, and A. Poutska, A strategy for the selection of transcribed sequences in the Xq28 region, *Hum. Molec. Genet.* 1:235:1992.
24. D.A. Tagle, M. Swaroop, M. Lovett, and F.S. Collins, Magnetic bead capture of expressed sequences encoded within large genomic segments, *Nature* 361:751 (1993).

25. G.M. Duyk, S. Kim, R.M. Myers, D.R. Cox, Exon trapping: A genetic screen to identify candidate transcribed sequences in cloned mammalian genome DNA, *Proc. Natl. Acad.Sci. USA* 87:8995 (1990).

26. D. Auch and M. Reth, exon trap cloning: Using PCR to rapidly detect and clone exons from genomic DNA fragments, *Nucl. Acids Res.* 18:6743 (1991).

27. A.J. Buckler, D.D. Chang, J.D. Brook, D.A. Harber, P.A. Sharp, and D.E. Housman, Exon amplification: A stategy to isolate mammalian genes based on RNA splicing, *Proc. Natl. Acad. Sci. USA* 88:4005 (1991).

28. M. Hamaguchi, H. Sakamoto, H. Tsuruta, H. Sasaki, T. Muto, T. Sugimura, and M. Terada, Establishment of highly sensitive and specific exon-trapping system, *Proc. Natl. Acad. Sci. USA* 89:9779 (1992).

29. D.M. Church, L.T. Banks, A.C. Rogers, S.L. Graw, D.E. Housman, J.F. Gusella, and A.J. Buckler, Identification of human chromosome 9 specific genes using exon amplification, *Hum. Molec. Genet.* 2:1915 (1993).

30. Delabar, J.M., N. CrÄau, P.M. Sinet, O. Ritter, S.E. Antonorakis, M. Burmeister, A. Chakravarti, D. Nizetic, D. Ohki, D. Patterson, M.B. Petersen, R.H. Reeves, and C. Van Broeckhoven, Report of the fourth international workshop on human chromosome 21 (April 1993, Paris, France); Genomics: In press.

31. I. Chumakov, P. Rigault, S. Guillou, P. Ougen, A. Billaut, G. Guasconi, P. Gervy, I. LeGall, P. Soularue, L. Grinas, L. Bougueleret, C. BellanÄ-Chantelot, B. Lacroix, E. Barillot, P. Gesnouin, S. Pook, G. Vaysseix, G. Frelat, A. Schmitz, J.L. Sambucy, A. Bosch, X. Estivill, J. Weissenbach, A. Vignal, H. Riethman, D. Cox, D. Patterson, K. Gardiner, M. Hattori, Y. Sakaki, H. Ichikawa, M. Ohki, D. LePaslier, R. Heilig, S. Antonorakis, and D. Cohen, Continuum of overlapping clones spanning the entire human chromosome 21q, *Nature* 359:380 (1992).

32. J. D. Hoheisel, G. Lennon, G. Zehetner, and H. Lehrach, Use of high coverage reference libraries of Drosophila melanogaster for relational data analysis; a step towards mapping and sequencing the genome, *J. Mol. Biol.* 220:903 (1991).

33. M.K. McCormick, A. Schinzel, M.B. Petersen, G. Stetten, D.J. Driscoll, E.S. Cantu, L. Tranebjaerg, M. Mikkelsen, P.C. Watkins, and S.E. Antonorakis, Molecular genetic approach to the characterization of the "Down syndrome region" of chromosome 21, *Genomics* 5:325 (1989).

34. Z. Rahmani, J.L. Blouin, N. CrÄau-Goldberg, P.C. Watkins, J.F. Mattei, M. Poissonnier, M.Prieur, Z. Chettouh, A. Nicole, A. Aurias, P.M. Sinet, and J.M. Delabar, Critical role of the D21S55 region on chromosome 21 in the pathogenesis of Down syndrome, *Proc. Natl. Acad. Sci. USA* 86:5958 (1989).

35. D. Nizetic, G. Zehetner, A.P. Monaco, L. Gellen, B.D. Young, and H. Lehrach, Construction, arraying and high-density screening of large insert libraries from human chromosomes X and 21: Their potential use as reference libraries, *Proc. Natl. Acad. Sci. USA* 88:3233 (1991).

36. D. Nizetic, R. Drmanac, and H. Lehrach, An improved colony lysis procedure enables direct DNA hybridizastion using short (10, 11 bases) oligonucleotides to cosmids, *Nucl. Acids Res.* 19:182 (1991).

37. A.P.Feinberg, and B. Vogelstein, A technique for radiolabelling DNA restriction endonuclease fragments to high specific activity, *Anal. Biochem.* 132:6 (1984).

38. M.L. Yaspo, M.A. North, and H. Lehrach, Exon-enriched probe derived from a human chromosome 21 YAC by exon-amplification, *Nucl. Acids Res.* 21:2271 (1993).

39. M.T. Ross, D. Nizetic, C. Nguyen, C. Knights, R. Vatcheva, N. Burden, C. Douglas, G. Zehetner, D.C. Ward, A. Baldini, and H. Lehrach, Selection of a human chromosome 21 enriched YAC sub-library using a chromosome-specific composite probe, *Nature Genetics* 1:284 (1992).

40. M.C. Potier, W.L. Kuo, A. Dutriaux, J. Gray, M. Goedert, Construction and characterization of a yeast artificial chromosome library containing 1.5 equivalents of human chromosome 21, *Genomics* 14:481 (1992).

41. H.M. Albersten, H. Abderrahim, H.M. Cann, J. Dausset, D. Le Paslier, and D. Cohen, Construction and characterization of a yeast artificial chromosome library containing seven haploid human genome equivalents, *Proc. Natl. Aca. Sci. USA* 87:4256 (1990).

42. D. Patterson, Report of the Second International Workshop on Human Chromosome 21, *Cytogenet. Cell Genet.* 57:167 (1991).

43. S. Baxendale, G.P. Bates, M.E. McDonald, J.F. Gusella, and H. Lehrach, The direct screening of cosmid libraries with YAC clones, *Nucl. Acids Res.* 19:6651 (1991).

44. E.S.P. Reddy, V.N. Rao, and T.S. Papas, The erg gene: A human gene related to the ets oncogene, *Proc. Natl. Acad. Sci. USA* 84:6131 (1987).

45. K. Gardiner, M. Horisberger, J. Kraus, U. Tantravahi, J. Korenberg, V. Rao, S. Reddy and D.

Patterson, Analysis of human chromosome 21: correlation of physical and cytogenetic maps; gene and CpG island distributions, *EMBO J.* 9:25 (1990).

46. S. Saccone, A. De Sario, G. Della Valle, and G. Bernardi, The highest gene concentrations in the human genome are in telomeric bands of metaphase chromosomes, *Proc. Natl. Acad. Sci. USA* 89:4913 (1992).

47. F. Tassone, S. Cheng, and K. Gardiner, Analysis of chromosome 21 yeast artificial chromosome (YAC) clones, *Am. J. Hum. Genet.* 51:1251 (1992).

48. L. Corbo, J.A. Maley, D.L. Nelson, and C.T. Caskey, Direct cloning of human transcripts with hnRNA from hybrid cell lines, *Science* 249:652 (1990).

49. P. Liu, R. Legersky, and M.J. Sicialono, Isolation of human transcribed sequences from human-rodent somatic cell hybrids, *Science* 246:813 (1989).

50. A.E. Lehesjoki, M. Koskiniemi, P. Sistonen, J. Miao, J. Hastbacka, R. Norio, and A. De La Chapelle, Localization of a gene for progressive myoclonus epilepsy to chromosome 21q22, *Proc. Natl. Acad. Sci. USA* 88:3696 (1991).

51. D. Patterson, Integrating maps of chromosome 21, *Current Opinion in Genetics and Development* 2:400 (1992).

52. B. Aissani, and G. Bernardi, CpG islands, genes and isochores in the genomes of vertebrates, *Gene* 106:185 (1991).

SHALLOW SHOTGUN SEQUENCING AS A STRATEGY FOR FINDING CODING EXONS

Jean-Michel Claverie

National Center for Biotechnology Information
National Library of Medicine
National Institutes of Health
8600 Rockville Pike, Bethesda MD 20894

ABSTRACT

Narrowing candidate genomic segments down to a size compatible with contig assembly by shotgun sequencing is a limiting step in the identification of mammalian genes by positional cloning. A 6 to 8 fold redundancy of sequencing (coverage) is usually required and, most often, does not alleviate the need for directed approaches for closing the remaining gaps. Once a complete sequence is obtained, computer analyzes are used to locate candidate exons, usually spanning less than 10% of the genomic sequence. Here, I propose an alternative strategy in which the need for contig assembly - and the high coverage it imposes - is removed. This strategy takes advantage of the fact that: i) mammalian coding exons are on average much smaller than individual sequencing runs, and ii) computer methods to identify coding regions only depend on local sequence information. In this context of "exon hunting" (opposed to genomic sequencing per se), I show that a 2 to 3 fold sequencing coverage is indeed sufficient to locate most candidate exons within genomic fragments, the size of which is now limited only by the available sequencing power. This strategy can be fully automated as it involves a single experimental technique and real-time computer analyzes of the data.

INTRODUCTION

The identification of a human disease gene by positional cloning typically involves the subcloning of the candidate region, contained in a Yeast Artificial Chromosome (YAC), into a set of λ clones prior to shotgun sequencing. For instance, in the case of the Kallmann syndrome gene (1) (our example throughout this article), the final 67 kb candidate region was subcloned into 7 overlapping λ clones with 15 to 20 kb inserts. While direct shotgun sequencing of the original genomic insert would have been possible, its fragmentation into pieces of 15-20 kb in size was required in order for a single contig to be obtained from random sequencing with a reasonable redundancy (6 to 8 fold coverage).

The subcloning and fine mapping of 100 kb-500 kb genomic fragment is a time-consuming process which cannot be automated. It also adds a 1.5 to 2 fold redundancy to the sequencing because of the necessary overlap of the λ subclones. For instance, identifying the exons in the 67 kb Kallmann region actually involved sequencing 117 kb, hence the determination of approximately 750,000 nucleotides (coverage >6).

Once reduced to less than 10 contigs, the 67 kb sequence was then analyzed by computers programs for potential coding regions and/or similarity with known proteins. Two exons, totalling less than 400 coding nucleotides (220 nt + 140 nt) were found to be located in this region and later confirmed by the isolation of the corresponding cDNA (1).

It is often not realized that none of the popular exon finding computer methods use sequence information over a span greater than 200 nucleotides. Furthermore, the average size for a mammalian coding exon is less than 150 nucleotides. Since, according to usual protocols, the sequence is obtained from individual runs of 400 to 600 nucleotides, the identification of candidate exons is largely independent of the contig assembly process. Hence the large sequencing redundancy and the lengthy λ subcloning and mapping it requires also become dispensable.

In the context of "exon hunting", I analyse here a more efficient strategy that would allow most coding exons to be discovered by direct shotgun sequencing of a much larger insert (40-500 kb) with only 2 to 3 fold coverage. Coupled with a real-time analysis of the raw sequence data, this strategy could be fully automated with existing machines and publicly available software.

METHODS

Statistical Overview of Shotgun Sequencing

Given a perfectly representative shotgun library, with all inserts of equal size, we define the following symbols (see Ref. 2, for a more complete treatment):

L = total insert length,
r = average usable length of individual sequencing runs
n = number of runs
c = average sequencing redundancy or "coverage' = $\dfrac{nr}{L}$

Expected Fraction of the Insert Sequenced at Least Once

the probability of a given position in the insert to be determined by one run is:

$$p = \frac{r}{L} \tag{1}$$

the probability of a given position NOT to be determined after n runs is

$$q_n = (1-p)^n = (1 - \frac{r}{L})^n \tag{2}$$

Thus the expected fraction F(c) of the insert for which at least one-pass sequencing information is available after n runs is

$$F(c) = 1 - q_n = 1 - (1 - \frac{r}{L})^{c\frac{L}{r}} \tag{3}$$

or approximately, for r/L small :

$$F(c) \approx 1 - e^{-c}$$ (3')

Standard Deviation of the Expected Fraction Sequenced

For each position, we can use an indicator variable X with values:

X = 0 , when the position is determined
X = 1 , when the position is not determined.

After n runs, the expected value of X is $<X> = q_n$
the second moment of X is also $<X^2> = q_n$
Thus, the variance $\sigma^2 = <X^2> - <X>^2 = q_n(1-q_n)$

Now, consider the sum of L indicator variables X, one for each position, this sum correspond to the number of undetermined residues. Its variance Σ^2 (with the assumption that all positions are independent, valid for small $\frac{r}{L}$ ratios), is now:

$$\Sigma^2 = L \, q_n \, (1 - q_n)$$

The standard deviation, relative to the insert size is:

$$\frac{\Sigma}{L} = \sqrt{[\frac{q_n(1-q_n)}{L}]}$$ (4)

or approximately, for r/L small :

$$\frac{\Sigma}{L} = \sqrt{[\frac{e^{-c}(1-e^{-c})}{L}]}$$ (4')

Figure 1 plots F(c), the expected fraction of determined sequence, and the relative standard deviation in function of the coverage c, for various $\frac{r}{L}$ ratios.

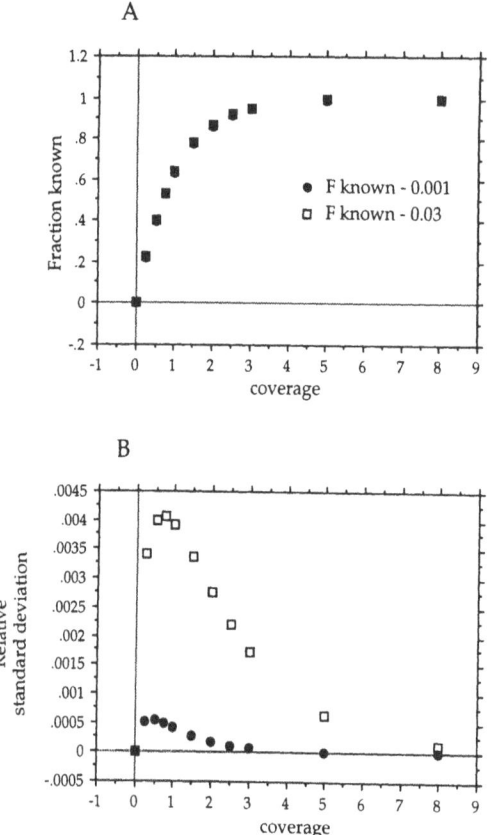

Figure 1. Expected fraction of determined sequence and relative standard deviation. A) The progression of sequencing as a function of the coverage is largely independent of the r/L ratio (or of the total length of the insert for a given average run size). F known-0.001 and F known-0.03 correspond respectively to 500 kb and 15 kb inserts for 500-nucleotide runs. B) Relative standard deviation of the functions shown in A.

Expected Number of Contigs in Function of the Coverage

The probability for a run to start at each of the L positions of the insert is

$\frac{n}{L}$ (2). The probability for the same run not to overlap with any other is :

$$\frac{n}{L} \left[\left(1 - \frac{n}{L} \right) \right]^{r(1-\theta)}$$

where θ is the fraction of the run length r required to assess overlap (i.e. 25 nucleotides over 500, thus $\theta = 0.05$). In function of the coverage $c = \frac{nr}{L}$, the expected number of contigs is:

$$N(c) = L\frac{c}{r}\left[1 - \frac{c}{r}\right]^{r(1-\theta)} \tag{5}$$

or approximately, for $\frac{c}{r}$ small :

$$N(c) \approx L\frac{c}{r}e^{-c(1-\theta)} \tag{5'}$$

Figure 2 plots the expected number of contig N(c) in function of the coverage for various insert length L, r = 500 and θ = 0.05.

Equation 3' and Fig. 1 show that the fraction of sequenced nucleotides is independent of the run size and of the length of the insert. On the other hand, Equation 5 and Fig. 2 show that the number of remaining "islands" for a given coverage is directly proportional to the insert length. For instance, a 5-fold coverage theoretically should lead to a unique island for a 15 kb insert, but leaves us with nine islands for a 100 kb segments, and more than 40 for a 500 kb YAC (Fig. 2). This is an optimistic estimate, not taking into account the presence of repeated sub-sequences or of non-clonable segments eventually delaying further (or making impossible) the contig process.

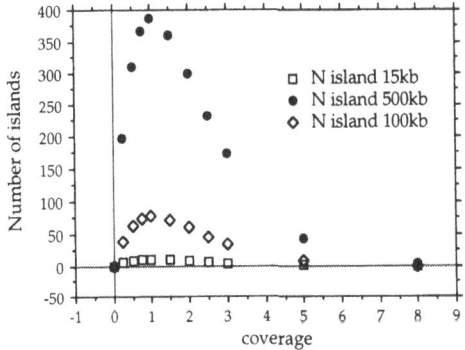

Figure 2. Number of contigs as a function of coverage and insert length. The three distributions are computed for individual sequencing runs of 500 nucleotides. The expected numbers of island for a 2.5-fold coverage are : 7 (15 kb), 46 (100 kb) and 232 (500 kb).

Computer Methods to Locate Coding Exons

Methods assessing the local coding potential of nucleotide sequences. Coding nucleotide sequences are constrained by the relative abundance of amino-acids as well as the differential usage of synonymous codons. This leads to a recognizable periodicity of the nucleotide frequencies at position +1, +2, +3 in the correct frame of coding exons. This is the principle behind the popular Testcode algorithm (3). This constraint, combined with other, more subtle effects such as the elimination of CpG in non-coding regions due to methylation, leads to a recognizable bias in the frequency of hexamers found in coding exons (4). 6-tuple frequency analysis can thus be applied to the detection of exons in genomic nucleotide sequence (4,5). The Grail program (6) combines these two main methods with others into a neural network, and results in improved performances (typically 80-90% correct predictions, with a low rate of false positive, after GC repeats, Alu and Line-1 elements are filtered out). All these methods only take advantage of the sequence information available within a local window of 100 nucleotides or less. Thus, a single 500-nucleotide run including about 100 nucleotides of a coding exon should score positive, rending the detection of candidate exon sequence independent of contig assembly.

Methods assessing the local similarity with known protein sequences. The blastx program (7) is used to scan all six reading frames of a nucleotide sequence query against the protein databases. The algorithm finds statistically significant local alignments irrespective of their length. Most of the time, an ungapped stretch of 30 to 40 similar amino-acid (90 - 120 coding nucleotides) is sufficient to characterize a statistically significant local alignment. The blastx algorithm is much more sensitive than direct nucleotide sequence comparison (because similar peptides can be encoded by highly divergent nucleotide sequences) and more tolerant of frame shift errors (the combination of two non-significant sub-alignments in different frames can become significant). Up to 1% insertion/deletion error, a realistic frequency for single pass raw data, is easily tolerated when the alignments are scored with the usual PAM120 matrix (7). Blastx can also be used to compare nucleotide sequences against the 6-frame translation of the rapidly growing Expressed Sequence Tag (EST) dataset. Here again, querying the databases with a fully assembled nucleotide sequence offers no benefit over using the individual 500-nucleotide runs as they become available. The limiting factor is more likely the actual size of the exon and its similarity with a known protein or a translated EST. However scanning databases with random genomic sequences requires special precautions in order to avoid an overwhelming number of false positive matches. Nucleotide sequences which translate into low-entropy amino-acid sequences have to be masked with the XNU program (8,9). Other ubiquitous repeated elements like Alu, line-1, etc, also produce misleading databases matches (10) or are positively rated by Grail. The elimination of those problematic sequences by a series of blast/Xblast procedures has been described elsewhere (8,9).

RESULTS AND DISCUSSION

An alternative exon-scanning strategy consists of submitting each individual single pass sequencing run to the usual battery of coding potential tests as soon as they are determined. Each time a promising candidate exon is detected (either as scoring "good" or "excellent" with Grail, or achieving a significant $P < <0.01$ similarity score with Blast, or both) the local island containing this run is examined for putative splice sites. In the meantime, nucleotide sequences from the candidate exons can be used to probe cDNA libraries, Zoo blots or expand the local contig (e.g. Alu-PCR). Of course, the real-time analysis of the sequencing runs can (and should) be implemented in parallel with any classical shotgun sequencing scheme (i.e. trying to achieve full contig assembly). In this context, it is a useful tool for quality control (putative contaminations, repeat content, small insert size, etc).

Using single pass sequencing runs as query sequences for Grail or Blastx requires a number of preliminary filtering steps in order to avoid a large number of false positive predictions. For instance, we want to eliminate any significant piece of cloning or vector (e.g. λ or M13) sequence prior to analysis. We also want to eliminate ubiquitous repeated elements (e.g. Alu) which induce misleading similarity scores (10) or are known to score as candidate exons with Grail (e.g. Line-1). Those filtering steps are automated as a series of blast/Xblast searches against a user-defined database of "junk" sequences depending on the detail of the cloning/sequencing protocol and the species being studied.

Another problem is caused by the low-entropy (e.g. doublet or triplet repeats, AT rich regions,...) segments often found in genomic sequences. These regions tend to produce uninformative high scores in similarity searches (8,9). Again pre-filtering provides an adequate answer, this time with the program XNU (8,9). The complete UNIX-based software tool box necessary to implement this strategy as been described elsewhere (8,9).

Once the suitable battery of filters has been set up, the real-time analysis of the sequencing runs can proceed automatically, with the occasional good candidates being detected from time to time. A flow chart of the protocol is shown in Fig. 3.

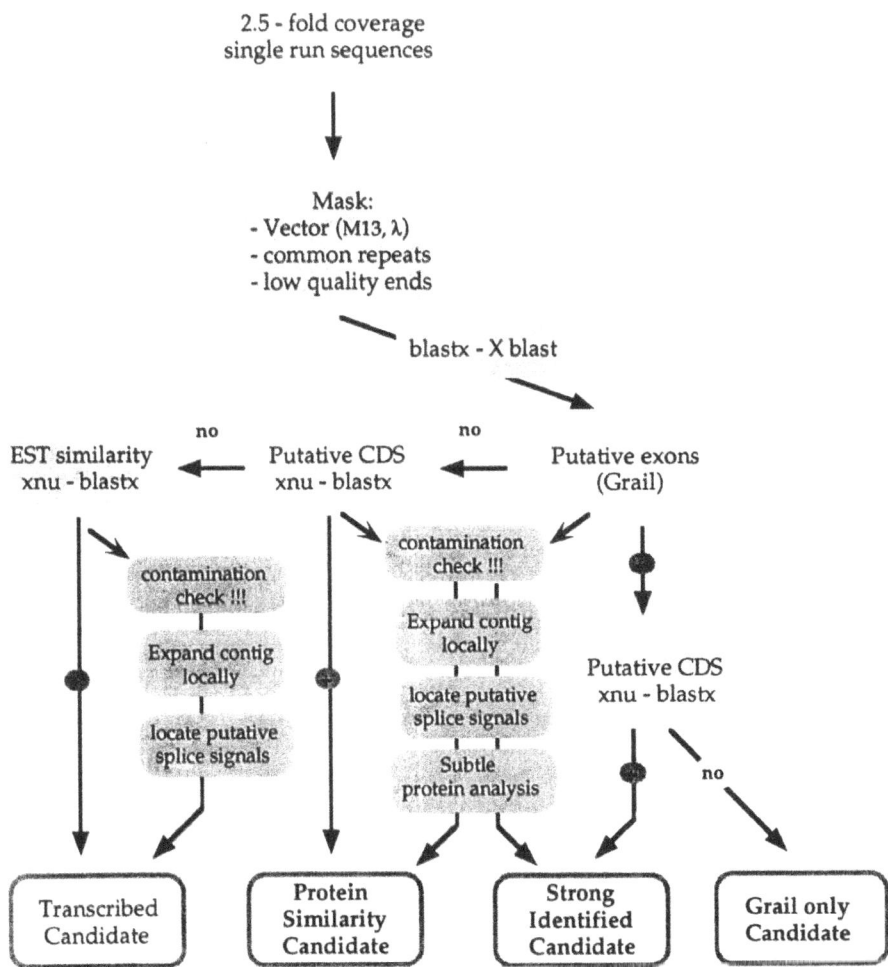

Figure 3. Real-time exon scanning procedures of the individual sequencing runs. After filtering for vector sequences, and known repeats, the individual sequencing runs are tested for coding potential with Grail, and similarity against protein or public databases of transcribed sequences.

Because total contig assembly is not the goal of this strategy anymore, direct shotgun sequencing can be applied to larger inserts from 40 kb to total YACs, the limit only being the sequencing power at hand. The project can stop as soon as suitable exon candidates have been found. At this point, the determined sequences (together with the genetic and physical location), although far from forming a single contig, could be stored in Single Pass Genomic Sequence (SPGS) public repository. This database could be used for statistical purpose, similarity search, or in the context of a subsequent full genome sequencing project.

This exon-scanning strategy is a method of choice for exploring genomic regions with a high density of (or unusually long) repeats, *a priori* impossible to contig. The statistical theory (Equation 3) shows that a 2.5-fold coverage is sufficient to achieve a probability better than 90% to sequence all the exons within a given insert. For a 40 kb cosmid, with the present technology, this corresponds to 200 runs and 100 kb of raw data, a task that can be performed in less than 10 days on a single sequencing machine. Meanwhile, the corresponding real-time data analysis can be performed overnight on a standard UNIX workstation. With two automated sequencers, 100 kb (P1) insert becomes accessible for exon scanning on the same time scale. For even better equipped laboratories, the complete repertoire of putative exon sequences could be obtained from YACs in a comparable time scale. A computer experiment was performed with the 67 kb genomic sequence of the Kallmann gene to verify the feasibility of the strategy. Using the genfrag (11) tool, I generated five independent sets of fragments corresponding to a 2.5-fold coverage of this sequence, including from 0 to 0.5% random mutations (with twice as much random insertion/deletion than substitution). In each case, the only two short coding exons present in this interval were correctly detected (one by Grail, the other by similarity search, removing the cognate protein hits) among a set of no more than 10 candidates. This exon scanning strategy is now being used in the context of a large scale sequencing project of a region of the murine X chromosome.

ACKNOWLEDGEMENTS

I thank Drs. J. Spouge and C. Abergel for helpful discussions and advice.

REFERENCES

1. R. Legouis, J-P. Hardelin , J. Levilliers, J.-M. Claverie, S. Compain, V. Wunderle, P. Millasseau , D. Le Paslier, D. Cohen, D. Caterina, L. Bougueleret, G. Lutfalla, J. Weissenbach, and C. Petit, The candidate gene for the X-linked Kallmann syndrome encodes a protein related to adhesion molecules, *Cell* 67:423 (1991).
2. E.S. Lander and M.S. Waterman, Genomic mapping by fingerprinting random clones: a mathematical analysis, *Genomics* 2:231 (1988).
3. J.W. Fickett, Recognition of protein coding regions in DNA sequences, *Nucl. Acids Res.* 10:5303 (1982).
4. J.-M. Claverie, J.-M. and L. Bougueleret, L. Heuristic informational analysis of sequences. *Nucl. Acids Res.* 14:179 (1986) .
5. J.-M. Claverie, I. Sauvaget, and L. Bougueleret, k-tuple frequency analysis: from intron/exon discrimination to T-Cell epitope mapping, *Meth. Enzymol.* 183:237 (1990).
6. E.C. Uberbacher and R.J. Mural, Locating protein-coding regions in human DNA by a multiple sensor neural network approach, *Proc. Natl. Acad. Sci. USA* 88:11261 (1991).
7. W. Gish and D.J. States, Identification of protein coding regions by database similarity search, *Nature Genetics* 3:266 (1993).
8. J.M. Claverie and D. States, Information enhancement methods for large scale sequence analysis. *Computers and Chemistry* 17:191 (1993).

9. J.M. Claverie, Large scale sequence analysis, *in* "Automated DNA Sequencing and Analysis Techniques," J.C. Venter, ed., Academic Press, New York, (in press).

10. J.-M. Claverie, Identifying coding exons by similarity search: Alu-derived and other potentially misleading protein sequences, *Genomics* 12:838 (1992).

11. M.L. Engle and C. Burks, Artificially generated data sets for testing DNA sequence assembly algorithms. *Genomics* 16:286 (1993).

REQUIREMENTS IN SCREENING cDNA LIBRARIES FOR NEW GENES AND SOLUTIONS OFFERED BY SBH TECHNOLOGY

R. Drmanac, S. Drmanac, I. Labat, and N. Stavropoulos

Integral Genetics Group
Center for Mechanistic Biology and Biotechnology
Argonne National Laboratory
9700 South Cass Avenue
Argonne, Il 60439-4833

ABSTRACT

Under different assumptions about the total number of genes, the number of housekeeping and tissue-specific genes, and the difference in the number of mRNAs per cell for functional and nonfunctional genes, significantly different results can be expected from screening random cDNA clones. We have developed gene expression models as a guide for interpretation of experimental results. For statistical, biological, and technical reasons, the search for 100,000 plus genes and discrimination between nonfunctional, housekeeping, and tissue-specific genes requires the analysis of up to 10 million clones from 20 to 50 tissues. Oligonucleotide hybridization of dense clone blots is an inexpensive and fast way to screen such large clone sets. Our preliminary results on control clones and thousands of cDNA clones from an infant brain library demonstrate the feasibility of the method.

INTRODUCTION

The number of mammalian genes is usually estimated to be 50,000 to 100,000 (or sometimes as high as 150,000) (1). Studies of gene expression have found that the number of mRNAs transcribed from a gene can vary a few thousand times and that there is an exponential increase of the number of genes having a small number of mRNA copies (2-4). Also, the existence of 200−300 cell types each expressing several hundred specific (usually highly expressed) genes is estimated. There is no clear-cut definition of what has to be considered as functional gene or whether there is transcription leakage of nonfunctional genes. These factors can seriously influence the identification of new or tissue-specific genes via the screening of cDNA libraries.

Partial sequencing by hybridization (SBH) of random cDNA clones as a way to catalog and define tissue specificity of genes was proposed four years ago (5). Instead of

Identification of Transcribed Sequences, Edited by
U. Hochgeschwender and K. Gardiner, Plenum Press, New York, 1994

tens of thousands of probes necessary for complete sequencing if a single genome is analyzed (6) or 3000 to 4000 if the data from similar genomes are integrated (7,8), 100-1000 probes are sufficient for partial SBH. In the last three years, successes of single pass gel sequencing of cDNAs have strongly demonstrated the usefulness of incomplete sequences (9-12). Partial SBH is, in principle, a less expensive approach with the ability to analyze millions of clones. SBH has been proven in a blind experiment (13) and recently we have developed a hybridization data production line to score up to 32 probes on 30,000 dots per day (14,15). Similar facilities are under development at ICRF (16).

In this paper, we present several models of gene expression and analyze the main factors which can influence the hunt for new genes via the screening of random cDNA libraries. The basic steps in the preparation and use of dense DNA dot arrays are described, and some results that demonstrate the feasibility and efficiency of making gene inventory by oligonucleotide hybridization are presented. Furthermore, partial SBH and single-pass gel sequencing are compared and a gene analysis scheme that combines the two approaches is discussed.

RESULTS AND DISCUSSION

Quantitative Models of Gene Expression: Implications for Screening cDNA Libraries

There are several possibilities for the distribution of genes in terms of the number of their mRNAs per cell in a homogenous tissue. We defined four models based on six gene expression levels (Table 1). Six expression levels representing averages for related parts of the distribution are a crude approximation. The level with five messengers per cell represents genes with 1 to 10 messengers per cell, the 30 messengers level comprises genes having 11 to 90 mRNAs per cell and so on. Levels below 0.1 mRNA per cell have no significant influence on the screening. The models (except model 4) do not differ in the two highest expression levels. Furthermore, since there is a general agreement that a gene is not functional if there is less than one messenger per cell, the genes belonging to the levels of 0.1 and 0.5 mRNA per cell will be considered as transcription leakage. This is not true for complex tissues consisting of several cell types. If 10% of cells represent one type some functional genes may have only 0.1 mRNA per cell of whole tissue. The basic differences among the models are the level of expression (leakage) for genes that are not functional in the given cell type and partition of housekeeping and tissue-specific genes between low (five mRNAs on average) and moderate (30 mRNAs on average) levels of expression.

Model 1 represents the case with a gap between the expression level of the functional and the nonfunctional (leaky) genes. In models 2 through 4 the gap is progressively eliminated. Consequently, less than 1% (in model 1) or up to 15.2% (in model 4) of cDNA clones represents genes without function in the given cells, assuming a total of 112,000 genes, of which 12,000 are housekeeping and cell type specific genes. If the number of genes is 150,000 and if model 4 turns out to be correct, then up to 20% of clones may represent genes non-specific for the given cells.

Leakage prevents the discrimination of active and inactive genes, but it offers an opportunity to define the catalog of all genes without analysis of all tissues. If 100,000 clones are screened per tissue, then 79-86% of functional genes will be represented with at least one clone. (For 200,000 clones, the range is 91-97%. The percentages are calculated by summing number of genes in 10^5 or 2×10^5 clones for the four highest expression levels and dividing by 12,000). By analyzing 10 cell types (a total of one million clones), most of the housekeeping genes will be recognized by occurrence in two or more cell types. A fraction of cell type specific genes having a few mRNAs per cell (or, due to statistical reasons), will be represented with one clone and can not be discriminated from non-functional genes. Furthermore, 14% in model 1 (1 − [120 × 0.6 + 180 × 0.99 + 50]/350) and 16% in model 3 (1 − [240 × 0.76 + 60 + 50]/350) of tissue-specific genes will not be found in 100,000 clones. On the other hand, 30, 54 and 77% of the genes which do not function in any of the 10 analyzed tissues can be found in models 2, 3, and 4, respectively (the calculation is done by the formula described in the footnote to Table 1 where $g = 100,000$ and c is 10 times the total number of clones in the library of 100,000 clones, which are expected to represent genes from the 0.1 and 0.5 mRNA-per-cell levels).

Table 1. Model distributions of genes in six levels of expression in a cell type.

	Number of mRNAs per gene per cell					
	0.1	0.5	5	30	200	2000
Model 1						
Total genes	1,000	2,000	5,000	6,000	1,000	10
Cell type specific genes	0	0	120	180	45	5
Housekeeping genes	0	0	4,880	5,820	500	5
% Clones	0.0	0.2	6.0	42	47	5.0
% Genes in 10^5 clones[a]	2	10	60	99	100	100
No. Genes in 10^5 clones	20	200	3,000	5,940	1,000	10
Model 2						
Total genes	50,000	15,000	7,000	4,000	1,000	10
% Clones	1.3	1.9	9.0	31	52	5
% Genes in 10^5 clones	2.5	12	72	100	100	100
No. Genes in 10^5 clones	1,300	1,900	4,500	4,000	1,000	10
Model 3						
Total genes	80,000	40,000	9,000	2,000	1,000	10
Cell type genes	0	0	240	60	45	5
Housekeeping genes	0	0	8,760	1,940	500	5
% Clones	2.3	5.9	13	18	58	5.8
% Genes in 10^5 clones	2.9	14	76	100	100	100
No. Genes in 10^5 clones	2,300	5,500	7,000	2,000	1,000	10
Model 4						
Total genes	20,000	80,000	10,000	1,500	500	20
% Clones	0.7	14.5	18	16	36	14.5
% Genes in 10^5 clones	3	18	83	100	100	100
No. genes in 10^5 clones	700	14,000	8,300	1,500	500	20

[a]The percentages are calculated by equation $1-(1-1/g)^c$, where g is the total number of genes in the given level and c is the number of clones representing these genes.

If 10,000 clones are screened per cell type, only 27% in model 4 and 38% in model 1 of housekeeping and cell type specific genes will be found in one library. Most of the genes from the 5- and 30-mRNA levels will be represented by a single clone and can not be discriminated from inactive genes. Furthermore, the number of clones representing one gene will be a very inaccurate measure of its expression level due to statistical factors. Cell type specific and housekeeping genes can be potentially distinguished if 30-40 cell types are analyzed. Thus, by screening a small number of clones per cell type, the majority of cell type specific genes will not be found or recognized, and the expression level can not be determined accurately for most of the genes. To be able to obtain these two types of data, 200,000 or more clones have to be analyzed per cell type if any one of the models is correct and if ordinary libraries are used.

The only differences between cDNA libraries of a cell type and a complex tissue is the reduction of clone frequency for genes specific for one cell type in the tissue. In this case, tissue specific genes having a few mRNAs per cell of one cell type may not be possible to discriminate from the leaky genes. The models allow calculation of the expected clone redundancy and the expected number of genes, which will be represented by a certain number of clones if a given number of clones is analyzed per tissue of known complexity. We are planning to develop a program to test the influence of particular variables and to assess the agreement between experimental and expected results.

Biological, Technical and Statistical Reasons Impose Screening Several Million cDNA Clones

The data collected by screening random cDNA libraries allow, in principle, the identification of tissue-specific genes and an estimate of their expression levels. The accuracy of the findings depends on various biological, statistical, and technical factors that influence the preparation and screening of cDNA libraries. The impact of statistical factors and transcription leakage can be anticipated by the described models. A few other biological facts can be taken in consideration. Cryptic promoters or transcription termination sites probably exist. Furthermore, incomplete splicing (small introns remain in some mRNAs) or trans-splicing (17–19) can occur. More than 2 Gb of non-coding sequences and hundreds of thousands of primary transcripts per nucleus can produce enormous numbers of "new genes" by very rare transcription or splicing errors. It is not impossible that 20% of the mRNAs representing thousands of genes or "gene like" sequences can be present in a cell without function in that cell or in any other cell. The cells will waste much less energy for this level of error than for the transcription of intron sequences, and the number of proteins for any of these unnecessary mRNAs will be below the level which can influence cell functions.

Technical problems in library preparation (contamination with genomic DNA or external mRNAs or DNAs, chimeric clones, false primed clones) and in library screening (deletions or recombinations during clone amplification or PCR, cross-talk of the wells, external contamination, sequencing or hybridization errors) will add further uncertainty in the meaning of the data. By summing the expected levels of all these types of error, we estimate that up to 30% of the cDNA clones in a library can be artifacts or can represent genes nonfunctional in the given tissue. The screening of large sets of clones from various tissues is one possibility that will discriminate between artifacts and real genes or between functional and leaky genes. It may be necessary to consider only clones found at least twice. This requirement will increase the number of clones to be screened several-fold. Can normalized libraries help? The two highest expression levels represent more than 50% of the clones, and in normalized libraries these genes will be represented by a significantly smaller fraction of clones. The average clone redundancy can be reduced up to twofold. Our preliminary screening of 10,000 clones showed a reduction in the number of redundant clones from 60% in the ordinary library to 35% in the normalized infant brain library (12) constructed by Bento Soares (Columbia University). On the other hand, very rare transcripts (comprising 2–10% of the mRNAs, which may represent transcription leakage) can be found in the normalized libraries by screening 5- to 10-times-smaller number of clones.

The basic disadvantage of normalized libraries is the loss of information about the expression level of genes. This information can not be determined by genome sequencing. Because of this loss, and a relatively small saving in the number of clones, fully representative cDNA libraries can be more important than normalized libraries. Standard methods of preparation of cDNA libraries can introduce a bias for very short and very long mRNAs, because of a narrow cDNA size selection and a reduced transformation efficiency, respectively.

Even in a perfectly normalized library, 10 gene equivalents have to be screened to find the last percentage of genes. To find genes having low levels of expression, and especially to find them twice, as many as 100 gene equivalents (10 million instead of one million clones) may be necessary.

Massive Clone Screening by Oligonucleotide Hybridization

To implement DNA screening, mapping, and sequencing by oligonucleotide hybridization, we have developed facilities with a present capacity to score 8−32 probes (6−12 bases in length) per day on 30,000−120,000 DNA fragments spotted on nylon membranes (14,15,20). Procedures for the high throughput clone arraying, amplification, and spotting have been developed (S. Drmanac and R. Drmanac, in preparation). The procedures involve arraying genomic or cDNA clones (prepared in M13 or plasmid vectors) in multi-well plates (96, 384, or 864 wells) by picking plaques or colonies, or by dispensing in the wells an optimally diluted transformation mixture. Replica plates of the master plates are prepared by transferring 2 μl of the cultures in the wells filled with 100 μl of water. An array of metal pins is used to transfer liquid from all wells simultaneously (instead of row by row using a multichannel pipet).

In the next step, cloned inserts are amplified by PCR directly from the bacterial cultures without DNA isolation. The PCR mixture is dispensed in the wells and 1 μl of each of the diluted cultures is transferred by pin array into the corresponding wells. BioOvens (BioTherm, Fairfax, Virginia) are used for cycling six plates in parallel (14). Many parameters are optimized to be able to routinely produce 20 ng of 2-kb clones per 1 μl of PCR reaction in 90% of the cases (Fig. 1).

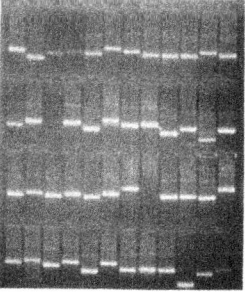

Figure 1. cDNA inserts amplified by PCR. Row A from four 96-well plates is tested on an agarose gel by loading 3 μl from each well. One to two inserts per row (15% on average) give very weak or invisible bands. Tfl polymerase (Epicentre Technologies, Madison, WI) which gives under our conditions better yield than AmplyTaq (Perkin Elmer, Norwalk, CN) was used. The superiority of Tfl has been demonstrated by D. Grujic and R. Crkvenjakov (personal communication).

Amplified DNA is used to prepare high-density dot blots. We have defined conditions to prepare well defined dots using an array of metal pins 0.3 mm in diameter. Interestingly, DNA can be spotted without removing the oil usually used to prevent the evaporation of PCR reaction mixtures. Membranes (15 × 23 cm, four 96-well plates) with 31,104 dots (4 × [(9×9×96)]) are routinely prepared using a Biomek1000 XYZ table (Beckman, Fullerton, CA). Several hardware modifications of the station have been made and specific software has been developed (I. Labat et al., unpublished results). Up to 100 replica membranes can be prepared from 15-μl PCR reactions.

Membranes are hybridized with $(N)_{0-2}(B)_{6-10}(N)_{0-2}$ probes (N, degenerated positions; B, specific base positions) at 12°C using 4 pM to 5 nM probe concentrations, and are washed for 20 min to 1 h at 2-20°C, depending on the length and GC content of the probes (15). Each filter is hybridized in a separate box with outlets for pumping out washing buffer. A setup which has four boxes fixed to a cooling plate and mounted on a shaker is presently in use. Filters can be reused over 50 times. The intensities of the hybridization signals are determined by our image analysis program (DOTS) (20) from the files of pixel values generated by a PhosphorImager (Molecular Dynamics, Sunnyvale, CA). One dot span is covered by 30 pixels. Examples of very different patterns obtained with two probes scored on one array of cDNA clones are shown on Fig. 2. The zinc finger consensus probe shown on this example demonstrated the possibility of defining a subset of genes that may encode specific protein motifs.

Recognition of Highly Similar cDNA Clones by Comparing Hybridization Signatures

Recently we developed user friendly software (SCORES) based on X windows (N. Stavropoulos and R. Drmanac, unpublished results) for data evaluation and normalization and for comparison of the generated oligonucleotide sequence signatures (OSS) (21). Instead of a less precise 0/1 scoring scheme, properly normalized hybridization intensities (scores) are used (R. Drmanac et al., in preparation). OSS of the pairs of dots prepared in two independent PCR reactions from the same master clones are presented in Fig. 3. The scores of a particular probe are very similar for the pairs of dots except in a few percent. For example, dot (7,75) has a score of 1 and corresponding dot (7,78) a score of 14 with probe F9. Possible reasons are dust or the shadow of the strong surrounding dots, especially if a dot has a small amount of DNA, which is the case for this dot (7,78); its relative mass is only 9.

Similarities of clone signatures are defined by calculating the distance parameter (Fig. 4). The distribution of the distance values for identical and random clones is shown. There is a significant separation of the two classes of pairs, which allows accurate identification of identical and similar pairs in a large set of clones. A. Milosavljevic (personal communication) motivated by the success of our score-based distance calculation procedure, has developed rank scaling of hybridization intensities and a different procedure for OSS comparisons. Also, we have defined an additional way to measure the similarity of the clone signatures (R. Drmanac et al., in preparation). In this approach, the number of probes which have significantly similar scores for the given pair of clones is defined. The evaluation of the advantages and limitations of these approaches is in the progress.

Clones with significantly small distances are grouped (clustered) together. The result obtained by the clustering procedure applied to the data from a test experiment is presented in Table 2. For this test every clone was spotted twice. Clones having low mass

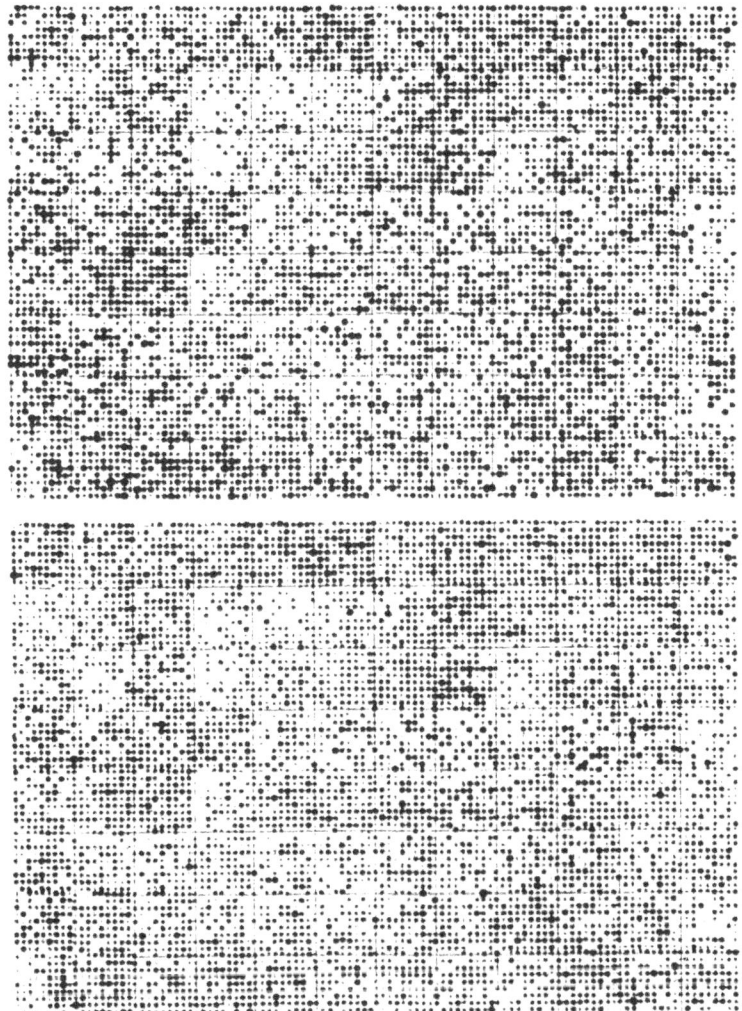

Figure 2. Hybridization patterns of two 7-mer probes with 7776 cDNA dots. Images represent one quarter of a filter (Gene Screen, NEN, Boston, MA) containing 31,104 dots. Grids are superimposed by our image analysis (DOTS) program. The first image is obtained by a coding-specific probe NNTGATGGTN, and the second by a zinc finger consensus probe (G,A)AAGCCNTTC. Hybridization conditions (15,26) are not full-match-specific in the case of the zinc finger probe.

Date:	Relative Mass	1 0415 7-25	2 0423 12N	3 0425 1I	4 0427 F24	5 0428 F3	6 0429 E10	7 0430 E30	8 0503 7-10	9 0504 7A	10 0505 F21	11 0507 8C	12 0508 C8	13 0511 E9	14 0512 F11	15 0514 5N	16 0517 E37	17 0519 7-31	18 0521 7N	19 0524 3N	20 0525 E21
Clone..																					
(7, 3)	19	3	1	1	2	5	5	17	9	1	1	3	1	1	2	1	2	1	3	1	9
(7, 6)	12	3	1	1	3	5	4	18	10	1	3	3	3	2	3	1	2	2	4	1	8
(7, 12)	42	7	9	1	7	3	1	4	9	1	4	2	2	1	1	1	4	1	7	2	1
(7, 15)	32	7	7	1	3	4	2	5	4	1	5	3	1	2	1	5	3	1	4	2	1
(7, 21)	16	5	6	2	1	5	1	11	3	3	1	2	3	1	1	8	3	1	1	7	10
(7, 24)	12	11	10	3	3	10	2	14	5	4	2	6	4	2	2	13	4	3	2	10	15
(7, 30)	20	1	1	4	4	1	2	1	5	1	2	3	1	3	2	1	1	1	1	5	2
(7, 33)	14	2	2	3	1	1	2	4	6	2	1	4	1	4	2	1	1	1	1	8	3
(7, 39)	33	3	6	1	2	2	1	3	6	1	1	5	1	1	1	2	1	1	3	3	5
(7, 42)	23	3	16	1	2	4	1	3	6	1	2	5	1	4	2	1	1	1	6	3	7
(7, 48)	75	3	1	1	1	1	2	1	1	1	2	3	1	1	8	1	1	1	2	4	2
(7, 51)	41	3	1	1	1	1	5	1	1	1	3	5	3	1	10	1	1	3	4	6	2
(7, 57)	38	1	4	1	1	1	1	3	3	6	1	3	3	1	2	2	2	2	3	2	3
(7, 60)	13	1	1	3	2	1	2	4	3	5	2	3	5	4	7	3	2	4	3	2	3
(7, 66)	65	1	4	2	2	1	1	1	8	1	1	7	8	1	1	1	2	2	24	2	1
(7, 69)	60	1	5	1	3	1	1	1	7	1	2	7	8	1	1	1	2	2	24	3	1
(7, 75)	24	2	2	1	3	13	2	1	5	2	1	1	2	1	4	3	1	2	1	1	10
(7, 78)	9	2	2	6	4	7	1	4	5	2	1	1	2	14	3	4	1	4	1	2	6
(7, 84)	63	1	1	1	1	2	2	1	1	1	1	2	14	7	2	1	3	1	1	4	1
(7, 87)	66	1	3	1	2	2	2	1	1	1	2	5	3	8	1	1	1	1	1	5	3

Figure 3. Oligonucleotide sequence signatures. The column labeled "Relative Mass" represents relative hybridization intensities obtained by a probe complementary to the coamplified vector sequence (mass probe) with dots containing DNA in comparison to the average signal of intentionally created empty dots. Columns 1–20 represent a subset of probes hybridized to one filter. Below the column number, the hybridization date and probe name are specified. Each row represent hybridization score values obtained with one dot specified by the row and column number on the filter. Pairs of dots separated by horizontal lines are generated from the same master clone by repeated PCR. Score values are adjusted for the difference in the amount of DNA using relative mass values (5). A score value of 1 represents dots with no detectable match.

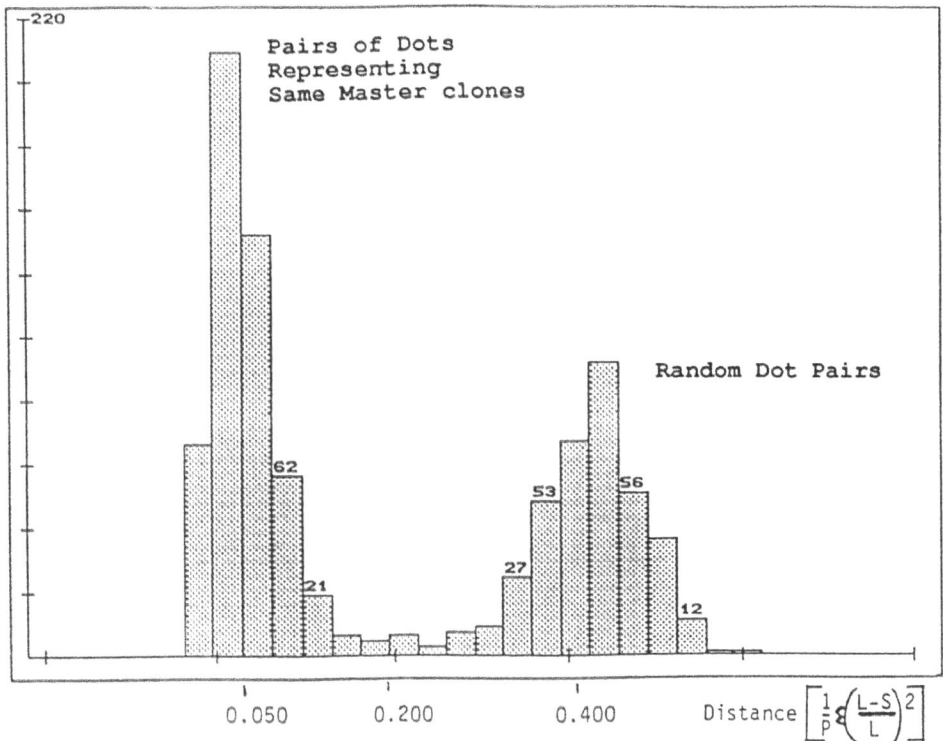

Figure 4. Distribution of OSS distances for identical and random clone pairs. The histogram is generated by Histo program (J. Jarvis, unpublished information). The formula for the distance calculation is written under the histogram. P is number of probes; L and S are the larger and smaller score values obtained by a particular probe with the given pair of clones. Bars represent how many (specified in some cases) clone pairs have distances which fall in the given range.

and an insignificant number of positive scores (in this case, less than 4) were eliminated from the comparisons. In this small set of 876 clones, 744 genes are represented (15% of the clones are redundant). We confirmed by restriction mapping (Fig. 5) that clusters consist of similar (not identical) clones in the majority of cases. The differences among the similar clones can not be explained by variation in clone size only. These clones represent either highly similar members from the gene families or alternatively spliced (or maybe trans-spliced) messengers. In the clustering procedure used, clones which differ up to 30% in size (or have 30% — i.e., a few exons of non-corresponding sequences) are recognized as significantly similar.

Alu Hae

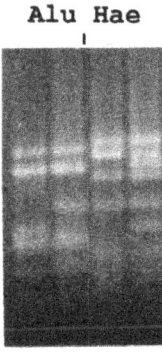

Figure 5. Restriction mapping of two clones with highly similar hybridization signatures. Inserts are amplified by PCR, digested by Alu I and Hae III restriction enzymes, and separated on 2% agarose gel.

Table 2. Profile of a cDNA-screening experiment.

Probes	51	
Dots:	3456	
PCR samples	1728	
Control clones	96	
Low mass	590	(36%)
Analyzable	1042	(64%)
Less than four hits	166	(16%)
For pairwise comparison	876	(54%)
Single clone clusters	687	(78%)
Multiple clone clusters	57	(3.3 clones/cluster)
Distinct clusters	744	(85% of 876 clones)

A Scheme for Gene Analysis by Combining Partial SBH and Single-Pass Gel Sequencing

Approximately 100,000 human cDNA clones have been sequenced from one or both ends by single-pass gel sequencing (at a cost of about $3 million). Screening one million clones probably can not be done in less than five years. If the reads are 300 bp on average, 600 Mb will be produced (twofold coverage of the expected 300 Mb of expressed sequences). By a very rough estimate based on the described models, we expect that less than 80% of the genes will be found. For most of them, only one piece will be determined, and it will be difficult to define the relationships of these clones and to identify artifacts. It is unlikely that more than half a million cDNA clones from the all other organisms will be end-sequenced in the next five years using presently available resources.

The described partial SBH approach is very cost- and time-effective. With our existing facilities, half a million clones can be analyzed in one year with 200 probes for less than $1 per clone of total cost. Further automation of the clone-managing and probe-labeling procedures, and particularly hybridization steps can increase the screening throughput to two million clones per year (Intelligent Automation Systems, Boston, MA, is constructing a machine to automatically operate 24 boxes which would have a daily throughput of 3 million clone/probe scores). The types of data expected from large-scale screening and the types of probes planned to be used are listed in the Tables 3 and 4, respectively. We recently have screened 20,000 cDNAs from the mentioned human brain libraries with 260 probes, and data collection for an additional 40,000 clones is in progress.

Table 3. Benefits from a large-scale cDNA screening by oligonucleotide hybridization.

1.	Catalogs of genes and gene families
2.	Tissue and time expression pattern for most genes
3.	Compositional features (G+C, Alu, coding capacity, motifs)
4.	Identification of the longest clone for each mRNA
5.	Clone contigs for long mRNAs (random priming)

Table 4. Composition of an oligonucleotide set suitable for cDNA analysis.

Type	No. specified bases	No. probes
Coding specific	7−8	40
Alu repeat	8−7	24
Gene domains	7−8	20
Extreme G/C or A/T	7−10	16
Protein motifs	7−9	40
Overlapped	7−8	34
Exceptional	5−6 and 9−15	26
		200

Partial SBH information is distributed over the whole insert and allows an estimate of the overall similarity of clone sequences. A smaller number of false positive or false negative clone matches is expected than if the comparisons are based on end sequences only. Also, 200 probes provide enough information to match the signatures with corresponding known gene sequences (unpublished results, 22). Because it is possible to screen several millions of clones in a few years, comprehensive gene catalogues with estimated expression levels for most of the genes in the analyzed tissues can be established. The enormous throughput can allow the elimination of artifacts by counting only cases represented by at least two clones. The main disadvantage of partial SBH based on a small number of probes is the impossibility of translating OSSs into protein sequence. Significantly long stretches of amino acids can be defined if 1000 or more probes are scored.

Data collected by SBH screening can be useful in several ways. Representative clones can be used as probes for gene mapping by FISH or by screening YAC, BAC, or cosmid libraries. In addition, a minimal set of representative clones (one gene equivalent) can be spotted on membranes for screening by genomic probes (e.g., cosmid or YAC clones) or by mRNA populations expressed under various physiological conditions. This can simplify the identification of genes for genetic diseases.

Complete gene sequencing can be rationalized and significantly accelerated. First of all, representative clones from the new families can be selected for complete sequencing. The molecular genetics will benefit enormously from the studies of the new gene families. The family of homeobox genes is a suggestive example. The complete sequencing of long mRNAs will be facilitated by selecting displaced clones from the defined contigs.

In addition to cDNA screening, partial SBH with less than 1000 probes (5) and single- pass gel sequencing can be combined to provide inexpensive overviews of genomic sequences (15,20). Resolution of genome fine structures and the identification of genes and their probable functions are anticipated (5,8,15,20). Partial cDNA and genomic sequences from a few species can be generated to the end of this century for half of the costs required to sequence the human genome completely. A "partial sequences first" approach gives an opportunity to practice "sequenetics" on the present level of development of sequencing techniques. An inexpensive full sequencing (15,20) and routine individual resequencing by next-generation methods (multiplex sequencing (23), directed sequencing by modular primers (24), capillary electrophoresis, mass spectrometry (25), and fast SBH by compact arrays called "sequencing chips", (reviewed in Refs. 8 and 15) will further improve the accuracy of the predictions and extend the field of the genetic discoveries achievable by comparative sequence analysis.

ACKNOWLEDGMENTS

In a discussion, Dr. George Church (Harvard Medical School, Boston, MA) brought to our attention the transcription leakage problem. We are grateful for the thoughtful comments of Drs. Radomir Crkvenjakov and Frank Collart. We thank David E. Nadziejka for technical editing and Kay Bexson for technical assistance. Work supported by the U.S. Department of Energy, Office of Health and Environmental Research, under Contract No. W-31-109-ENG-38.

REFERENCES

1. I. Labat, N.A. Stavropoulos, and R. Drmanac, Search for re-estimated 150,000−200,000 human genes in unsequenced genomic fragments, in: "Genome Mapping and Sequencing," Cold Spring Harbor Laboratory, Cold Spring Harbor, NY (1993).

2. D.M. Chikaraishi, Complexity of cytoplasmic polyadenylated and nonpolyadenylated rat brain ribonucleic acids, Biochem. 18:3249 (1979).

3. N. Chaudhari and W.E. Hahn, Genetic expression in the developing brain, Science 220:924 (1983).

4. R.J. Milner and J.G. Sutcliffe, Gene expression in rat brain, Nucl. Acid Res. 11:5497 (1983).

5. R. Drmanac., G. Lennon, S. Drmanac, I. Labat, R. Crkvenjakov, and H. Lehrach, Partial sequencing by oligohybridization: concept and applications in genome analysis, in "Electrophoreses, Supercomputing and the Human Genome," C.R. Cantor and H.A. Lim, eds., World Scientific, Singapore (1991).

6. R. Drmanac, I. Labat, I. Brukner, and R. Crkvenjakov, Sequencing of megabase plus DNA by hybridization: theory of the method, Genomics 4:114 (1989).

7. R. Drmanac, How 1000 base pairs with 10% error become 10,000 base pairs of correct sequence: a felicitous marriage of the gel and hybridization sequencing methods. Abstracts of Genome Mapping and Sequencing Meeting, Cold Spring Harbor Laboratory, Cold Spring Harbor, NY (1992).

8. R. Drmanac and R. Crkvenjakov, Sequencing by hybridization (SBH) with oligonucleotide probes as an integral approach for the analysis of complex genomes, Int. J. Genome Res. 1:59 (1992).

9. M.D. Adams, J.M. Kelley, J.D. Gocayne, M. Dubnick, M.H. Polymeropoulos, H. Xiao, C.R. Merril, A. Wu, O. Olde, R.F. Moreno, A.R. Kerlavage, W.R. McCombie, and J.C. Venter, Complementary DNA sequence: expressed sequence tags and human genome project, Science 252:1651 (1991).

10. A.S. Wilcox, A.S. Khan, J.A. Hopkins, and J.M. Sikela, Use of 3' untranslated sequences of human cDNAs for rapid chromosome assignment and conversion to STSs: implications of an expression map of the genome, Nucl. Acids Res. 13:1837 (1991).

11. K. Okubo, N. Hori, R. Matoba, T. Niiyama, A. Fukushima, Y. Kojima, and K. Matsubara, Large scale cDNA sequencing for analysis of quantitative and qualitative aspects of gene expression, Nature Genetics 2:173 (1992).

12. M.D. Adams, M.B. Soares, A.R. Kerlavage, C. Fields, and J.C. Venter, Rapid cDNA sequencing (expressed sequence tags) from a directionally cloned human infant brain cDNA library, Nature Genetics 4:373 (1993).

13. R. Drmanac, S. Drmanac, Z. Strezoska, T. Paunesku, I. Labat, M. Zeremski, J. Snoddy, W.K. Funkhouser, B. Koop, L. Hood, and R. Crkvenjakov, DNA sequence determination by hybridization: a strategy for efficient large-scale sequencing, Science 260:1649 (1993).

14. R. Drmanac, S. Drmanac, I. Labat, R. Crkvenjakov, A Vicentic, and A. Gemmell, Sequencing by hybridization: towards an automated sequencing of one million M13 clones arrayed on membranes, Electrophoresis 13:566 (1992).

15. R. Drmanac, S. Drmanac, J. Jarvis, and I. Labat, Sequencing by hybridization, in "Automated DNA Sequencing and Analysis Techniques," J.C. Venter, ed., Academic Press, London, in press.

16. S. Meier-Ewert, E. Maier, A. Ahmadi, J. Curtis, and H. Lehrach, An automated approach to generating expressed sequence catalogues, Nature 361:375 (1993).

17. T. Blumenthal, Mammalian cells can trans-splice? But do they?, BioEssays 15:347 (1993).

18. J.P. Bruzik and T. Maniatis, Spliced leader RNAs from lower eukaryotes are trans-spliced in mammalian cells, Nature 360:692 (1992).

19. P.G. Zaphiropoulos, Differential expression of cytochrome P450 2C450 2C24 transcripts in rat kidney and prostate: evidence indicative of alternative and possibly trans splicing events, *Biochem. Biophys. Res. Comm.* 192:778 (1993).

20. R. Drmanac, S. Drmanac, I. Labat, A. Vicentic, A. Gemmell, N. Stavropoulos, and J. Jarvis, SBH and the integration of complementary approaches in the mapping, sequencing, and understanding of complex genomes, *in* "Proceedings of Second International Conference on Bioinformatics, Supercomputing and Complex Genome Analysis," H.A. Lim, J.W. Fickett, C.R. Cantor, and R.J. Robbins, eds., World Scientific, Singapore (1992).

21. G.S. Lennon and H. Lehrach, Hybridization analyses of arrayed cDNA libraries. *Trends Genet.* 7:314 (1991).

22. A. Milosavljevic, Discovering sequence similarity by the algorithmic significance method, *in* "Proceedings of the First International Conference on Intelligent Systems for Molecular Biology," H. Hunter, D. Searls, and J. Shavlik, eds., AAAI Press, Menlo Park, CA (1993).

23. G.M. Church and S. Kieffer-Higgins, Multiplex DNA sequencing, *Science* 240:185 (1988).

24. J. Kieleczawa, J.J. Dunn, and F.W. Studier, DNA sequencing by primer walking with strings of contiguous hexamers, *Science* 258:1787 (1992).

25. L.M. Smith, The future of DNA sequencing, *Science* 262:530 (1993).

26. R. Drmanac, Z. Strezoska, I. Labat, S. Drmanac, and R. Crkvenjakov, Reliable hybridization of oligonucleotides as short as six nucleotides, *DNA and Cell Biology* 9:527 (1990).

ESTABLISHING CATALOGUES OF EXPRESSED SEQUENCES BY OLIGONUCLEOTIDE FINGERPRINTING OF cDNA LIBRARIES

Sebastian Meier-Ewert, Joachim Rothe, Richard Mott, and Hans Lehrach

Genome Analysis Laboratory, Imperial Cancer Research Fund
44 Lincoln's Inn Fields, London WC2A 3PX, United Kingdom

ABSTRACT

The number of DNA clones to be manipulated and analysed in a variety of projects dealing with the analysis of complex genomes exceeds the potential of current methodology by orders of magnitude. By focussing first on the transcribed parts of the genome, i.e., by using cDNA or exon-trap libraries, it is possible to reduce the volume of the task considerably, presumably without sacrificing too much information. We present here an integrated series of mostly automated processes which together allow the isolation, amplification, arrayed spotting and analysis by oligonucleotide fingerprinting of > 100,000 cDNA clones in a few months with little operator involvement. The sequence information thus derived will be used to search databases for related sequences and to establish catalogues of expressed sequences. The technique is currently being applied to the analysis of cDNA derived from various human and mouse tissues and developmental stages. An expanded version of the process will allow us to analyse cDNA libraries from a range of representative human tissues, thereby giving us access to a significant fraction of the human genome.

INTRODUCTION

The detailed analysis of the genomes of a whole range of organisms, ranging in complexity from bacteria to mammals, including mouse and man, for the first time gives us the opportunity to access much of the information that cells draw upon for the process of development and maintenance of the organism (for an update of the Human Genome Project, see 1 and 2). Significant progress has been made recently in physical mapping (3-6), expressed sequence analysis (7-12) and the integration of these two sets of data by mapping 'expressed sequence tags' (ESTs) onto specific chromosomes (13,14). Expressed genes do not represent more than a few percent of the mouse and human genomes, i.e., in the range of 100 million bp of DNA out of a total of 3 billion bp. This small fraction, however, represents a very large proportion of the information contained in the genome. As long as the experimental tools of genome analysis are not powerful enough as to allow

Identification of Transcribed Sequences, Edited by
U. Hochgeschwender and K. Gardiner, Plenum Press, New York, 1994

for high resolution whole-genome approaches, there is a strong incentive to focus research efforts first on the components of the genome with the highest informational content. This aim may best be achieved by large scale analysis of cDNA- or exon-trap libraries. It has therefore been one of the aims of the Human Genome Project to identify new genes by 'tag sequencing' random cDNA clones. Cells of any tissue express a rather invariable set of housekeeping genes in addition to a selection of specific genes which define the phenotypic characteristics of a particular cell type. In one particular analysis of cDNAs derived from brain, a tissue with high transcriptional complexity, it was found that between 30% and 50% of the estimated 30,000 mRNA species represent messages specific for brain (15). Consequently, particular transcripts are very unevenly represented in cDNA libraries derived from different sources, mainly due to the varying expression levels in the source tissues. Since tissue-specific genes are usually expressed at low levels or are even transcriptionally silent in tissues other than the appropriate one, cDNAs from multiple organs and tissues must be analysed to obtain a close to complete complement of all expressed sequences. A random approach towards the analysis of a large proportion of transcribed sequences will require the characterisation of a much larger number of clones than 100,000 - 150,000, which is the estimated number of genes in mammalian genomes.

METHODS AND RESULTS

We have developed a hybridisation-based system which allows the characterisation of all clones in an arrayed cDNA library (16). Using short radiolabeled oligonucleotides as probes and a hybridisation technique that allows the discrimination of a perfect match (for example, 8 nucleotides out of 8 in an octamer) from a mismatch, it is possible to identify all clones in a library that contain a given oligonucleotide sequence (17, 18). A fingerprint of all the clones present in a cDNA library can essentially be derived by accumulation of data from many such hybridisations, by assigning any clone as either positive or negative for any oligonucleotide in a given set of probes. Using octanucleotides as probes and based on an average cDNA insert size of 1 kb, it will be necessary to perform approximately 200 hybridisations to yield sufficient information to unambiguously discriminate more than 100,000 sequences. Any fingerprint represents discontinuous sequence information for one clone. Hence, oligonucleotide-fingerprinting does not have an inherent bias for non-translated regions as the EST approach mentioned earlier. By simply increasing the number of hybridisations, more and more sequence information can be obtained and more and more gaps can be closed. The informational content of these data is at least equivalent to that produced by 'tag sequencing' in that it identifies new genes and homologies to previously identified genes. As with conventional gel sequencing data it is possible to compare clones to sequences stored in databases such as Genbank, EMBL and dbEST (19).

For the large scale analysis of cDNA libraries, new automated high throughput procedures had to be developed, and many different techniques had to be combined in order to form an efficient analysis system (20). A future extension of this analysis technology generating at least one order of magnitude more hybridisation data will allow the reconstruction of the complete sequence of all the target DNAs (21), this has been successfully demonstrated in small test systems (22,23). Simple calculation shows that the rate of sequence data acquisition in such a system, based on 100,000 gridded cDNA clones on a nylon membrane and several hybridisations per day can be expected to be between 10 and 100 times higher than what can be readily achieved with current gel-based technology.

Automated, High-Throughput cDNA Analysis

The outline of a procedure developed in our laboratory over the past few years which allows simultaneous analysis of one hundred thousand cDNA clones of any library of interest is shown in Fig. 1. Most parts of the protocol are highly automated and, in fact, being performed by robotic devices which require little or no control by a human operator. Automation includes the initial picking of the clones into quadruple-density microtiter plates, amplification of the inserts using linker specific primers, high-density spotting of the amplified DNA onto standard nylon membranes and data acquisition and analysis. The actual hybridisations and washes are currently the only major remaining manual steps. The following paragraphs will describe the specifications of the system in some more detail.

Picking, Insert Amplification and Spotting

The analysis of arrayed cDNA libraries requires the manipulation of 10 times the number of clones used in YAC genomic DNA libraries and chromosome specific cosmid libraries. Of the order of 10^5 clones per cDNA library are needed to achieve a meaningful representation of the expressed sequences within the source tissue. For this purpose we have developed an automatic picking robot which can array more than 3000 clones per hour into 384 well plates (Fig. 2a). A CCD camera identifies individual colonies plated at low density (between 3000 and 6000 clones) on 22 cm x 22 cm agar dishes and an image analysis program calculates the x-y co-ordinates of every colony. There are several parameters that can be adjusted, taking into consideration roundness factors, size, colour and texture. When the list with the co-ordinates of all suitable clones is established, the robot addresses them with 96 individually controlled picking pins. After inoculation of prefilled microtitre plate wells, all 96 pins are sterilised in an ethanol bath and a new cycle of 96 pickings can begin.   The use of short oligonucleotides as hybridisation probes requires the clone DNA to be purified in order to suppress the high background signals caused by even small amounts of *E. coli* DNA. For this purpose, the polymerase chain reaction (PCR) is a suitable technique that lends itself both to automation and scale up. Using a water bath cycling system in which DNA amplification is carried out directly in microtitre plate wells, we are able to perform up to 46,000 reactions in one experiment (Fig. 2b). In practice, a rack containing the microtiter plates is being transferred back and forth between a high-temperature (95°C) and a low-temperature (73°C) water bath, thereby effectively performing a two-step PCR amplification. The purity and yield of the cDNA products are such that the PCR mix can be arrayed directly onto nylon membranes. Arraying again is performed by a robotic device that transfers 36,000 samples from 384 well microtitre plates onto 15 separate membranes in under 2 h (Fig. 2c).

Hybridisation and Data Analysis

Taking advantage of expressed sequences in the EMBL database, a computer algorithm was designed that chooses a set of oligonucleotides to best partition the expressed sequences. This set of oligonucleotides is likely to generate more informative fingerprints than randomly designed probes since it takes into account some of the sequence bias expected to be found in transcribed DNA. Additionally, a number of motif-specific oligonucleotides have been designed to target clones encoding members of gene families of general interest. In order to assess the quality of all hybridisation data, a set of 2,000 previously sequenced clones are included in each experiment. The correlation between

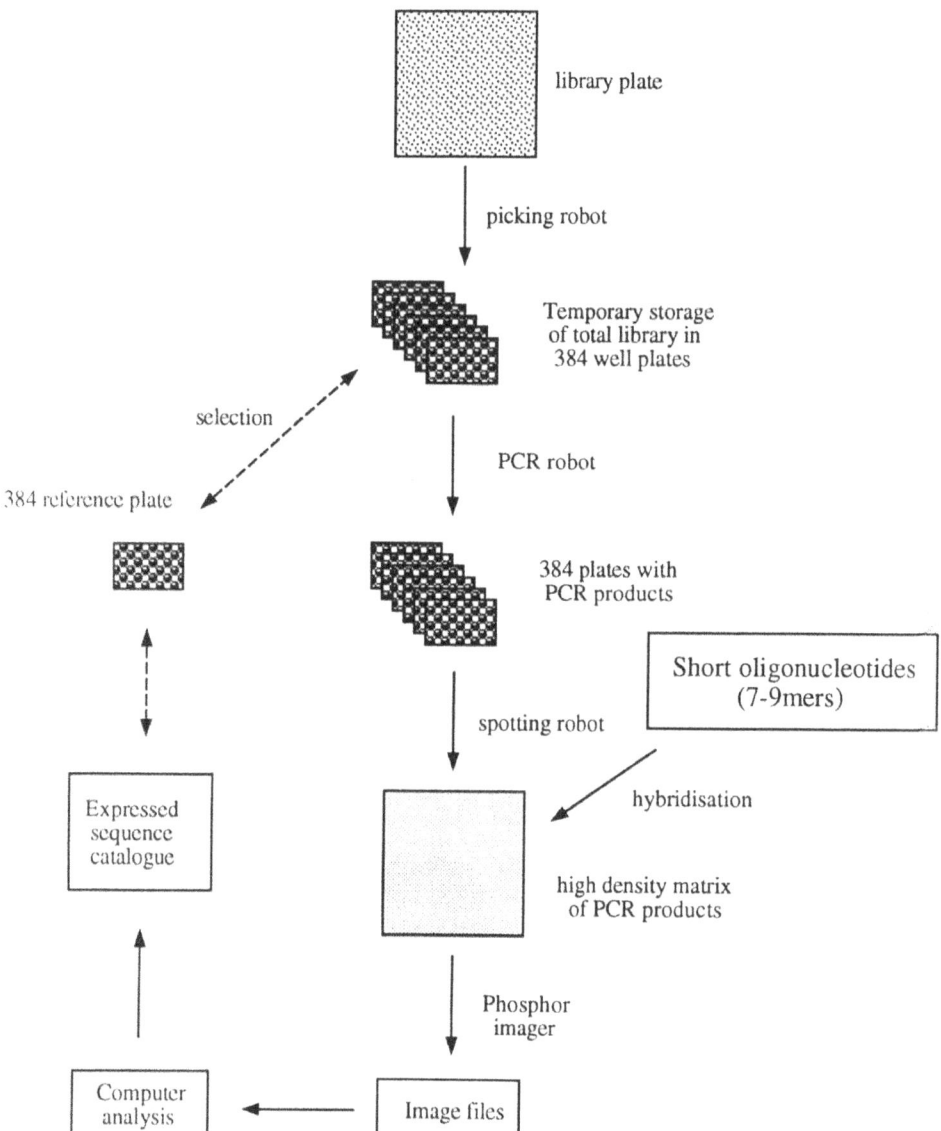

library plate

picking robot

Temporary storage
of total library in
384 well plates

selection

PCR robot

384 reference plate

384 plates with
PCR products

Short oligonucleotides
(7-9mers)

spotting robot

hybridisation

Expressed
sequence
catalogue

high density matrix
of PCR products

Phosphor
imager

Computer
analysis

Image files

Figure 1. Outline of the sequence fingerprinting procedure. Individual bacterial colonies are transferred by a picking robot from a low-density library plate into 384-well reference plates and grown over night. Aliquots of these cultures are then used as templates in a large scale PCR which is performed in polypropylene 384-well plates by a PCR robot. As the primers used flank the plasmids' inserts, the PCR results in a specific amplification of the cDNAs. A spotting robot then transfers the PCR products onto nylon membranes, creating high density gridded arrays of DNAs. Radioactively labelled short oligonucleotides are hybridised to these filters and hybridisation results are acquired via a Phosphorimager. Computer analysis of the accumulated sequence data allows the generation of an expressed sequence catalogue of the tissue under investigation. Databases can be searched for matches or closely related sequences. It is anticipated that a considerable portion of the cDNA clones can be identified at this stage. Clones of interest can be retrieved without delay from the reference plate for functional analysis.

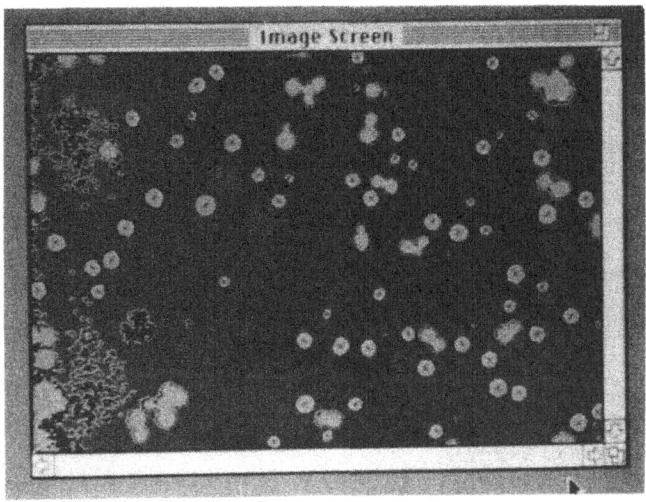

Figure 2a. Picking robot. Automatic on-screen evaluation of bacterial colonies grown on a library plate. The first step of the picking procedure consists of a stepwise scanning of the library plate with a CCD camera. The parameters of the colonies with a black mark in their center are within the limits set by the operator of the picking robot and their coordinates are written into the computer's memory. The second part of the procedure is the actual picking.

Figure 2b. A 96-pin picking manifold is mounted on a precision xyz-drive. The pins of this device successively address the positions of the selected colonies. Once at the right position, the pin is driven down by a pneumatic pusher and it's tip contacts the bacterial colony (this is the moment shown in b). When all 96 pins have addressed their target clones, the manifold inoculates them into 384-well plates. A new round of picking begins after the sterilisation of the pins in an ethanol bath.

Figure 2c. Spotting robot. A 384-well plate containing amplified plasmid inserts is shown on the left hand side in front of a bar-code reader which identifies the bar-code on the far side of the plate. The 384-pin manifold on the right hand side of the figure is transferring DNA onto a nylon membrane. In the next step, the manifold returns to the plate, picks up new DNA samples and transfers DNA to the next membrane. When all the membranes are spotted, the pins are sterilised, a new plate is inserted and the second round of spotting begins.

observed and expected results for these clones is used to estimate the reliability of the cDNA fingerprint data. To date, overall error rates (false positive plus false negative hybridisations) have been approximately 10%.

Oligonucleotides are labelled with ^{33}P-ATP at the 5' end by T4 polynucleotide kinase. Hybridisations and washings are carried out in the same buffer (high-salt, sarcosyl; Drmanac *et al.*, 1990b)) at low temperatures ($0°C$ - $5°C$). The membranes are then exposed to phosphor storage screens. After a few hours, these screens are scanned by a Phosphor Imager. The resulting images are analysed in a specifically developed UNIX based package that is run on Sun workstations. A typical example of the hybridisation of an octanucleotide to an array of approximately 20000 cDNA clones is shown in Fig. 3.

Due to the arrayed structure of the libraries, clones which display interesting features like informative fingerprints or unusual expression patterns are immediately accessible for further analysis without the need to isolate or subclone them.

CONCLUSIONS AND PROSPECTS

The first step towards understanding how the genome controls cellular processes involves the identification of the components that code for the information required to exert this control. Oligonucleotide fingerprinting is a technique that will present the possibility to obtain both sequence and expression data on a very large proportion of the informational content of mammalian genomes within a very reasonable time frame: 1 person can

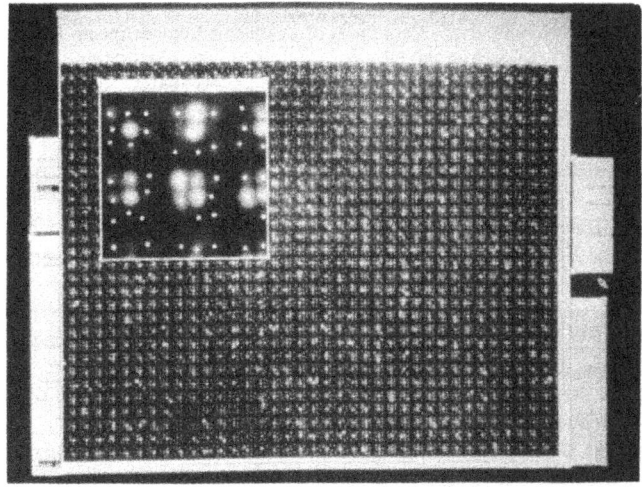

Figure 3. Oligonucleotide hybridisation. A ^{33}P-labelled octamer is hybridised to approximately 20,000 cDNA clones arrayed on a nylon membrane in a 3x3 pattern. The digital image of this hybridisation is shown along with an enlarged view (insert at the upper left hand corner). For normalisation and aligning purposes, the central position of each 3x3 square contains salmon sperm DNA which gives a positive hybridisation signal with any oligonucleotide. Signals deriving from positions other than the central one represent hybridisation events to cDNA clones.

fingerprint a cDNA library of roughly 110,000 clones with 200 oligonucleotides in about 2 weeks (2 filters of 55,000 clones each, 10 sets of filters used in parallel, 1-2 hybridisations per day). Oligonucleotide fingerprints of cDNA libraries from various tissue sources and stages of development will allow the parallel determination of the location and the level of expression of many thousands of transcripts, producing an expressed sequence catalogue of any particular organ, or even, by combining data derived from many independent cDNA libraries, of an entire genome. Such catalogues doubtless will be excellent resources for biological research in general and the characterisation of genes involved in genetic disorders in particular, especially in view of the increasing potential of gene therapy.

REFERENCES

1. F. Collins, and D. Galas, A new five-year plan for the U.S. Human Genome Project, *Science* 262: 43 (1993).
2. A.J. Cuticchia, M.A. Chipperfield, C.J. Porter, W. Kearns, and P.L. Pearson, Managing all those bytes: The human genome project, *Science* 262:47 (1993).
3. C. Bellanne Chantelot, B. Lacroix, P. Ougen, A. Billault, S. Beaufils, S. Bertrand, I. Georges, F. Glibert, I. Gros, G. Lucotte, and *et al.*, Mapping the whole human genome by fingerprinting yeast artificial chromosomes, *Cell* 70:1059 (1992).

4. I. Chumakov, P. Rigault, S. Guillou, P. Ougen, A. Billaut, G. Guasconi, P. Gervy, I. LeGall, P. Soularue, L. Grinas, and et al., Continuum of overlapping clones spanning the entire human chromosome 21q, *Nature* 359:380 (1992).

5. S. Foote, D. Vollrath, A. Hilton, and D.C. Page, The human Y chromosome: overlapping DNA clones spanning the euchromatic region, *Science* 258:60 (1992).

6. D. Vollrath, S. Foote, A. Hilton, L.G. Brown, P. Beer Romero, J.S. Bogan, and D.C. Page, The human Y chromosome: a 43-interval map based on naturally occurring deletions, *Science* 258:52 (1992).

7. M.D. Adams, M. Dubnick, A.R. Kerlavage, R. Moreno, J.M. Kelley, T.R. Utterback, J.W. Nagle, C. Fields, and J.C. Venter, Sequence identification of 2,375 human brain genes, *Nature* 355:632 (1992).

8. M.D. Adams, J.M. Kelley, J.D. Gocayne, M. Dubnick, M.H. Polymeropoulos, H. Xiao, C.R. Merril, A. Wu, B. Olde, R.F. Moreno, and et al., Complementary DNA sequencing: expressed sequence tags and human genome project, *Science* 252:1651 (1991).

9. M.D. Adams, A.R. Kerlavage, C. Fields, and J.C. Venter, 3400 new expressed sequence tags identify diversity of transcripts in human brain, *Nature Genetics* 4:256 (1993).

10. M.D. Adams, B.M. Soares, A.R. Kerlavage, C. Fields, and J.C. Venter, Rapid cDNA sequencing (expressed sequence tags) from a directionally cloned human infant brain cDNA library, *Nature Genetics* 4:373 (1993).

11. A.S. Kahn, A.S. Wilcox, M.H. Polymeropoulos, J.A. Hopkins, T.J. Stevens, M. Robinson, A.K. Orpana, and J.M. Sikela, Single pass sequencing and physical and genetic mapping of human brain cDNAs, *Nature Genetics* 2:180 (1992).

12. K. Okubo, N. Hori, R. Matoba, T. Niiyama, A. Fukushima, Y. Kojima, and K.Matsubara, Large scale cDNA sequencing for analysis of quantitative and qualitative aspects of gene expression, *Nature Genetics* 2:173 (1992).

13. M.H. Polymeropoulos, H. Xiao, A. Glodek, M. Gorski, M.D. Adams, R.F. Moreno, M.G. Fitzgerald, J.C. Venter, and C.R. Merril, Chromosomal assignment of 46 brain cDNAs, *Genomics* 12:492 (1992).

14. M.H. Polymeropoulos, H. Xiao, J.M. Sikela, M.D. Adams, J.C. Venter, and C.R. Merril, Chromosomal distribution of 320 genes from a brain cDNA library, *Nature Genetics* 4:381 (1993).

15. R.J. Milner, and G.J. Sutcliffe, Gene expression in rat brain, *Nucl. Acids Res.* 11: 5497 (1983).

16. S. Meier Ewert, E. Maier, A. Ahmadi, J. Curtis, and H. Lehrach, An automated approach to generating expressed sequence catalogues, *Nature* 361:375 (1993).

17. R. Drmanac, Z. Strezoska, I. Labat, S. Drmanac, and R. Crkvenjakov, Reliable hybridization of oligonucleotides as short as six nucleotides, *DNA Cell. Biol.* 9:527 (1990b). 18. R. Drmanac, G.G. Lennon, S. Drmanac, I. Labat, R. Crkvenjakov, and H. Lehrach, Partial Sequencing by Oligo-Hybridization: Concept and Applications in Genome Analysis, *in* "The First International Conference on Electrophoresis, Supercomputing, and the Human Genome", C. Cantor and H.A. Lim, eds., World Scientific (1990a).

19. M.S. Boguski, T.M.J. Lowe, and C.M. Tolstoshev, dbEST - database for "expressed sequence tags", *Nature Genetics* 4:332 (1993).

20. E. Maier, S. Meier Ewert, and H. Lehrach, Automation of partial sequence analysis by hybridisation, *J. Biotech.* in press, (1993).

21. R. Drmanac, I. Labat, I. Brukner, and R. Crkvenjakov, Sequencing of megabase plus DNA by hybridization: theory of the method, *Genomics* 4:114 (1989).

22. Z. Strezoska, T. Paunesku, D. Radosavljevic, I. Labat, R. Drmanac, and R. Crkvenjakov, DNA sequencing by hybridization: 100 bases read by a non-gel-based method, *Proc. Natl. Acad. Sci. USA* 88:10089 (1991).

23. R. Drmanac, S. Drmanac, Z. Strezoska, T. Paunesku, I. Labat, M. Zeremski, J. Snoddy, W.K. Funkhouser, B. Koop, L. Hood, and R. Crkvenjakov, DNA sequence determination by hybridization: A atrategy for efficient large-scale sequencing, *Science* 260:1649 (1993).

PCR-BASED TECHNOLOGIES TO STUDY DIFFERENTIAL GENE EXPRESSION IN RAT BRAIN

Mark G. Erlander, Ana Dopazo, Pamela E. Foye, and J. Gregor Sutcliffe

Department of Molecular Biology, The Scripps Research Institute
La Jolla, CA 92037

ABSTRACT

We are interested in cloning/identifying virtually all mammalian brain mRNAs which are differentially expressed in response to various chemical, electrical, pharmacological and behavioral challenges to the central nervous system. To reach this ambitious objective, we propose that a PCR-based differential display methodology which is digital, portable, systematic, quantitative and database-friendly is required. To develop this technology, we first examined previously reported PCR-based differential display methodologies; we report that although current methods are sufficient for identifying some differentially expressed mRNAs, a more rigorous technique is needed to fit the necessary criteria. To this end, we report preliminary work which is aimed at the development of a modified method of PCR-based differential display.

INTRODUCTION

For biochemically minded scientists, an ultimate goal of research ought to be a complete characterization of the protein molecules that make up an organism. This would include their identification, sequence determination, demonstration of their anatomical sites of expression, elucidation of their biochemical activities, and understanding of how these activities determine organismic physiology. In pursuit of this ultimate goal, this laboratory is interested in developing gene discovery technologies which lead to the cloning of brain mRNAs which are expressed discretely within the central nervous system or whose expression is differentially regulated in response to different challenges.

There are three conceptually different approaches for cloning the products of differentially expressed genes: classical differential screening (plus-minus screening), subtractive hybridization and PCR-based differential display. Classical differential screening allows for isolation of clones of only relatively abundant (greater than 0.06%) mRNAs (1) and is thus usually not the method of choice. In contrast, subtractive hybridization technologies can lead to the cloning of differentially expressed mRNAs of relatively rare prevalence (0.001%). In this workshop, this laboratory presented a directional tag PCR subtraction method which is technically easy, efficient and reproducible (2,3).

Despite the obvious power of this refined subtraction methodology for identifying clones of mRNAs that have a particular selective pattern of expression, the overall patterns of expression emerge one gene at a time. Thus, the process is slow and laborious if one wants to examine differential gene expression of an entire mRNA population of interest. Therefore, we have examined PCR-based differential display methodologies in order to develop a technology that integrates information about sequence and differential expression patterns of mRNAs within the context of a digital, portable, systematic and database-friendly methodology. Here, we present our progress in obtaining this objective.

RESULTS AND DISCUSSION

Current PCR-Based Differential Display Methods

In discussing ways in which to combine information about sequence and expression pattern, we realized that PCR offered a potentially rapid way in which a combined database could be established. Williams *et al.* (4) and Welsh and McClelland (5) showed that single 10-mer PCR primers of arbitrarily chosen sequence (that is, any 10-mer primer off the shelf), when used for PCR with complex DNA templates such as human, plant, yeast or bacterial genomic DNA, gave rise to an array of PCR products. The priming events were demonstrated to involve incomplete complementarity between the primer and the template DNA. Presumably, partially mismatched primer-binding sites are randomly distributed throughout the genome. Occasionally, two of these sites in opposing orientation were located closely enough together to give rise to a PCR product band. There were on average 8-10 products, which varied in size from about 0.4 - 4 kb and had different mobilities for each primer. The array of PCR products exhibited differences among individuals of the same species. These authors proposed that the single arbitrary primers could be used to produce RFLP-like information for genetic studies, and others have since proven this (6-8).

Two groups (9,10) adapted the method to compare mRNA populations. In particular, Liang and Pardee (9) adapted this method, called mRNA differential display, to compare the populations of mRNAs expressed by two related cell types, normal and tumorigenic mouse A31 cells. For each experiment, they used one arbitrary 10-mer as the 5' primer and an oligonucleotide complementary to a subset of poly A tails as the 3' anchor primer, performing PCR amplification in the presence of ^{35}S-dNTPs on cDNAs prepared from the two cell types. The products were resolved on sequencing gels and 50-100 bands ranging from 100-500 nucleotides were observed. The bands presumably result from amplification of cDNAs corresponding to the 3' ends of mRNAs that contain the complement of the 3' anchor primer and a partially mismatched 5' primer site, as had been observed (4,5) on genomic DNA templates. For each primer pair, the pattern of bands amplified from the two cDNAs was similar, with the intensities of about 80% of the bands being indistinguishable. Some of the bands were more intense in one or the other PCR samples; a few were detected in only one of the two samples. Further studies (11) have demonstrated that the procedure works with low concentrations of input RNA (although it is not quantitative for rarer species), and that specificity resides primarily in the last nucleotide of the 3' anchor primer. At least a third of identified differentially detected PCR products correspond to differentially expressed RNAs, with a false positive rate of at least 25%. A few technical obstacles have been identified, but generally the method is suited for a high throughput screen of the most prevalent mRNAs (11).

We repeated this experiment (example in Fig. 1), comparing cDNAs made from liver and various brain regional mRNA samples, and obtained generally similar results, although a greater percentage of the amplified bands differ between brain and liver than between types of A31 cells, as would be expected from previous studies. We found that the appearance of bands i) was reproducible in that duplicate reactions gave rise to the same set of bands, ii) differed in the majority of bands when liver and brain were compared and in only a few bands per gel when brain regions such as cerebral cortex, cerebellum and hypothalamus were compared, and iii) was indicative of the mRNAs expressed in the tissues in that when primer sets corresponding to known mRNAs were used, PCR products whose sequences corresponded to those of the known mRNAs were detected with their expected tissue distributions. Furthermore, 36 of the PCR-cDNAs of mRNAs with overtly differential patterns of expression were excised from the gels, amplified by a further round of PCR using the same primer pairs, and nucleotide sequences determined. These data have confirmed the identities of the knowns and provided new sequences. Most of the new sequences have been novel and a few have been used to design specific pairs of PCR primers which have shown by RT-PCR that most of the novel sequences correspond to mRNAs with the distributions indicated by the initial PCR gels. The few that did not are believed to correspond to comigrating PCR species.

If all of the 50,000 to 100,000 mRNAs of the mammal were accessible to this arbitrary-primer PCR approach, then about 80-90 5' arbitrary primers and 12 3' anchor primers would be required in approximately 1000 PCR panels and gels to give a likelihood (calculated by the Poisson distribution) that about 2/3 of these mRNAs would be identified. It is unlikely that all mRNAs are amenable to detection by this method for the following reasons. First, for an mRNA to surface in such a survey, it must be prevalent enough to produce a signal on the autoradiograph and contain a sequence in its 3' 500 nucleotides capable of serving as a site for mismatched primer binding and priming. The more prevalent an individual mRNA species, the more likely it would be to generate a product, thus prevalent species may give bands with many different arbitrary primers. Because this latter property would contain an unpredictable element of chance based on selection of the arbitrary primers, it would be difficult to approach closure by the arbitrary primer method.

Second, for the information to be portable from one lab to another, the mismatch priming must be highly reproducible under different laboratory conditions using different PCR machines. As the basis for mismatch priming is poorly understood, we thought that this might be a drawback of building a database from data obtained by the exact Liang and Pardee differential display method. Although we have been successful with the method, as discussed above, to avoid these anticipated problems, we have modified the method so as to reduce greatly the uncertain aspect of 5' end generation and allow data to be absolutely reproducible in different settings.

Modified Method of Differential Display

In considering a methodology for surveying brain mRNA expression, we were concerned about several points. First, the method should be able reproducibly and portably to identify large numbers of mRNAs and compare their expression not only within brain but in other tissues and in response to various physiological paradigms. Second, it should allow individual mRNAs that are identified by the method to be pursued easily at deeper levels of analysis, such as sequence determination, *in situ* hybridization and genetic mapping. Third, it should be more dependent upon some digital feature of the mRNA than upon its prevalence in a given tissue. Finally, the method should, in principal, allow for the accounting of all mRNAs (it should have closure) and, at any time during its application, should allow the portion of mRNAs that have been measured to be defined.

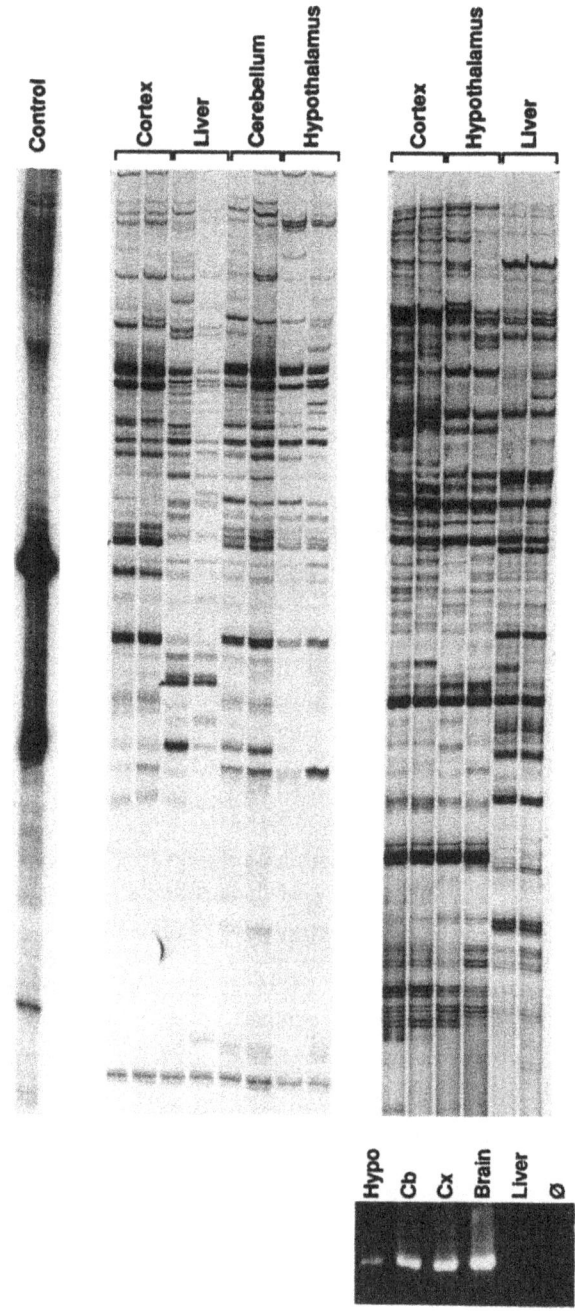

Figure 1.

We have developed a new PCR-based method, as described in Fig. 2, which allows for the visualization of nearly every mRNA expressed by a tissue as a distinct band on a gel whose intensity corresponds roughly to the concentration of the mRNA. The method is conceptually related to the differential display methodologies discussed above, but has attributes that allow precise accounting of all species to be achieved and differentially expressed mRNAs to be recognized.

Our method operationally begins with double-stranded cDNA syntheses using an equimolar mixture of the 12 primers of the set $ANCHOR1T_{18}VN$ (ANCHOR1 = AACTGGAAGAATTCGCGGCCGCAGGAA; V = A,C,or G; N = A,C,G,or T) to initiate reverse transcription. One member of this mixture of 12 primers initiates synthesis at a fixed position at the 3' end of all copies of a given mRNA species in the sample, thereby defining a 3' endpoint for each species. Following this, the cDNA sample is cleaved with the restriction endonucleases MspI and NotI (since MspI recognizes a 4 nucleotide sequence, it cleaves at multiple sites in most cDNAs and the NotI cleaves at a single site in the ANCHOR1 sequence but not within most cDNAs), and subsequently subcloned directionally into ClaI-, NotI-cleaved plasmid pBC SK$^+$ in an orientation that is antisense with respect to the vector's T3 promoter. Libraries with in excess of 5×10^5 recombinants are then generated.

To survey for differentially expressed cDNAs, an aliquot of each library is divided into two pools, one of which is cleaved with XhoI, the second with SalI. The pools of linearized plasmids are then combined, mixed, then divided into thirds. The thirds are digested with HindIII, BamHI and EcoRI, respectively, recombined and incubated with T3 RNA polymerase, which generates antisense cRNA transcripts of the cloned inserts that contain known vector sequences (tags) abutting the MspI and NotI sites from the original cDNAs. Plasmid DNA is removed by incubation with RNase-free DNase.

At this stage, each cRNA preparation is subdivided into 16 subpools for further processing in a two-step reaction. First-strand cDNA is made from each subpool using each of the 16 primers of the form 5PRIMERNN (N = A,C,G,or T; 5PRIMER = AGGTCGACGGTATCGG) and the r*Tth* thermostable reverse transcriptase (annealing at 60°C, reaction at 70°C) to promote high fidelity complementarity between primer and cRNA. The product of each reaction is used as template for PCR with the universal 3' primer 3PRIMER = GAACAAAAGCTGGAGCTCCACCGC and the

---

Figure 1. (See facing page.) Example of PCR-based differential display. In the middle panel of the figure, samples of cerebral cortex, liver, cerebellum and hypothalamus mRNA were used in duplicate for differential display by the Liang-Pardee method. In the far left lane, as a control template, we used a cDNA clone of the RNA encoding RC3, an mRNA encoding a 78 residue protein kinase C substrate enriched in forebrain dendritic spines, present as 0.01% of cortex mRNA but not detectable in liver, cerebellum or hypothalamus. The selected primers were chosen so as to amplify RC3. The PCR survey reactions with the RC3 clone produce a major band, indicated by an asterisk. The survey reactions produce patterns that are almost identical in the duplicate samples, but differ substantially between brain and liver, and to a lesser extent among brain regions. A band comigrating with the control RC3 major band was detected in cortex but much less or not at all in the other samples, as expected from the known RC3 distribution. The cDNA was excised from the cortex lane, amplified, and the sequence determined: it corresponded exactly to that of the known sequence of RC3. The panel on the far right side of the figure was produced with a different primer pair (that does not recognize the RC3 sequence). Again, duplicates are faithful and there are several differences between cortex and hypothalamus, but many more between brain and liver. The band indicated by the asterisk was excised from the gel, amplified by PCR, and its sequence determined and found to be novel. A primer pair, based on the novel sequence, was synthesized and used in a PCR with cDNA from hypothalamus (hypo), cerebellum (cb), cortex (cx), whole brain, liver or no cDNA as template. The amplification pattern is consistent with that observed in the survey gel. Cumulatively, our data indicate that the appearance of a band in the survey gels corresponds qualitatively and quantitatively to the presence of an mRNA in the tissue sample, and show that a large number of differentially expressed mRNAs are encountered.

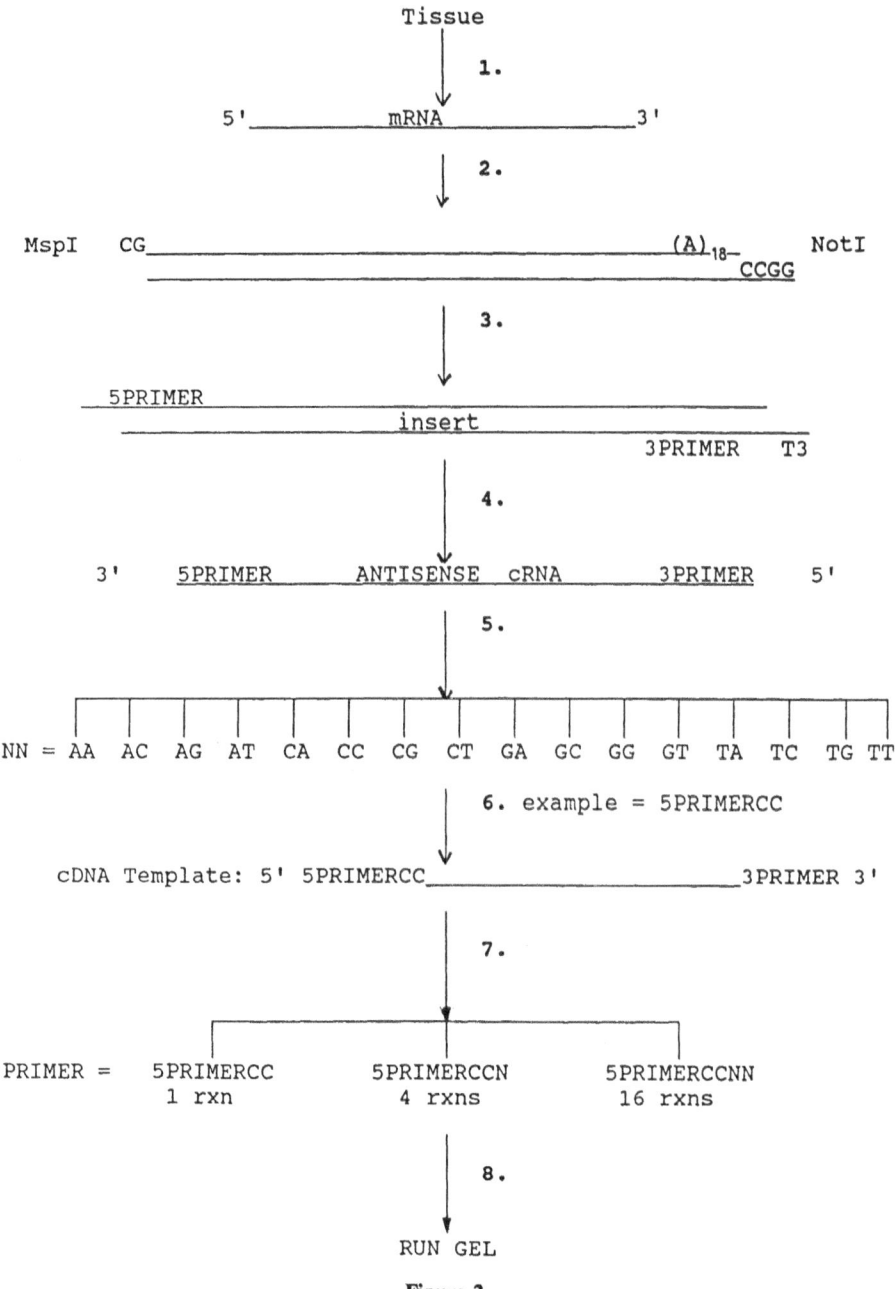

Tissue

1.

5'_____mRNA_____3'

2.

MspI   CG_____(A)₁₈— NotI
       _____CCGG

3.

___5PRIMER_____
_____insert_____
                              3PRIMER   T3

4.

3'   5PRIMER   ANTISENSE   cRNA   3PRIMER   5'

5.

NN = AA   AC   AG   AT   CA   CC   CG   CT   GA   GC   GG   GT   TA   TC   TG TT

6. example = 5PRIMERCC

cDNA Template: 5' 5PRIMERCC_____3PRIMER 3'

7.

PRIMER =   5PRIMERCC   5PRIMERCCN   5PRIMERCCNN
           1 rxn   4 rxns   16 rxns

8.

RUN GEL

Figure 2.

266

appropriate primer of the form 5PRIMERNN, 5PRIMERNNN, or 5PRIMERNNNN as the second primer, in the presence of ^{35}S-dATP using a PCR program (94°C 15",60°C 15",72°C 30" on a Perkin-Elmer 9600 apparatus) that includes a high temperature annealing step that minimizes artifactual mispriming by the 5PRIMER at its 3' end and promotes high fidelity copying. This series of reactions produces 16, 64 and 256 product pools for the 5PRIMERNN, NNN and NNNN primers, respectively, for each cRNA sample used.

According to this scheme, the cDNA libraries produced from each of the mRNA samples should contain copies of the extreme 3' ends [from the most distal site for MspI to the beginning of the poly(A) tail] of all poly(A)$^+$ mRNAs in the starting RNA sample approximately according to the initial relative concentrations of the mRNAs. Because both ends of the inserts for each species are exactly defined by sequence, their lengths will be uniform for each species allowing their later visualization as discrete bands on a gel, regardless of the tissue source of the mRNA, an important fundamental concept of the approach. [mRNAs containing no MspI recognition sequences would not be represented, but these are quite rare. Also rare are species corresponding to internal NotI-MspI fragments, since NotI has an 8 bp recognition sequence.]

To generate antisense transcripts of the inserts with T3 RNA polymerase, the template must first be cleaved with a restriction endonuclease that cuts within flanking sequences but not within the inserts themselves. Given that the average lengths of the 3'-terminal MspI fragments is 256 bp, approximately 6% of the inserts are expected to contain sites for any enzyme with a hexamer recognition sequence. Those inserts would be lost to further analysis were only a single enzyme utilized. Here we have divided the reaction so that only one of either of two enzymes (XhoI and SalI) is used for linearization of each half reaction. Only inserts containing sites for both enzymes (approximately 0.4%) are lost from both halves of the samples. Similarly, each cRNA sample would be contaminated to a different extent with transcripts from insertless plasmids, which could lead to variability in the efficiency of the later PCRs for different samples because of differential competition for primers. Cleavage of thirds of the samples with one of 3 enzymes that have single targets in pBC SK$^+$ between its ClaI and NotI sites eliminates the production of transcripts containing 5PRIMER binding sites from insertless plasmids. The use of 3 enzymes on thirds of the reaction reduces the loss of insert-containing sequences that also contain sites for the enzyme (10% were only one enzyme used) to an acceptable 0.1%, while neatly solving the no insert problem. Thus, after transcription by T3 RNA polymerase, cRNA pools representing the 3' ends of greater than 99% of the 0.001% or greater prevalence mRNA population (assuming libraries within excess of 10^5 recombinants) with primer binding sites attached at both ends of the cRNAs are present for future PCR amplification.

Figure 2. (See facing page.) Modified differential display method. **1.** Each tissue source is dissected and poly(A) RNA extracted. **2.** Double-stranded cDNA is synthesized by reverse transcription followed by second strand syntheses. ANCHOR1-(T)$_{18}$-VN (V = ACG, N = ACGT) is the primer used in first strand synthesis. The double-stranded cDNA is then cleaved with MspI and NotI, thus defining both ends precisely for each mRNA species. **3.** The digested cDNA is then cloned into compatible ends of ClaI/NotI-cleaved pBC SK$^+$. The CsCl-purified cDNA library is then linearized in two separate reactions with SalI or XhoI and subsequently mixed in equal molar amounts. **4.** Antisense cRNA is synthesized via T3 RNA polymerase. **5.** Antisense cRNA is divided into 16 equal aliquots and cDNA template is synthesized by using in each tube one primer of the possible 16 permutations of 5PRIMERNN. **6.** Each on the 16 cDNAs is then utilized as a template in subsequent PCRs for differential display. As an example, the cDNA template synthesized by using PRIMERCC is indicated in subsequent steps. **7.** The PRIMERCC cDNA template is amplified in one reaction using primers 5PRIMERCC and 3PRIMER, 4 reactions using all possible permutations of 5PRIMERCCN and 3PRIMER and 16 reactions using all possible permutations of 5PRIMERCCNN and 3PRIMER (1+4+16 = 21 PCRs per 1/16 pop of cDNA). **8.** PCR products are then separated on a denaturing gel (Urea-PAGE). In an actual experiment, several tissue sources are compared amongst each other.

Up until this stage the procedure is compact, thus the quality of the samples is easily assessed.

A second fundamental aspect of the approach is the use of successive steps to survey the cDNAs. These steps, which essentially act like a nested PCR, enhance quality control and diminish the background that potentially could result from amplification of untargeted cDNAs. The second reverse transcription step subdivides each cRNA sample into 16 subpools, utilizing a primer that anneals with sequences derived from pBC SK$^+$ but extends across the CGG of the non-regenerated MspI site and including two nucleotides (NN) of the insert. This step segregates the starting population of potentially 50,000-100,000 mRNAs into 16 subpools of approximately 3000-6000 members each. In serial iterations of the subsequent PCR step, in which radioactive label is incorporated into the products for their autoradiographic visualization, those pools are further segregated by division into 4 or 16 subsubpools by using progressively longer 5PRIMERs of the 5PRIMERNNN and 5PRIMERNNNN forms. By first demanding by high temperature annealing a high fidelity 3'-end match at the reverse transcription step in the NN positions, and subsequently demanding again at the NNN or NNNN iterations, bleed-through from mismatch priming at the NN positions is drastically minimized. The autoradiogram resolves the molecules into distinct bands.

Example of Modified Method

We have applied this methodology using 5PRIMERNNNNs corresponding to the sequences of known brain mRNAs of different concentrations [for example, neuron-specific enolase (NSE) roughly 0.5%, RC3 0.01%, somatostatin 0.001%] to compare cDNAs made from libraries constructed from cerebral cortex, striatum, cerebellum and liver RNAs, made as described above (examples in Fig. 3). 50-100 bands were obtained on short autoradiographic exposures from any particular RNA sample, and bands were absolutely reproducible in duplicate samples. Approximately 2/3 of the bands differed between brain and liver samples, including the bands of the correct lengths corresponding to the known brain-specific mRNAs (confirmed by excision from the gels, amplification and sequencing),

Figure 3. (See facing page.) Specificity controls. To demonstrate that the appearance of a PCR product is indicative of the presence of an mRNA in the tissue that corresponds to the 5PRIMERNNNN used in its detection, poly(A)$^+$-enriched RNA samples isolated from cerebral cortex and liver were used to construct cDNA libraries and tagged antisense cRNA preparations as described in text and Fig. 2 (steps 1-4). The cRNAs were then used as substrates for 2-stage reactions (steps 5-8). In the 5 gel lanes on the left, cortex cRNA was substrate for reverse transcription with 5PRIMERNNs, where NN = CT (primer 118), GT (primer 116) or CG (primer 106). The PCR amplification used 3PRIMER and 5PRIMERNNNNs, where NNNN = CTAC (primer 128), CTGA (primer 127), CTGC (primer 111), GTGC (primer 134) and CGGC (primer 130), as indicated in figure. Primers 118 and 111 match the sequence of the two and four nucleotides, respectively, downstream from the most 3' MspI site in the NSE mRNA sequence. Primer 127 is mismatched with the NSE sequence in the last (-1) position, primer 128 in the next-to-last (-2) position, primers 106 and 130 in the -3 position, and primers 116 and 134 in the -4 position. Primer 134 extended 2 nucleotides further upstream than the others shown here, hence its PCR products are 2 nucleotides longer relative to the products in other lanes. In each lane 50-100 bands were visible in 15' exposures (^{32}P-dCTP was used in this experiment; ^{35}S-dATP gives higher resolution), and these bands were apparently distinct for each primer pair, with the exception that a subset of the 118-111 bands appeared more faintly in the 116-134 lane (trailing by two nucleotides), indicating bleed through in the -4 position. The 118-111 primer set was used again on separate cortex (CX) and liver (LV) cRNAs. The cortex pattern was identical to that in lane 118-111 demonstrating reproducibility, the liver pattern differed from CX in the majority of species. The asterisk indicates the position of the NSE product. Analogous primer sets detected RC3 and somatostatin (somat.) products (asterisks) in CX but not LV lanes. The relative band intensities of a given PCR product can be compared within lanes using the same primer set, but not different sets.

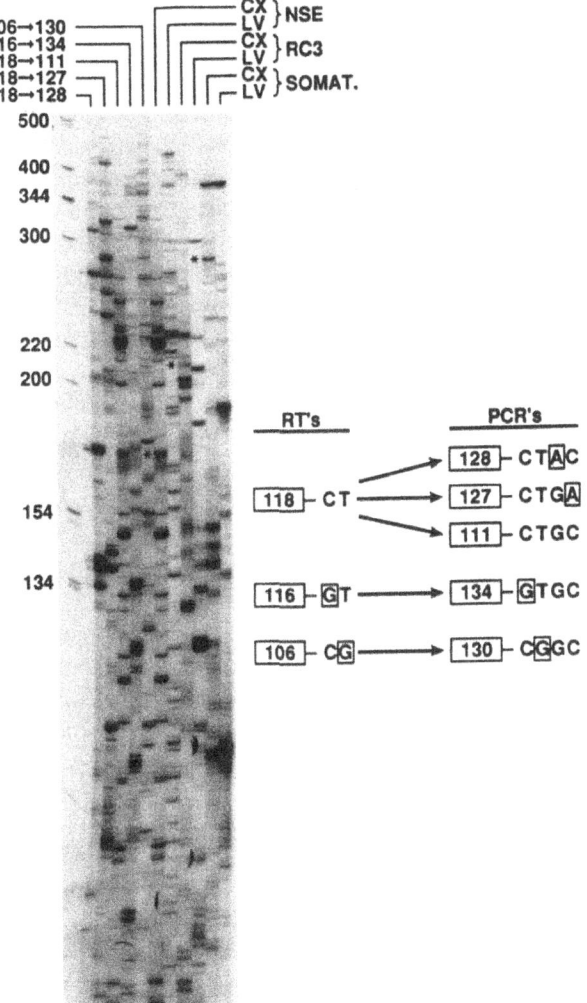

Figure 3.

but only a few bands differed among samples from various brain regions for any particular primer (not shown), although some band intensities differed. The band corresponding to NSE, a relatively prevalent mRNA species, appeared in all of the brain but not liver samples, but was not observed when any of the last 3 single nucleotides within NNNN was changed in the synthetic 5PRIMER. When the first N was changed, a small amount of bleed through was detected. For the known species, the intensity of the autoradiographic signal was roughly proportional to mRNA prevalence, and mRNAs with concentrations of one part in 10^5 or greater of the poly(A)$^+$ RNA were visible, with the occasional problem that cDNAs that migrate close to more intense bands were obscured. Dilution of the cRNA samples had no effect on gel pattern.

These data demonstrate that the appearance of a particular band with a given 5PRIMER is evidence that a particular mRNA specified discretely by the sequence of that primer (with a slight bleed through of prevalent species with single-nucleotide differences at the -4 position) is present within the mRNA population at a prevalence roughly proportional to band intensity. Thus the method simultaneously gives information about a large number of mRNAs, some of which are differentially expressed. Because of the separation method, the assay allows mRNAs of both high and low prevalence to be detected simultaneously (hence, it is much less subject to mRNA prevalence than methods that rely on random clone isolation or arbitrary-primer differential display and in this regard resembles a normalization) and is especially diagnostic of all or none differences in mRNA expression between tissue samples. Each mRNA appears only once (except for low level bleed through) unless it has alternative 3' ends.

Thus, a survey of patterns of mRNA expression could be performed with as many cDNA samples as conveniently fit on a gel and requiring only a relatively small amount of cDNA from each tissue. Each of the individual amplified PCR products can be excised from the separation gel and characterized further by rapid sequence analysis. The excised fragments can also be radiolabeled for use in probing a Northern blot or for *in situ* hybridization to verify the mRNA distribution and learn the size and prevalence of the corresponding mRNA; the probe can be used to screen a cDNA library to isolate clones for more reliable and complete sequence determination, since the PCR-derived material would be of partial length and no doubt be subject to PCR-introduced mutations. Such a program would allow systematic and rapid accumulation of data about expression ofmRNAs under several circumstances coupled with a small amount of single-pass sequence. Because it is more dependent upon a digital feature of the mRNA (its sequence) than upon its prevalence in a given tissue, the method should, in principal, allow for the accounting of all mRNAs (it should have closure) and, at any time during its application, should allow the portion of mRNAs that have been measured to be defined.

There are many ways this survey method could be applied to approach fundamental issues; in the neuroscience arena, for example, identification of genes whose expression is altered during neuronal development, in models of plasticity and regeneration, in response to chemical or electrophysiological challenges (neurotoxicity and long-term potentiation), in response to behavioral, viral, drug/alcohol paradigms, cell death/apoptosis, aging, pathological conditions, etc. In the long term, such an approach would lead to the recognition of regulated and regulatory genes along a multitude of investigative lines. As the databases of nucleic acid sequences expand the sequence- and length-based method of mRNA recognition described here may lend itself to assignment of band identity with species compiled in the databases, thus eliminating the necessity for excising PCR products and determining their sequences. This may lead to a high throughput mechanism by which investigators can examine the expression of all mRNAs and quickly recognize those that are regulated in response to a particular challenge.

CONCLUSIONS

We have examined PCR-based differential display methodologies with the goal of implementing a technology which enables the identification, cloning and sequencing of differentially expressed mRNAs in a high throughput manner. Our preliminary studies suggest that a method can be developed. Studies such as these will potentially provide a baseline for a technology that is certain to become, because of its digital approach to total gene expression , a major tool for identifying mRNAs differentially expressed in a multitude of biological paradigms.

ACKNOWLEDGEMENTS

We thank Frank Burton, Monica Carson and Frank Mercurio for participating in scientific discussions related to differential display technologies. We gratefully acknowledge the assistance of Linda Elder in the preparation of this manuscript. This work was supported in part by NIH grants NS22347 and GM32355.

REFERENCES

1. M.B. Dworkin and I.B. Dawid, Construction of a cloned library of expressed embryonic gene sequences from Xenopus laevis. *Dev. Biol.* 76:435 (1980).
2. J.D. Falk, H. Usui, and J.G. Sutcliffe, Identification of expressed sequences on human chromosome 9q32-34, *Proceed. Transcribed Seq. Workshop*, in press (1994).
3. H. Usui, J. Falk, A. Dopazo, L. de Lecea, M.G. Erlander, and J.G. Sutcliffe, Isolation of clones of rat striatum-specific mRNAs by directional tag PCR subtraction, *J. Neurosci.* In Press, (1994).
4. J.G.K. Williams, A.R. Kubelik, K.J. Livak, J.A. Rafalski, and S.V. Tingey, DNA polymorphisms amplified by arbitrary primers are useful as genetic markers, *Nucl. Acid Res.* 18:6531 (1990).
5. J. Welsh and M. McClelland, Fingerprinting genomes using PCR with arbitrary primers, *Nucl. Acids Res.* 18:7213 (1990).
6. J. Welsh and M. McClelland, Genomic fingerprinting using arbitrarily primed PCR and a matrix of pairwise combinations of primers, *Nucl. Acids Res.* 19:5275 (1991).
7. S.R. Woodward, J. Sudweeks, and C. Teuscher, Random sequence oligonucleotide primers detect polymorphic DNA products which segregate in inbred strains of mice, *Mammal. Genome* 3:73 (1992).
8. J.H. Nadeau, H.G. Bedigian, G. Bouchard, T. Denial, M. Kosowsky, R. Norberg, S. Pugh, E. Sargeant, R. Turner, and B. Paigen, Multilocus markers for mouse genome analysis: PCR amplification based on single primers of arbitrary nucleotide sequence, *Mammal. Genome* 3:55 (1992).
9. P. Liang, and A.B. Pardee, Differential display of eukaryotic messenger RNA by means of the polymerase chain reaction, *Science* 257:967 (1992).
10. J. Welsh, K. Chada, S.S. Dalal, R. Cheng, D. Ralph, and M. McClelland, Arbitrarily primed PCR fingerprinting of RNA, *Nucl. Acids Res.* 20:4965 (1992).
11. P. Liang, L. Averboukh, and A.B. Pardee, Distribution and cloning of eukaryotic mRNAs by means of differential display: refinements and optimization, *Nucl. Acids Res.* 21:3269 (1993).

259 HUMAN BRAIN EXPRESSED SEQUENCE TAGS (ESTs): CHROMOSOME LOCALIZATION, SUBREGIONAL ASSIGNMENT, AND SEQUENCE ANALYSIS

Donna R. Maglott[1], A. Scott Durkin[2], and William C. Nierman[2]

Departments of [1]Bioinformatics and [2]Molecular Biology
American *Type Culture* Collection
12301 Parklawn Drive
Rockville, MD 20852-1776

ABSTRACT

Using PCR with automated product analysis, 259 human brain cDNA sequences have been assigned to individual human chromosomes, 99 with subregional localizations. Primers were designed from single-pass cDNA sequences (expressed sequence tags or ESTs) and tested for specific amplification from human genomic DNA. Primers were then used in PCR reactions with DNA from somatic cell hybrid mapping panels as templates, often with multiplexing. Amplification products were identified using automated fluorescence detection and chromosomal assignments were made by discordancy analysis.

Several mapped ESTs were similar to previously reported sequences, including other ESTs. These matches have provided information on the chromosome assignments of previously unmapped genes or members of known gene families, and have also identified unknown genes that appear to be commonly expressed.

INTRODUCTION

Genes expressed in human tissues are rapidly being identified by partially sequencing cDNA clones (1-10). The number of these sequences that have been generated (16,227 human ESTs in dbEST version 1.28) far exceeds the number that have been mapped. Several groups are using PCR and somatic cell hybrid mapping panels to localize the sequences to individual human chromosomes (3,6,7,11-14). Other approaches to identifying and mapping expressed sequences include fluorescence *in situ* hybridization of cDNA clones to metaphase spreads of human chromosomes (15), hybridization selection of cDNAs using genomic lambda, cosmid or YAC clones of known map position (16), and trapping exons from genomic clones (17).

The focus of our approach is to expedite mapping ESTs by multiplexing PCR reactions and by automating the sensitive detection of products using an ABI 373A sequencer configured with 672 GENESCAN software (11,14,18-19). The instrumentation

Identification of Transcribed Sequences, Edited by
U. Hochgeschwender and K. Gardiner, Plenum Press, New York, 1994

permits high resolution of multiple products separated by denaturing polyacrylamide gel electrophoresis in combination with the four color discrimination of the fluorescence detection system. Because different primers can be labeled with different fluorescent dyes, products can be identified by a combination of size and dye label. To save time and reduce cost, multiple pairs of PCR primers have been combined in single PCR reactions (multiplexing), and products from several PCR reactions have been pooled in a single lane for electrophoretic analysis.

MATERIALS AND METHODS

EST sequences were obtained in electronic format and primers were designed that met the criteria of 50% GC content, primer Tm of 55-59°C, and product length of 80-150 bp (11, 14). To expedite processing, all source sequence files were analyzed without regard to the correspondence between the predicted PCR product and any potential coding sequence. Thus, for the ESTs that were mapped because of similarities to previously reported sequences, the amplified products were not restricted to lie within the region of matching.

One of each primer pair was synthesized with 5' Aminolink II and conjugated to a fluorescent dye using procedures provided by ABI (20). Primer pairs were first tested in three single-template (uniplex) PCR reactions (25 cycles) with total human, total mouse, and total hamster genomic DNA using the following standard conditions: each 15 μl PCR reaction contained 50 ng template DNA, 40 ng of each PCR primer, 0.6 U Amplitaq AS, 1.5 mM MgCl$_2$, and 200 μM of each dNTP. The thermal profile and gel loading procedure are described in Durkin *et al.*, 1992 (11). Amplification products were analyzed by denaturing polyacrylamide gel electrophoresis using an ABI 373A sequencer and GENESCAN software. The products were detected by the fluorescent label conjugated to one of the primers, with the size determined automatically by the GENESCAN software based on internal size standards (GS-2500, ABI). In multiplexed reactions (using somatic cell hybrid mapping panels only), the reaction conditions were not changed except for the addition of 40 ng of each additional primer.

Templates were derived from several rodent-human cell hybrid mapping panels and a hamster-mouse cell hybrid mapping panel: NIGMS human/rodent somatic cell hybrid mapping panels 1 and 2, Coriell Institute for Medical Research, Camden, NJ (21,22); PCRable DNA from BIOS Corporation, New Haven, CT (23); a chromosome 3 panel (NIGMS Mutant Cell Repository, Coriell Institute for Medical Research, Camden, NJ); a chromosome 6 panel (24,25); a chromosome 10 panel (26-29); a chromosome 11 panel (30,31); a chromosome 15 panel (32,33); a chromosome 18 panel (34,35); an X chromosome panel (36-38); and a mouse/hamster panel (39). The NIGMS panel 1 was supplemented with a human 21-only cell line (GM10323 from NIGMS panel 2).

For sequences mapping to chromosomes 4, 7, 8, and 19, the EST identifications, the primer sequences, and/or cDNA clones were sent to collaborating laboratories for higher resolution mapping. In the case of chromosome 8, DNAs from cell lines CL17, 9HL10, 1HL33, 1SHL3, 1HL12, 20XPO435-2, XVIII-23Ha, MC-2F, 3;8/4-1, GM-1F, MC-1E, and NH-1B were used (40,41).

ESTs were assigned to a human chromosome when the discordancy value was less than or equal to 9% for the assigned chromosome. Cell lines reported as containing a particular chromosome in fewer than 12% of metaphases were not included in the discordancy analysis. Each primer pair was tested on one of the mapping panels, and if that panel did not permit resolution by at least 2 discordant results from the chromosome with the next fewest discordancies, selected cell lines from the other panel were used. For subregional localizations based on small mapping panels (fewer than 10 cell lines), only results with no discordancies were reported. Those panels were not sufficiently redundant to require the next most likely assignment to have 2 or more discordant values.

ESTs mapped in this study were retrieved from dbEST (42) using the NCBI e-mail server (search date by dbEST 19-Oct-1993). ESTs showing a nucleotide match with a BLAST (43) score greater than 200 or a protein match with a BLAST score greater than 50 were investigated in more detail. The sequences were submitted for BLASTN (nucleotide sequence matching) and BLASTX (translation/protein sequence matching) analysis using the NCBI e-mail server to determine whether the sequence used for PCR amplification lay within the region of overlap between the EST and the matching sequence. The mask used by dbEST, the BLAST score, and the Poisson probability values for the best matched sequences were tabulated (Table 1). In some cases where the best match was a non-human sequence but the human sequence record existed, both matches were included.

The sequences of the primers used for mapping and the chromosome localizations were submitted to the Genome Data Base (GDB) for assignment of D-segment numbers.

RESULTS AND DISCUSSION

Mapping and Multiplexing

Chromosomal assignments, EST identifiers, ATCC numbers of the clones from which the sequences were derived, D-segment identifiers, and information on matching sequences are presented in Table 1. Except for EST00858, EST00884, EST01699, and EST01683, all primers directed amplification of a product from human genomic DNA that was within 5 base pairs of the size predicted by the cDNA sequence. For EST00858, the product was 450 bp instead of the predicted 143 bp. When the genomic amplification product was sequenced, it was apparent that the primers (nt 93-110 and 214-235 of the EST) bracketed an intron. This intron occurs after nt 190 of the EST record. For EST00884, the primers correspond to positions 117-136 (exon 22) and 478-499 (exon 23) of GenBank accession X12923 for an Na+, K+-activated adenosine triphosphase (86). The primers apparently amplified across intron 22 to yield a product of 382 bp. Because EST00844 maps to chromosome 19, these data suggest this EST is the ATP1A3 isoform, which has already been mapped to 19q13.1-q13.2 (94). Similarly, the primers for EST01699 generated a product containing an intron of 100 bp near nt 43 of the EST, and the primers for EST01683 generated a product including an intron of 140 bp near nt 185. The presence of these introns in the genomic sequence was confirmed by sequence analysis (data not shown).

Some primers amplify a reproducible product from rodent templates. The primers derived from EST00903 (mapping to human chromosome 1) amplified products of similar sizes from the human and mouse genomes and were therefore used against a Chinese hamster X mouse hybrid cell mapping panel (39). There were no discordancies with mouse chromosome 4. This is consistent with a syntenic relationship between human chromosome 1p and mouse chromosome 4 (95), and therefore suggests that EST00903 may be more precisely localized to human chromosome 1p (Table 1).

It is interesting that primers derived from EST00903 were successful in amplifying a mappable product in the mouse genome. The mouse product is presumed to be related, because the product size (133 bp) was similar to that in the human (128 bp), and because of the confirmation of the comparative map. EST00903 did not match a sequence database record, although a GRAIL (96) analysis suggested a high probability of a coding region in the sequence used for amplification. It is thus likely the primers detected a conserved coding region in both genomes.

Early in the project, only "uniplex" reactions were used with the mapping panel templates. More recently, we have successfully multiplexed these reactions and consider an EST to have been mapped by multiplexing when more than 50% of the mapping PCR reactions were multiplexed. Primer pairs have been combined for multiplexing based on the predicted resolution of all amplification products--products differing by at least 10 bp if the primers are labeled with the same fluorescent dye or at least 5 bp if different dyes are used.

Of the 162 mapped sequences for which multiplexing was attempted, 97 of 108 (90%) ESTs for which duplex reactions were used, and 43 of 59 (73%) ESTs for which triplex reactions were used, were mapped as a result of those reactions. The 5 extra duplex reactions resulted when 5 sequences that failed to be mapped in triplex reactions were mapped in duplex reactions. No attempt was made to identify causes for multiplexing failures, but the rate of success has justified making duplex mapping reactions a standard component of our protocol.

Mapped ESTs Matching Sequence Database Records

Many ESTs mapped in this project match sequence database records (Table 1). To evaluate the significance of these matches, the position of the primers relative to the matching region (In), the BLAST scores and the Poisson probability values of the EST/sequence record match were determined. Those ESTs for which at least one primer was derived from a region of the EST that matched the target sequence are indicated by a (+) in Table 1. Because of the possibility of chimerism in the clones from which the ESTs were derived (97,98), the genes matched by the ESTs were assigned to chromosomes with more confidence when the primer sequences lay within the region of sequence matching.

In some cases, the gene product identified by the sequence match had already been mapped. In all cases when the sequence match was sufficiently stringent, the map location determined from the EST was consistent with the previous report: EST02446, neutral amino acid transporter, chromosome 2 (51); EST01471, microtubule associated protein 1B, chromosome 5q13, (99); EST02092, protein tyrosine phosphatase beta or zeta, 7q31-q32 (100); EST01806, prohibition, chromosome 17 (101); IB781, myelin basic protein, 18q22-qter (102); EST00884, Na, K-ATPase, chromosome 19q (94). This evidence strongly supports the validity of assigning unknown genes to human chromosomes based on the EST sequence matches.

For at least 8 ESTs, even though the actual sequence used to design primers came from the EST, the percent identity of the match was such that the primers could have been derived from the matched sequence instead. In these cases, therefore, mapping the ESTs to a chromosome can also be considered to have mapped the known gene product: rolipram-sensitive cAMP phosphodiesterase, chromosome 1; protein phosphatase 2Aß, chromosome 4; alpha-catenin, chromosome 5; the ELE1 oncogene, chromosome 10q11.2 or q22.1-q23; MXI1 protein, chromosome 10q24-qter; ribosomal protein L18a homologue, chromosome 14; ribosomal protein L3, chromosome 17; and moesin, Xq11.2-q12 (103). There were also ESTs mapped in this laboratory that were less closely related to human sequence records or that were closely related to non-human sequence records. Table 1 describes these matches, which can therefore be considered to identify human counterparts of known gene products, or members of known gene families. Examples of these include membrane proteins, translation-associated proteins, structural proteins, and enzymes.

These analyses have also identified ESTs for which the only database match is another single-pass sequence (Table 1). Since most of the libraries from which these were derived were not normalized, these ESTs may be considered to represent moderately expressed, unknown genes.

In summary, these data demonstrate many of the strengths of the EST approach to genomic analysis (104). Single pass sequence information has been sufficient to design PCR primers useful for assigning cDNA sequences to human chromosomes. With template DNA from additional mapping panels, those primers have also been used to make subregional localizations, or even to map a related sequence in another genome. When the

Table 1. Summary of ESTs mapped.

Map location	EST	ATCC#	D seg	Other EST	Sequence	Gene Product (HGM symbol)	Mask	In	Blast Score	Poisson Prob.	Ref.
1	EST00100	61888	D1S269E				no				
1	EST00503	77651	D1S299E				no				
1	EST00552	77718	D1S314E				no				
1	EST00745	77897	D1S325E	IB1415	GB:X53529	dog signal sequence receptor	no	+	303	2.4e-16	44
1	EST00751	77995	D1S326E		GB:X53591	dog glycoprotein gp25H	no	+	347	4.6e-20	45
1	EST00767	78014	D1S327E				no				
1	EST00883	78158	D1S316E				no				
1	EST00885	78160	D1S345E				no				
1p	EST00903	78181	D1S328E				no				
1	EST00904	78183	D1S329E				no				
1	EST00919	78203	D1S356E		PIR:A35308	bovine ARPP-16	no	-	98	2.4e-7	46
1	EST01041	78364	D1S330E		PIR:B35308	bovine ARPP-19	no	-	98	3.1e-7	46
	EST01699*	78532	D1S357E		GP:L04538	mouse amyloid precursor-like protein	no	+	277	1.7e-33	47
	EST01764	78742	D1S341E		PIR:A36427	chicken lamin B receptor	no	+	204	5.8e-23	48
	EST01919	78981	D1S331E				no				
	EST01950	81011	D1S342E	s16f03			no				
	EST01988	81049	D1S343E				no				
	EST01996	81057	D1S362E				no				
	EST01997	81058	D1S344E				no				
	EST02015	81076	D1S346E				no				
	EST02064	81125	D1S347E				no				
	EST02107	81168	D1S348E				no				
	EST02139	81199	D1S367E				no				
	EST02164	81224	D1S350E				no				
	EST02218	81277	D1S351E				no				
	EST02228	81286	D1S352E				no				
	EST02239	81297	D1S353E				no				
	EST02245	81303	D1S354E				no				
	EST02270	81328	D1S530E				no				
	EST02301	81357	D1S527E	EST02288, FB10G10			no				
	EST02308	81364	D1S528E				no				
1	EST02311	81367	D1S529E				no				
1	EST02447	81500	D1S349E		GB:M97515	human rolipram-sensitive cAMP phosphodiesterase	no	+	796	3.6e-143	49
1	EST02679	81728	D1S532E				no				
1	EST06713	85318	D2S217E				no				
2	EST00587	77762	D2S204E				no				
2	EST00707	77927	D2S204E				no				
2	EST00876	78147	D2S212E				no				

Map location	EST	ATCC#	D seg	Other EST	Sequence	Gene Product (HGM symbol)	Mask	In	Blast Score	Poisson Prob.	Ref.
2	EST00892	78166	D2S236E				no				
2	EST00915	78199	D2S232E	EST01013, EST01030			no				
2	EST01870	78933	D2S214E				no				
2	EST01880	78943	D2S215E				no				
2	EST01911	78974	D2S216E				no				
2	EST01972	81033	D2S233E				no				
2	EST02013	81074	D2S234E				no				
2	EST02044	81105	D2S235E				no				
2	EST02132	81192	D2S247E				no				
2	EST02178	81238	D2S237E				no				
2	EST02240	81298	D2S241E				no				
2	EST02303	81359	D2S420E				no				
2	EST02446	81499	D2S421E		GB:L24595	human alanine/serine/cysteine/threonine transporter	no	+	1405	9.3e-110	50
					GB:L19444	human neutral amino acid transporter	no	+	1405	9.3e-110	51
2,10	EST02530	81582	D2F118S1E / D10F118S2E		GB:L13689	human proto-oncogene (BMI1)	no	+	747	1.2e-55	52
3q27(28)-qter	EST00094	37920	D3S1342E	EST00857			no				
3pter-p24.2	EST00476	77618	D3S1343E				no				
3p21.2-p13	EST00514	77668	D3S1356E				no				
3q26-q27(28)	EST00605	77785	D3S1355E	EST00094	GB:L09675	human glucose transporter 2 (GLUT2)	no	-	362	2.5e-21	53
3q27(28)-qter	EST00857	78122	D3S1342E				no				
3p21.2-p13	EST01894	78957	D3S1357E				no				
3q21-q25	EST01990	81051	D3S1492E				no				
3q11-q13.2	EST02067	81128	D3S1493E				no				
3pter-p24.2	EST02072	81133	D3S1506E				no				
3	EST02093	81154	D3S1494E				no				
3p24.2-p21.1	EST02101	81162	D3S1495E				no				
3q26-q27(28)	EST02155	81215	D3S1500E				no				
3q21-q25	EST02196	81255	D3S1501E				no				
3q11-q13.2	EST02229	81287	D3S1502E				no				
3q27(28)-qter	EST02236	81294	D3S1503E				no				
3p21.2-p13	EST02284	81340	D3S1751E				no				
3pter-p24.2	EST02307	81363	D3S1750E				no				
3p21.2-p13	EST06732	85336	D3S1748E				no				
4q21-q25	EST00038	77690	D4S814E		GB:X63465	human smg GDS	no	+	618	8.0e-44	54
4q12-q21	EST00530	77998	D4S813E				no				
4p16.1-p15.1	EST00754	78121	D4S808E				no				
4q25-q35	EST00863	78365	D4S815E		GB:M64930	human protein phosphatase 2A beta	no	+	842	1.8e-153	55
4p16.1	EST01650	78936	D4S816E	EST00945			no				
4	EST01873	81037	D4S1495E				no				
4q25-q35	EST01976	81190	D4S1496E				no				
4q25-q35	EST02129	81204	D4S1497E				no				
	EST02144										

Map location	EST	ATCC#	D seg	Other EST	Sequence	Gene Product (HGM symbol)	Mask	In	Blast Score	Poisson Prob.	Ref.
4q25-q35	EST02147	81207	D4S1492E				no				
4q35-qter	EST02370	81423	D4S1490E	AAACMEA	PIR:S25063	D. melanogaster diff6 homologue	no	+	139	4.6e-26	56
4	EST02533	81585	D4S1491E				no				
5	EST00104	61896					no				
5	EST00478	77620	D5S521E		GB:M62510	Alu repeat sequence	no		168	7.5e-05	
5	EST00488	77633	D5S516E				no				
5q	EST00835	78093	D5S517E		GB:D14705	human alpha-catenin (CTNNA1)	no	+	1592	3.3e-125	57
5	EST00911	78194	D5S566E				no				
5	EST00922	78206	D5S569E				no				
5	EST01471	77689	D5S522E		GB:X16623	rat neuraxin	no	+	1298	1.3e-101	58
5	EST01744	78681	D5S523E		GB:J03534	bovine nicotinamide nucleotide trans-hydrogenase	no	+	1027	5.4e-79	59
5	EST01943	81004	D5S562E		SP:P23960	Alu class B	no	+	56	0.4	
5	EST01977	81038	D5S563E	EST02653			no				
5	EST01986	81047	D5S572E	HUMRTPGER			mer4				
5	EST01998	81059	D5S573E		GB:J05158	human carboxypeptidase N	no	+	255	3.8e-12	60
5	EST02033	81094	D5S564E				Alu				
5	EST02054	81115	D5S565E				no				
5	EST02204	81263	D5S567E				no				
5	EST02222	81281	D5S568E				no				
6q21-qter	EST00770	78018	D6S350E				no				
6q21-qter	EST00833	78091	D6S345E				no				
6p21.3-21.1	EST00874	78145	D6S346E				no				
6	EST00914	78198	D6S371E				no				
6	EST01915	78978	D6S351E				no				
6q21-qter	EST01920	78982	D6S352E				no				
6q21-qter	EST01975	81036	D6S374E				no				
6cen-q14	EST02029	81090	D6S367E				no				
6	EST02060	81121	D6S368E				L1				
6q21-qter	EST02111	81172	D6S369E				no				
6q14-q21	EST02188	81247	D6S370E				no				
6pter-p23	EST02205	81264	D6S484E				no				
6	EST02280	81336	D6S485E				no				
7cen-q21.1	EST00548	77712	D7S551E	EST02419			no				
7pter-cen	EST00601	77779	D7S552E				no				
7q22-q34	EST00654	77856	D7S556E				no				
7cen-q21.1	EST00838	78096	D7S553E				no				
7q22-q34	EST01888	78951	D7S557E				Alu				
7q35-q36	EST02091	81152	D7S600E		GB:M93426	human protein tyrosine phosphatase (PTPRZ)	no	+	1725	3.5e-137	61
7q22-q34	EST02092	81153	D7S601E		GB:M58318	human ala gene	no	-	316	2.9e-17	62
7q22-q34	EST02113	81174	D7S602E		GB:M76425	human intron 2 Alu repetitive element	no		287	1.3e-15	
7pter-cen	EST02120	81181	D7S603E				no				
7q	EST02143	81203	D7S610E				no				

Map location	EST	ATCC#	D seg	Other EST	Sequence	Gene Product (HGM symbol)	Mask	In	Blast Score	Poisson Prob.	Ref.
7pter-cen	EST02207	81266	D7S604E				no				
7q35-q36	EST02224	81283	D7S605E				no				
7q22-q34	EST02268	81326	D7S810E				no				
8q22	EST00582	77752	D8S361E				no				
8q24.13-qter	EST00614	77799	D8S295E				no				
8q13-q22.1	EST00680	77890	D8S340E				no				
8	EST02105	81166	D8S362E				no				
8	EST02157	81217	D8S364E				Alu				
8	EST02297	81353	D8S581E				no				
9	EST00510	77662	D9S191E				no				
9	EST00895	78170	D9S220E				no				
9	EST00917	78201	D9S221E				no				
9	EST01906	78969	D9S204E				no				
9	EST02055	81116	D9S218E		GB:M94172	human N-type calcium channel, alpha-1	no	+	511	6.4e-36	63
9	EST02125	81186	D9S219E				no				
9	EST06927	85519	D9S308E				no				
10	EST00577	77751	D10S409E	EST02128			no				
10q11.2-q22.1	EST00684	77895	D10S406E				no				
10q11.2-q22.1	EST00695	77906	D10S407E				no				
10q11.2-q22.1	EST00856	78121	D10S250E				no				
10pter-cen	EST00916	78200	D10S488E				no				
10q24-qter	EST01876	78939	D10S408E				no				
10q11.2 or q22.1-q23	EST01945	81006	D10S474E				no				
10q11.2 or q22.1-q23	EST01947	81008	D10S490E		GB:X71413	human ELE1	no	+	1053	1.0e-81	64
10pter-cen	EST01958	81019	D10S475E		SP:P19146	yeast ARF2 ADP-ribosylation factor	no	+	150	2.8e-15	65
10	EST01982	81043	D10S476E				no				
10q11.2 or q22.1-q23	EST02019	81080	D10S477E		GB:M2934	human *Alu* repeat	no		117	0.77	66
10q24-qter	EST02041	81102	D10S478E		GB:L07648	human MXI1	no	+	1597	1.1e-125	
10q24-qter	EST02043	81104	D10S479E				no				
10	EST02089	81150	D10S480E				no				
10pter-cen	EST02097	81158	D10S481E				no				
10pter-cen	EST02099	81160	D10S482E				no				
10q11.2 or q22.1-q23	EST02122	81183	D10S483E				Alu				
10q11.2 or q22.1-q23	EST02127	81188	D10S500E				no				
10, 14	EST02146	81206	D10F117S1E, D14F117S2E				no				
10pter-cen	EST02191	81250	D10S485E		GB:M31178	rat calbindin D28	no	+	320	1.5e-17	67
					GP:L09190	human trichohyalin	no	+	62	0.0002	68
					GP:218361	ovine trichohyalin	*Alu*	+	59	8.8e-06	69
10q23-q24	EST02269	81327	D10S669E								

Map location	EST	ATCC#	D seg	Other EST	Sequence	Gene Product (HGM symbol)	Mask	In	Blast Score	Poisson Prob.	Ref.
11	EST00109	61906	D11S955E				no				
11p15.3-p13	EST00501	77649	D11S973E				no				
11p15.3-p13	EST00532	77693	D11S981E				no				
11	EST00547	77711	D11S979E				no				
11	EST00776	78027	D11S980E				no				
11	EST00901	78178	D11S1271E	EST00861			no				
11	EST00926	78212	D11S1272E				no				
11	EST02084	81145	D11S1273E				*Alu*				
11	EST02088	81149	D11S1274E				no				
11	EST02145	81205	D11S1275E				no				
11q11-q13.4	EST02209	81268	D11S1381E	EST01499			*Alu*				
11p15.3-p13	EST02264	81322	D11S1380E				no				
11q13.4-q23.2	EST02300	81356	D11S1379E	EST01301, EST07054			no				
11q13.4-q23.2	IB753	86323			GB:M73512	bovine cGMP-stimulated cyclic nucleotide phosphodiesterase	no	+	655	5.9e-46	70
12	EST00867	78135	D12S275E				no				
12	EST00899	78174	D12S180E				no				
12	EST00908	78189	D12S280E				no				
12	EST01973	81034	D12S282E				no				
12	EST02006	81067	D12S276E				*Alu*				
12	EST02011	81072	D12S277E				no				
12	EST02066	81127	D12S278E				no				
12	EST02076	81137	D12S279E				no				
13	EST00527	77684	D13S184E				no				
13	EST00537	77698	D13S224E				no				
13	EST00920	78204	D13S237E		GB:X15723	human fur	no	-	198	2.3e-06	71
13	EST01889	78952	D13S229E				no				
13	EST01896	78959	D13S230E				no				
13	EST02037	81098	D13S236E				no				
14	EST00795	78053	D14S103E		GB:M87634	rat brain factor 1	no	+	832	9.7e-63	72
14	EST00845	78107	D14S102E				no				
14	EST00902	78180	D14S104E		GB:L05093	human ribosomal protein L18a homolog	no	+	859	1.5e-67	73
14	EST01583	78131	D14S105E	tb031, S445	GB:X05216	Xenopus laevis ribosomal protein L1a	no	+	403	5.3e-25	74
14	EST01627	78285	D14S106E	a33	GB:X05217	Xenopus laevis ribosomal protein L1b	no	+	412	8.8e-26	74
14	EST01868	78931	D14S107E				no				
14	EST01875	78938	D14S108E				no				
14	EST02182	81241	D14S248E				no				
15q15(21.1)-q22	EST00780	78036	D15S167E				no				
15pter-q15(21.1)	EST00918	78202	D15S179E				no				
15q22-qter	EST01678	78455	D15S168E				no				
15q15(21.1)-q22	EST01683*	78473	D15S180E	EST01865	GB:S38337	mouse milk fat globule protein	no	+	359	5.0e-21	75
15pter-q15(21.1)	EST02279	81335	D15S224E		GB:M83196	rat microtubule-associated protein 1A	no	+	1218	7.0e-95	76
15q22-qter	EST02292	81348	D15S223E				no				

281

Map location	EST	ATCC#	D seg	Other EST	Sequence	Gene Product (HGM symbol)	Mask	In	Blast Score	Poisson Prob.	Ref.
15q22-qter	EST02550	81602	D15S177E		SP:P03360	avian reticuloendotheliosis virus pol	no	+	141	8.8e-14	77
16	EST00566	77737	D15S445E				no				
16	EST00831	78089	D16S442E				no				
16	EST00889	78163	D16S469E	EST01281, IB1433	GB:M80244	human E16 mRNA	no	+	1049	7.9e-81	78
16	EST01953	81014	D16S471E				mer13				
16	EST01954	81015	D16S472E				no				
16	EST01969	81030	D16S473E	EST01990			no				
16	EST02246	81304	D15S470E				no				
16	EST06702	85307	D16S532E				no				
17	EST00483	77628	D17S853E	EST00972	SP:P23960	Alu class B	no				
17	EST00675	77784	D17S864E	FIA015, AAEICW H28D06	PIR:S21636	rice GOS2 protein	no	-	129	2.3e-12	79
17	EST00854	78118	D17S851E				no				
17	EST00869	78137	D17S852E	EST00949	GB:M24105 (VAMP-2)	rat vesicle associated membrane protein	no	+	239	2.3e-11	80
17	EST01667	78408	D17S891E		GB:M90054	human ribosomal protein L3 (RPL3)	no	+	1520	1.7e-120	81
17	EST01806	78834	D17S895E		GB:S85655	human prohibitin (PHB)	no	+	1820	3.3e-146	82
17	EST02042	81103	D17S896E				no				
17	EST02085	81146	D17S897E				no				
17	EST02119	81180	D17S892E				Alu				
17	EST02254	81312	D17S893E				no				
17	EST02287	81343	D17S961E				no				
17	EST02291	81347	D17S962E	FB10F7	GB:L05367	human oligodendrocyte myelin glycoprotein	no	+	286	5.4e-15	83
18q12.3-q21.1	EST00542	77704	D18S105E				no				
18p11.21-p11.1	EST00775	78026	D18S360E				no				
18	EST00893	78167	D18S361E				no				
18pter-p11.21	EST00906	78187	D18S362E				no				
18q21.1-q21.3	EST01784	78787	D18S373E	EST00250	SP:P10723	Brugia malayi 63 kDa antigen	no	+	275	7.4e-34	84
18p11.1-q11.2	EST02112	81173	D18S374E	IB742, IB1444, EST01814, IB934, IB941			no				
18q23-qter	IB781	86344	D18S540E				no				
19p13.3	EST00708	77928	D19S382E		GB:M13577	human myelin basic protein (MBP)	no	+	1891	3.0e-151	85
19q13.3-q13.4	EST00875	78146	D19S264E				no				
19	EST00884	78159	D19S265E	HE0022, HEI028	GB:X12923	human Na K ATPase (ATP1A3)	no	+	456	9.3e-30	86
19q13.1-q13.2	EST00894	78169	D19S383E				no				
19q13.1-q13.2	EST01791	78798	D19S387E		GB:X54938	human inositol 1,4,5-triphosphate 3-kinase	no	+	487	5.1e-32	87
19	EST02086	81147	D19S390E				Alu				
19	EST02247	81305	D19S388E				no				
20	EST00113	61922	D20S125E				no				
20	EST00890	78164	D20S143E				no				
20	EST00909	78190	D20S134E				no				
20	EST00925	78210	D20S140E	EST00669, EST01078			no				
20	EST01992	81053	D20S141E				no				

Map location	EST	ATCC#	D seg	Other EST	Sequence	Gene Product (HGM symbol)	Mask	In	Blast Score	Poisson Prob.	Ref.
20	EST02017	81078	D20S142E				no				
20	IB820	86369	D20S203E	EST01220			no				
20	EST02285	81341	D20S205E				no				
21	EST00541	77703	D21S363E				no				
21	EST00591	77767	D21S413E				no				
21	EST02022	81083	D21S1236E				no				
22	EST00101	61890									
22	EST00924	78208	D22S412E		GB:M58561	bovine rhodanese	no		807	8.6e-61	88
22	EST01601	78192	D22S317E		GB:S61764	human cyanide sulfurtransferase	no	-	461	1.0e-29	89
22	EST02116	81177	D22S413E				Alu				
22	EST02150	81210	D22S410E				no				
22	EST02219	81278	D22S411E				no				
Xq26-q27.1	EST00574	77748	DXS1119E				no				
Xpter-p21.3	EST00737	77974	DXS1115E				no				
Xp21.3-p11	EST00858	78124					no				
Xp21.3-p11	EST00887	78162	DXS1116E		GP:M33553	rat spot 14	no	+	89	3.1e-05	90
Xq11.2-q12	EST00896	78171	DXS1117E		GB:M69066	human moesin (MSN)	no	+	433	7.1e-113	91
Xpter-p21.3	EST01879	78942	DXS1118E		GB:M18332	rat protein kinase C zeta	no	+	954	5.6e-73	92
Xq21-q24	EST02087	81148	DXS1179E		GB:Z15108	human protein kinase C zeta	no	+	931	4.9e-71	93

Legend: **Other EST**: other ESTs with a sequence match to the mapped EST; **Sequence**: identifier for the sequence record matched: GB = GenBank; GB = GenBank, PIR = Protein Identification Resource, GP = GenPept; SP = SwissProt; **Mask** - mask used by dbEST in sequence analysis; EST* - ESTs demonstrated to span an intron.

EST sequence matches previous sequence database records, the chromosome assignments of the EST can be used to make preliminary assignments of the human gene to a human chromosome. When the human EST matches a sequence record from another genome, and when the gene in that other genome has already been mapped, then the human assignment can be used to expand or confirm comparative maps.

ACKNOWLEDGEMENTS

We wish to acknowledge M. Graham, J. Mao, and M. Lee for expert technical assistance; M. Adams, The Institute for Genomic Research, and J. Sikela, University of Colorado for sharing EST sequence information; NCBI for IRX access to Medline and sequence databases; Drs. K. H. Grzeschik, D. Ledbetter, T. Mohandas, D. Nelson, J. Overhauser, B. Ponder, R. Schultz, A. Tunnacliffe, M. Wapenaar, and J. J. Wasmuth for providing valuable cell lines; our collaborators R. Myers and R. Goold (chromosome 4), H. Zoghbi (chromosomes 6 and X), L. Tsui, S. Scherer, K. H. Grzeschik and J. Kunz (chromosome 7), J. Parrish (chromosomes 8 and X), C. Jones (chromosome 11), G. Lennon (chromosome 19), D. Nelson (X chromosome) for sub-regional assignments of ESTs on the indicated chromosomes; and K. Meyer, ATCC, for laboratory database implementation, including automation of discordancy analysis. Supported by DE-FG05-91ER61232 from the Department of Energy.

REFERENCES

1. M.D. Adams, J.M. Kelley, J.D. Gocayne, M. Dubnick, M.H. Polymeropoulos, H. Xiao, C.R. Merril, A. Wu, O. Olde, R.F. Moreno, A.R. Kerlavage, W.R. McCombie, and J.C. Venter, Complementary DNA sequencing: expressed sequence tags and human genome project, *Science* 252:1651 (1991).
2. C. Hoog, Isolation of a large number of novel mammalian genes by a differential cDNA library screening strategy, *Nucl. Acids Res.* 19:6123 (1991).
3. A.S. Wilcox, A.S. Khan, J.A. Hopkins, and J.M. Sikela, Use of 3' untranslated sequences of human cDNAs for rapid chromosome assignment and conversion to STSs: implications for an expression map of the genome, *Nucl. Acids Res.* 19:1837 (1991).
4. M.D. Adams, M. Dubnick, A.R. Kerlavage, R. Moreno, J.M. Kelley, T.R. Utterback, J.W. Nagle, C. Fields, and J.C. Venter, Sequence identification of 2,375 human brain genes, *Nature* 355:632 (1992).
5. K. Okubo, N. Hori, R. Matoba, T. Niiyama, A. Fukushima, Y. Kojima Y., and K. Matsubara, Large scale cDNA sequencing for analysis of quantitative and qualitative aspects of gene expression, *Nature Genetics* 2:173 (1992).
6. L. Gieser and A. Swaroop, Expressed sequence tags and chromosomal localization of cDNA clones from a subtracted retinal pigment epithelium library, *Genomics* 13:873 (1992).
7. A.S. Khan, A.S. Wilcox, M.H. Polymeropoulos, J.A. Hopkins, T.J. Stevens, M. Robinson, A.K. Orpana, and J.M. Sikela, Single pass sequencing and physical and genetic mapping of human brain cDNAs, *Nature Genetics* 2:180 (1992).
8. M.D. Adams, A.R. Kerlavage, C. Fields, and J.C. Venter, 3400 expressed sequence tags identify diversity of transcripts from human brain, *Nature Genetics* 4:256 (1993).
9. M.D. Adams, M.B. Soares, A.R. Kerlavage, C.F. Fields, J.C. and Venter, Rapid cDNA sequencing (expressed sequence tags) from a directionally cloned human infant brain cDNA library, *Nature Genetics* 4:373 (1993).
10. J. Takeda, H. Yano, S. Eng, Y. Zeng, and G.I. Bell, A molecular inventory of human pancreatic islets: sequence analysis of 1000 cDNA clones, *Hum. Mol. Genet.* 2:1793 (1993).
11. A.S. Durkin, D.R. Maglott, and W.C. Nierman, Chromosomal assignment of 39 human brain expressed sequence tags (ESTs) by analyzing fluorescently-labeled PCR products from hybrid cell panels, *Genomics* 14:808 (1992).
12. M.H. Polymeropoulos, H. Xiao, A. Glodek, M. Gorski, M.D. Adams, R.F. Moreno, M.G. Fitzgerald, J.C. Venter, and C.R. Merril, Chromosomal assignment of 46 brain cDNAs, *Genomics* 12:492 (1992).

13. M.H. Polymeropoulos, H. Xiao, J.M. Sikela, M. Adams, J.C. Venter, and C.R. Merril, Chromosomal distribution of 320 genes from a brain cDNA library, *Nature Genetics* 4:381 (1993).

14. A.S. Durkin, W.C. Nierman, H. Zoghbi, C. Jones, C.A. Kozak and D.R. Maglott, Chromosomal Assignment of Human Brain Expressed Sequence Tags (ESTs) by Analyzing Fluorescently-labeled PCR Products from Hybrid Cell Panels, *Cytogenet. Cell Genet.* 65:86 (1994).

15. W.J. Wood, A.A. Thompson, J. Korenberg, X-N. Chen, W. May, R. Wall, and C.T. Denny, Isolation and chromosomal mapping of the human immunoglobin-associated B29 gene (IGB), *Genomics* 16:187 (1993).

16. G. Pengue, V. Viola, P. Cannada-Bartoli, P. De Luca, T. Esposito, P. Taillon-Miller, S. LaForgia, T. Druck, K. Huebner, M. D'Urso, and L. Lania, YAC-assisted cloning of transcribed sequences from the human chromosome 3p21 region, *Hum. Mol. Genet.* 2:791 (1993).

17. A.J. Buckler, D.G. Chang, S.L. Graw, J.D. Brook, D.A. Haber, P.A. Sharp, P.A., and D.E. Housman, Exon amplification: a strategy to isolate mammalian genes based on RNA splicing, *Proc. Natl. Acad. Sci. USA.* 88:4005 (1991).

18. J.S. Ziegle, Y. Su, K.P. Corcoran, L. Nie, P.E. Mayrand, L.B. Hoff, L.J. McBride, M.N. Kronick, and S.R. Diehl, Application of automated DNA sizing technology for genotyping microsatellite loci, *Genomics* 14:1026 (1992).

19. C.P. Kimpton, P. Gill, A. Walton, A. Urquhart, E.S. Millican, and M. Adams, Automated DNA profiling employing multiplex amplification of short tandem repeat loci, *PCR Methods Applic.* 3:13 (1993).

20. Applied Biosystems User Bulletin, Synthesis of fluorescent dye-labeled oligonucleotides for use as primers in fluorescent-based DNA sequencing, Issue 11 (1989).

21. H.L. Drwinga, L.H. Toji, C.H. Kim, A.E. Greene, and R.A. Mulivor, NIGMS human/rodent somatic cell hybrid mapping panels 1 and 2, *Genomics* 16:311 (1993).

22. B.L. Dubois and S.L. Naylor, Characterization of NIGMS Human/Rodent Somatic Cell Hybrid Mapping Panel 2 by PCR, *Genomics* 16:315 (1993).

23. D. Fong, D.I. Smith, and W. Hsieh, The human kininogen gene (KNG) mapped to chromosome 3q26-qter by analysis of somatic cell hybrids using the polymerase chain reaction, *Hum. Genet.* 87:189 (1991).

24. H.Y. Zoghbi, A.E. McCall, and F. LeBorgne-Demarquoy, Sixty-five radiation hybrids for the short arm of human chromosome 6: their value as a mapping panel and as a source for rapid isolation of new probes using repeat element-mediated PCR, *Genomics* 9:713 (1991).

25. G.R. Cutting, S. Curristin, H. Zoghbi, B. O'Hara, M.F. Seldin, and G.R. Uhl, Identification of a putative γ-aminobutyric acid (GABA) receptor subunit rho_2 cDNA and colocalization of the genes encoding rho_2 (GABRR2) and rho_1 (GABRR1) to human chromosome 6q14-q21 and mouse chromosome 4, *Genomics* 12:801 (1992).

26. O.T. Mueller, W.M. Henry, L.L. Haley, M.G. Byers, R.L. Eddy, and T.B. Shows, Sialidosis and galactosialidosis: chromosomal assignment of two genes associated with neuraminidase-deficiency disorders, *Proc. Natl. Acad. Sci. USA.* 83:1817 (1986).

27. D.F. Callen, V.J. Hyland, E.G. Baker, A. Fratini, R.N. Simmers, J.C. Mulley, and G.R. Sutherland, Fine mapping of gene probes and anonymous DNA fragments to the long arm of chromosome 16, *Genomics* 2:144 (1988).

28. R.A. Norum, M.J. Worsham, D.L. Van Dyke, V.R. Babu, and C.E. Jackson, Chromosome rearrangements localizing the chromosome 10 loci RBP3 and D10S5. *Am. J. Hum. Genet.* 43 (Suppl.):A154 (1988).

29. A. Tunnacliffe, L. Kiu, J.K. Moore, M.A. Leversha, M.S. Jackson, L. Papi, M.A. Ferguson-Smith, H-J. Thiesen, and B.A.J. Ponder, Duplicated KOX zinc finger clusters flank the centromere of human chromosome 10: evidence for a pericentric inversion during primate evolution, *Nucl. Acids Res.* 21:1409 (1993).

30. T. Glaser, D. Housman, W.H. Lewis, D. Gerhard, and C. Jones, A fine-structure deletion map of human chromosome 11p: analysis of J1 series hybrids, *Somat. Cell Mol. Genet.* 15:477 (1989).

31. T. Tokino, E. Takahashi, M. Mori, A. Tanigami, T. Glaser, J.W. Park, C. Jones, T. Hori, and Y. Nakamura, Isolation and mapping of 62 new RFLP markers on human chromosome 11, *Am. J. Hum. Genet.* 48:258 (1991).

32. P. vanTuinen, D.C. Rich, K.M. Summers, and D.H. Ledbetter, Regional mapping panel for human chromosome 17: application to neurofibromatosis type 1, *Genomics* 1:374 (1987).

33. L.D. McDaniel and R.A. Schultz, Elevated sister chromatid exchange phenotype of Bloom syndrome cells is complemented by human chromosome 15, *Proc. Natl. Acad. Sci. USA.* 89:7968 (1992).

34. A.D. Kline, K. Rojas, D. Moshinsky, and J. Overhauser, A deletion mapping panel of human chromosome 18, *Genomics* 13:1 (1992).

35. A.D. Kline, M.E. White, R. Wapner, K. Rojas, J. Kamholtz, M. Muenke, and J. Overhauser, Molecular analysis of the 18q syndrome and correlation with phenotype, *Am. J. Hum. Genet.*, in press (1993).

36. D.L. Nelson, S.A. Ledbetter, L. Corbo, M.F. Victoria, R. Ramirez-Solis, T.D. Webster, D.H. Ledbetter, and C.T. Caskey, *Alu* polymerase chain reaction: a method for rapid isolation of human-specific sequences from complex DNA sources, *Proc. Natl. Acad. Sci. USA* 86:6686 (1989).

37. D.L. Nelson, A. Ballabio, M.F. Victoria, M. Pieretti, R.D. Bies, R.A. Gibbs, J.A. Maley, A.C. Chinault, T.D. Webster, and C.T. Caskey, *Alu*-primed polymerase chain reaction for regional assignment of 110 yeast artificial chromosome clones from the human X chromosome: identification of clones associated with a disease locus, *Proc. Natl. Acad. Sci. USA* 88:6157 (1991).

38. M.C. Wapenaar, T. Kievits, P. Meera Kahn, P.L. Pearson, and G.J. Van Ommen, Isolation and characterization of cell hybrids containing human Xp-chromosome fragments, *Cytogenet. Cell. Genet.* 54:10 (1991).

39. M.D. Hoggan, N.F. Halden, C.E. Buckler, C.A. Kozak, Genetic mapping of the mouse c-fms proto-oncogene to chromosome 18. *J. Virol.* 62:1055 (1988).

40. M.J. Wagner, Y. Ge, M. Siciliano, and D.E. Wells, A hybrid cell mapping panel for regional localization of probes to human chromosome 8, *Genomics* 10:114 (1991).

41. J.E. Parrish, M.J. Wagner, J.T. Hecht, C.I. Scott, Jr., and D.E. Wells, Molecular analysis of overlapping chromosome deletions in patients with Langer-Giedion syndrome, *Genomics* 11:54 (1991).

42. M.S. Boguski, T.M. Lowe, and C.M. Tolstoshev, dbEST - the database for "Expressed sequence tags", *Nature Genetics* 4:332 (1993).

43. S.F. Altschul, W. Gish, W. Miller, E.W. Myers, and D.J. Lipman, Basic local alignment search tool, *J. Mol. Biol.* 215:403 (1990).

44. D. Gorlich, S. Prehn, E. Hartmann, J. Herz, A. Otto, R. Kraft, M. Wiedmann, S. Knespel, B. Dobberstein, and T.A. Rapoport, The signal sequence receptor has a second subunit and is part of a translocation complex in the endoplasmic reticulum as probed by bifunctional reagents, *J. Cell Biol.* 111:2283 (1990).

45. I. Wada, D. Rindress, P.H. Cameron, W-J. Ou, J.J. Doherty II, D. Louvard, A.W. Bell, D. Dignard, D.Y. Thomas, and J.J.M. Bergeron, SSRα and associated calnexin are major calcium binding proteins of the endoplasmic reticulum membrane. *J. Biol. Chem.* 266:19599 (1991).

46. A. Horiuchi, K.R. Williams, T. Kurihara, A.C. Nairn, and P. Greengard, Purification and cDNA cloning of ARPP-16, a cAMP-regulated phosphoprotein enriched in basal ganglia, and of a related phosphoprotein, ARPP-19, *J. Biol. Chem.* 265:9476 (1990).

47. W. Wasco, K. Bupp, M. Magendantz, J.F. Gusella, R.E. Tanzi, and F. Solomon, Identification of a mouse brain cDNA that encodes a protein related to the Alzheimer disease-associated amyloid β protein precursor, *Proc. Natl. Acad. Sci. USA* 89:10758 (1992).

48. H.J. Worman, C.D. Evans, and G. Blobel G., The lamin B receptor of the nuclear envelope inner membrane: a polytopic protein with eight potential transmembrane domains, *J. Cell Biol.* 111:1535 (1990).

49. M.M. McLaughlin, L.B. Cieslinski, M. Burman, T.J. Torphy, and G.P. Livi, A low-K_m, rolipram-sensitive, cAMP-specific phosphodiesterase from human brain, *J. Biol. Chem.* 268:6470 (1993).

50. J.L. Arriza, M.P. Kavanaugh, W.A. Fairman, Y. Wu, G.H. Murdoch, R.A. North, S.G. Amara, Cloning and expression of a human neutral amino acid transporter with structural similarity to the glutamate transporter gene family, *J. Biol. Chem.* 268:15329 (1993).

51. S. Shafqat, B.K. Tamarappoo, M.S. Kilberg, R.S. Puranam, J.O. McNamara, A. Guadaño-Ferraz, and R.T. Fremeau, Jr., Cloning and expression of a novel Na$^+$-dependent neutral amino acid transporter structurally related to mammalian Na$^+$/glutamate cotransporters, *J. Biol. Chem.* 268:15351 (1993).

52. M.J. Alkema, J. Wiegant, A.K. Raap, A. Berns, and M. van Lohuizen, Characterization and chromosomal localization of the human proto-oncogene BMI-1, *Hum. Mol. Gen.* 2:1597 (1993).

53. J. Takeda, T. Kayano, H. Fukumoto and G.I. Bell, Organization of the human GLUT2 (pancreatic beta-cell and hepatocyte) glucose transporter gene, *Diabetes* 42:773 (1993).

54. A. Kikuchi, K. Kaibuchi, Y. Hori, H. Nonaka, T. Sakoda, M. Kawamura, T. Mizuno and Y. Takai, Molecular cloning of the cDNA for a stimulatory GDP/GTP exchange protein for c-Ki-ras p21 and smg p21, *Oncogene* 7:289 (1992).

55. R.E. Mayer, P. Hendrix, P. Cron, R. Matthies, S.R. Stone, J. Goris, W. Merlevede, J. Hofsteenge, and B.A. Hemmings, Structure of the 55 kDa regulatory subunit of protein phosphatase 2A: evidence for a neuronal specific isoform, *Biochemistry* 30:3589 (1991).

56. C. Nottenburg, W.M. Gallatin, and T. St. John, Lymphocyte HEV adhesion variants differ in expression of multiple gene sequences, *Gene* 95:279 (1990).

57. Y. Furukawa, S. Nakatsuru, A. Nagafuchi, S. Tsukita, T. Muto, Y. Nakamura, and A. Horii, Structure, expression and chromosomal assignment of the human alpha-Catenin gene, unpublished (1993).

58. A. Rienitz, G. Grenningloh, I. Hermans-Borgmeyer, J. Kirsch, U.Z. Littauer, P. Prior, E.D. Gundelfinger, B. Schmitt, and H. Betz, Neuraxin, a novel putative structural protein of the rat central nervous system that is immunologically related to microtubule-associated protein 5, *EMBO J.* 8:2879 (1989).

59. M. Yamaguchi, Y. Hatefi, K. Trach, and J.A. Hoch, The primary structure of the mitochondrial energy-linked nicotinamide nucleotide transhydrogenase deduced from the sequence of cDNA clones, *J. Biol. Chem.* 263:2761 (1988).

60. F. Tan, D.K. Weerasinghe, R.A. Skidgel, H. Tamei, R.K. Kaul, I.B. Roninson, J.W. Schilling, and E.G. Erdos, The deduced protein sequence of the human carboxypeptidase N high molecular weight subunit reveals the presence of leucine-rich tandem repeats, *J. Biol. Chem.* 265:13 (1990).

61. N.X. Krueger and H. Saito, A human transmembrane protein-tyrosine-phosphatase, PTPzetam is expressed in brain and has an N-terminal receptor domain homologous to carbonic anhydrases, *Proc. Natl. Acad. Sci. USA* 89:7417 (1992).

62. G.B. Price, ala gene, GenBank M58318 (1990).

63. M.E. Williams, P.F. Brust, D.H. Feldman, S. Patthi, S. Simerson, A. Maroufi, A.F. McCue, G. Veliçelebi, S.B. Ellis, and M.M. Harpold, Structure and functional expression of an omega-conotoxin-sensitive human N-type calcium channel, *Science* 257:389 (1992).

64. I. Bongarzone, Direct submission, (1993).

65. T. Stearns, R.A. Kahn, D. Botstein, and M.A. Hoyt, ADP ribosylation factor is an essential protein in *Saccharomyces cervisiae* and is encoded by two genes, *Mol. Cell. Biol.* 10:6690 (1990).

66. A.S. Zervos, J. Gyuris, and R. Brent, Mxil, a protein that specifically interacts with Max to bind Myc-Max recognition sites, *Cell* 72:223 (1993).

67. W. Hunziker, and S. Schrickel, Rat brain calbindin D28: Six domain structure and extensive amino acid homology with chicken calbindin D28, *Mol. Endocrinol.* 2:465 (1988).

68. S. Lee, I. Kim, L.N. Marekov, E.J. O'Keefe, D.A. Parry, and P.M. Steinert, The structure of human trichohyalin. Potential multiple roles as a functional EF-hand-like calcium-binding protein, a cornified cell envelope precursor, and an intermediate filament-associated (cross-linking) protein, *J. Biol. Chem.* 268:12164 (1993).

69. M.J. Fietz, C.J. McLaughlan, M.T. Campbell, and G.E. Rogers, Analysis of the sheep trichohyalin gene: potential structural and calcium-binding roles in the hair follicle, *J. Cell Biol.* 121:855 (1993).

70. W.K. Sonnenburg, P.J. Mullaney, and J.A. Beavo, Molecular cloning of a cyclic GMP-stimulated cyclic nucleotide phosphodiesterase cDNA, *J. Biol. Chem.* 266:17655 (1991).

71. A.M. Van den Ouweland, J.J. Van Groningen, A.J.M. Roebrock, C. Onnekink, and W.J. Van de Ven, Nucleotide sequence analysis of the human fur gene, *Nucl. Acids Res.* 17:7101 (1989).

72. W. Tao and E. Lai, Telencephalon restricted expression of BF-1, a new member of the HNF-3/ fork head gene family, in the developing rat brain, *Neuron* 8:957 (1992).

73. K.S. Bhat, Expressed sequence tags from a human cell line, unpublished (1992).

74. F. Loreni, I. Ruberti, I. Bozzoni, P. Pierandrei-Amaldi, and F. Amaldi, Nucleotide sequence of the L1 ribosomal protein gene of *Xenopus laevis*: remarkable sequence homology among introns, *EMBO J.* 4:3483 (1985).

75. J.D. Stubbs, C. Lekutis, K.L. Singer, A. Bui, D. Yuzuki, U. Srinivasan, and G. Parry, cDNA cloning of a mouse mammary epithelial cell surface protein reveals the existence of epidermal growth factor-like domains linked to factor VIII-like sequences, *Proc. Natl. Acad. Sci. USA* 87:8417 (1990).

76. A. Langkopf, J.A. Hammarback, R. Muller, R.B. Vallee, and C.C. Garner, Microtubule-associated proteins 1A and LC2, *J. Biol. Chem.* 267:16561 (1992).

77. K.C. Wilhelmsen, K. Eggleton, and H.M. Temin, Nucleic acid sequences of the oncogene v-rel in reticuloendotheliosis virus strain T and its cellular homolog, the proto-oncogene c-rel, *J. Virol.* 52:172 (1984).

78. H.W. Gaugitsch, E.E. Prieschl, F. Kalthoff, N.E. Huber, and T. Baumruker, A novel transiently expressed, integral membrane protein linked to cell activation. Molecular cloning via the rapid degradation signal AUUA, *J. Biol. Chem.* 267:11267 (1992).

79. B.S. de Pater and R.A. Schilperoort, submitted, EMBL Data Library (1990).

80. L.A. Elferink, W.S. Trimble, and R.H. Scheller, Two vesicle-associated membrane protein genes are

differentially expressed in the rat central nervous system, *J. Biol. Chem.* 264:11061 (1989).

81. I.H. Still, GenBank accession M90054.

82. T. Sato, H. Saito, J. Swenson, A. Olifant, C. Wood, D. Danner, T. Sakamoto, K. Takita, F. Kasumi, Y. Miki, M. Skolnick, and Y. Nakamura, The human prohibitin gene located on chromosome 17q21 is mutated in sporadic breast cancer, *Cancer Res.* 52:1643 (1992).

83. R.M. Cawthon, R.B. Weiss, G. Xu, D. Viskochil, M. Culver, J. Stevens, M. Robertson, D. Dunn, R. Gesteland, P. O'Connell, and R. White, A major segment of the neurofibromatosis type 1 gene: cDNA sequence, genomic structure and point mutations, *Cell* 62:193 (1990).

84. T.W. Nilsen, P.A. Maroney, R.G. Goodwin, K.G. Perrine, J.A. Denker, J. Nanduri, and J.W. Kazura, Cloning and characterization of a potentially protective antigen in lymphatic filariasis, *Proc. Natl. Acad. Sci. USA* 85:3604 (1988).

85. J. Kamholz, F. de Ferra, C. Puckett, and R. Lazzarini, Identification of three forms of human myelin basic protein by cDNA cloning, *Proc. Natl. Acad. Sci. USA* 83:4962 (1986).

86. E.D. Sverdlov, G.S. Monastyrskaya, N.E. Broude, Y.A. Ushkarev, A.M. Melkov, Y.V. Smirnov, I.V. Malyshev, R.L. Allikmets, M.B. Kostina, I.E. Dulubova, N.I. Kiyatkin, A.V. Grishin, N.N. Modyanov, and Y.A. Ovchinniko, Family of human Na+, K+-ATPase genes. Structure of the gene of isoform alpha III, *Dokl. Biochem.* 297:426 (1987).

87. K. Takazawa, J. Perret, J.E. Dumont, and C. Erneux, Human brain inositol 1,4,5-triphosphate-3-kinase cDNA sequence, *Nucl. Acids Res.* 18:7141 (1990).

88. D.M. Miller, R. Delgado, J.M. Chirgwin, S.C. Hardies, and P.M. Horowitz, Expression of cloned bovine adrenal rhodanese, *J. Biol. Chem.* 266:4686 (1991).

89. R. Pallini, G.C. Guazzi, C. Cannella, and M.G. Cacace, Cloning and sequence analysis of the human liver rhodanese: comparison with the bovine and chicken enzymes, *Biochem. Biophys. Res. Commun.* 180:887 (1991).

90. D.B. Jump, A. Bell, and V. Santiago, Thyroid hormone and dietary carbohydrate interact to regulate rat liver S14 gene transcription and chromatin structure, *J. Biol. Chem.* 265:3474 (1990).

91. W.T. Lankes and H. Furthmayr, Moesin: a member of the protein 4.1-talin-ezrin family of proteins, *Proc. Natl. Acad. Sci. USA* 88:8297 (1991).

92. Y. Ono, T. Fujii, K. Ogita, U. Kikkawa, K. Igarashi, and Y. Nishizuka, The structure, expression, and properties of additional members of the protein kinase C family, *J. Biol. Chem.* 263:6927 (1988).

93. G. Kochs, D. Meyer, H. Hug, D. Marme, and T.F. Sarre, Activation and substrate specificity of the human protein kinase C gamma and zeta isoenzymes, *Eur. J. Biochem.* 216:597 (1993).

94. H.G. Harley, J.D. Brook, C.L. Jackson, T. Glaser, K.V. Walsh, M. Sarafarzi, R. Kent, M. Lager, M. Koch, P.S. Harper, R. Levenson, D.E. Housman, and D.J. Shaw, Localization of a human Na+, K+-ATPase α subunit gene to chromosome 19q12 → q13.2 and linkage to the myotonic dystrophy locus, *Genomics* 3:380 (1988).

95. C.M. Abbott, R. Blank, J.T. Eppig, J.M. Friedman, K.E. Huppi, I. Jackson, B.A. Mock, J. Stoye, R. Wiseman, Mouse Chromosome 4, *Mammal. Genome* 3:S55 (1992).

96. E. Uberbacher, and R. Mural, Locating protein-coding regions in human DNA sequences by a multiple sensor-neural network approach, *Proc. Natl. Acad. Sci. USA* 88:11261 (1991).

97. T.R. Burglin, and T.M. Barnes, Introns in sequence tags, *Nature* 357:367 (1992).

98. M.D. Adams, C. Fields, and J.C. Venter, Introns in sequence tags, *Nature* 357:367(1992).

99. L.L. Lien, F.M. Boyce, P. Kleyn, L.M. Brzustowicz, J. Menninger, D.C. Ward, T.C. Gilliam, and L.M. Kunkel, Mapping of human microtubule-associated protein 1B in proximity to the spinal muscular atrophy locus at 5q13, *Proc. Natl. Acad. Sci. USA* 88:7873 (1991).

100. B. Levy, P.D. Canoll, O. Silvennoinen, G. Barnea, B. Morse, A.M. Honegger, J-T. Huang, L.A. Cannizzaro, S-H. Park, T. Druck, K. Huebner, J. Sap, M. Ehrlich, J.M. Masacchio, and J. Schlessinger, The cloning of a receptor-type protein tyrosine phosphatase expressed in the central nervous system, *J. Biol. Chem.* 268:10573 (1993).

101. J.J. White, D.H. Ledbetter, R.L. Eddy, Jr., T.B. Shows, D.A. Stewart, N.M. Nuell, V. Friedman, C.M. Wood, G.A. Owens, J.K. McClung, D.B. Danner, and C. Morton, Assignment of the human prohibitin gene (PHB) to chromosome 17 and identification of a DNA polymorphism, *Genomics* 11:228 (1991).

102. D.F. Saxe, N. Takahashi, L. Hood, and M.I. Simon, Localization of the human myelin basic protein gene (MBP) to region 18q22-->qter by in situ hybridization, *Cytogenet. Cell Genet.* 39:246 (1985).

103. J.E. Parrish and D.L. Nelson, Regional assignment of 19 X-linked ESTs, *Hum. Mol. Genet.* 2:1901 (1993).

104. J.M. Sikela and C. Auffray, Finding new genes faster than ever, *Nat. Genet.* 3:189(1993).

288

MAPPING cDNAS BY HYBRIDIZATION TO GRIDDED ARRAYS OF DNA FROM YAC CLONES

Donald T. Moir, Ron Lundstrom, Peter Richterich, Xiaohong Wang,
Maria Atkinson, Kathy Falls, Jen-i Mao, Douglas R. Smith, and Gerald
F. Vovis

Department of Human and Molecular Genetics
Collaborative Research, Inc.
Waltham, MA 02154

ABSTRACT

A physical map of overlapping clones covering the human genome will provide a substrate for rapid, high-throughput, high resolution mapping of genes. Such a map of megaYAC clones is currently under development in several laboratories. To permit utilization of this resource, we are developing technology for mapping of cDNAs by hybridization to gridded arrays of DNA from megaYAC clones. Included in this approach are methods and instrumentation for reducing false negatives and false positives by pooling of megaYAC DNAs, for reducing the number of hybridizations by pooling of cDNA probes, and for automating the hybridization and detection steps. Results from a pilot project involving megaYACs representing about one-quarter of the human genome are described. Total yeast DNA was prepared from 730 megaYACs, diluted to a uniform concentration, and pooled with a representation of three and a pool size of three. Various amounts of pooled DNAs from megaYAC clones were gridded onto nylon filters in a medium density array. PCR amplified inserts from cDNA libraries were radiochemically labeled by random priming and used as probes. The sensitivity of detection was adequate even at the lowest megaYAC quantity per dot, i.e., about 0.08 μg of total yeast DNA from each YAC in each dot. Most signals were triplets as expected from the pooling strategy, and the signal intensity of dots was quite uniform. Filters could be re-probed at least five times with no detectable degradation of signal to noise ratio. Application of a "two out of three" rule for validity of "real" signals appears adequate since it permitted accurate identification of all control YACs on the filters. These results suggest that such a hybridization-based approach will permit accurate, rapid, high throughput mapping of cDNAs to intervals in the emerging YAC contig map.

Identification of Transcribed Sequences, Edited by
U. Hochgeschwender and K. Gardiner, Plenum Press, New York, 1994

INTRODUCTION

Rapid, efficient sequencing methods have already provided partial sequences for tens of thousands of cDNAs (1-5). Complete sequences of these cDNAs and of the corresponding genes will eventually be determined by various laboratories, based on specific interests. Rapid sequencing methods are also being used to determine the array of genes expressed in specific tissues (6). Knowledge of both DNA sequence and tissue expression patterns of genes is a crucial part of efforts to identify genes responsible for specific phenotypes. However, the key feature in positional cloning strategies, namely the chromosomal location of genes, is not known for most of the sequenced cDNAs. Indeed, current cDNA mapping technology is inadequate to keep pace with the rapid rate of partial sequencing. For example, conversion of the DNA sequence information to STSs followed by PCR-based mapping to panels of hybrid cell lines can provide low resolution chromosomal location for about 500 cDNAs per person per year (7), but over 50,000 cDNAs have been partially sequenced so far. Large cDNAs of over 1-2 kb can be mapped with higher resolution by fluorescence *in situ* hybridization to human metaphase chromosome spreads, but the rate of mapping is similar to that of the PCR-based approach (8). PCR-based strategies can be scaled up to achieve adequate rates, but hybridization-based approaches appear capable of providing high mapping rates at much less cost. The resolution of mapping can be improved by applying these techniques to the highly redundant set of megaYACs which are being assembled into ordered contigs representing the entire human genome (9). Each end of a megaYAC provides a mapping interval; thus, 20,000 megaYACs (about 7 genome equivalents) will provide mapping intervals of under 100 kb, on average.

Alternative, top-down strategies for developing gene maps rely on identifying genes in genomic clones such as cosmids or YACs by methods such as exon-trapping or cDNA hybrid selection (10,11). While these approaches are useful, they are tedious to apply to the whole genome. A bottom-up approach of mapping cDNAs is complementary and adds considerable value to the already existing database of partial cDNA sequences (dbEST, ref.12). Through chromosomal location, specific cDNAs can become candidate genes for inherited diseases whose phenotypes have been localized by genetic linkage mapping and for cancers associated with known cytogenetic abnormalities.

A YAC hybridization approach to cDNA mapping could utilize either of two types of YAC substrates -- micro-colony blots or DNA dot blots. YAC micro-colony blots have been described previously (13). They are convenient because the DNA is prepared *in situ*, but signal to noise ratios are limited by the amount of DNA available in the micro-colony. We have undertaken a study of YAC DNA dot blots using total yeast DNA from megaYAC clones in order to assess the feasibility of mapping cDNAs by hybridization to several genome equivalents of megaYACs. The overall objective of this pilot study was to establish the feasibility of the approach by building and probing filters carrying a subset of megaYAC DNAs. Specific objectives were to investigate (a) rapid methods for preparing DNA from YAC clones, (b) pooling of YAC DNAs to reduce error rates, (c) methods for the rapid preparation of high quality cDNA probes, (d) pooling of probes to reduce the number of hybridizations, and (e) the signal strength and longevity of dot blot filters.

METHODS

Preparation of Total Yeast DNA from Cultures of MegaYAC Clones

MegaYAC clones were grown from frozen storage dishes on AHC agar plates and inoculated into 50 ml AHC liquid medium (14). After overnight growth at 30°C with shaking, cells were harvested by centrifugation in 50 ml disposable tubes. DNA was prepared in the same tubes by a bead-shear lysis method including treatment with RNAse (15). A heavy-duty multi-tube vortexer ("Big Vortexer" by Glas-Col, Terre- Haut, IN) was used to increase throughput. Up to 48 cultures could be handled per person per day. Yields were about 250 μg nucleic acid as determined by A_{260} for over 90% of the preparations. Judging from agarose gel electrophoretic analysis, the preparations consisted of about equal quantities of DNA and RNA. MegaYAC DNA is expected to represent about 8-10% of the total yeast DNA from the clones.

PCR Amplification and Labeling of cDNA Probes

The cDNA insert was excised from plasmids of a directionally cloned human infant brain (HIBB) or human fetal brain (FBL) library (Bento Soares, Columbia University) with EcoRI and HindIII, purified by agarose gel electrophoresis, and labeled with ^{32}P by random priming with hexamers to 10^9 cpm/μg. Alternatively, instead of restriction enzyme digestion, the cDNA insert was amplified from the plasmid DNA in a PCR using primers PR1 (5'AGCTATGACCATGATTACGCCA) and PR2 (5'GACGGCCAGTGAATTCCCCT) flanking the cloning site before gel purification and labeling as described. PCR conditions were as follows: 94°C 15 sec; 50°C 15 sec; 72°C 1.5 min for 34 cycles in a Perkin-Elmer Cetus Model 9600 thermal cycler with 2 mM MgCl$_2$ in a 40 μl volume.

Hybridization

Filters were prehybridized in hybridization buffer without probe for 1-6 hr at 65°C. Filters were hybridized overnight in 0.5 M NaPi, pH 7.2; 7% SDS; 1% BSA; 1 mM EDTA; 0.1 mg/ml yeast tRNA, and 50 μg/ml denatured and sheared human DNA. After hybridization, filters were washed (40 mM NaPi, pH 7.2; 0.1% SDS) at 65° twice for 20 min. Film exposures were made for 1-2 days with screens. Probes were stripped in 2 mM Tris-HCl, pH 7.5, 1 mM EDTA, 0.1% SDS, 65° for 15-30 min prior to reuse. Probes HIBBN14 (ATCC 85466) and HIBBN18 (ATCC 85469) were gifts from Mark Adams and Craig Venter (TIGR). Probe FBL40 was the gift of Bento Soares (Columbia University). Control YACs (ICRFy900-D115 and ICRFy900-C0828) known to carry sequences homologous to probe HIBBN14 were obtained by hybridization of HIBBN14 probes to high density micro-colony blots carrying two genome equivalents of YACs (gift of H. Lehrach, M. Ross, G. Zehetner, ICRF, London).

RESULTS

YAC DNA Preparation

Preparation of DNA from YAC clones permits the application of large amounts of YAC DNA to filters and should provide signal to noise ratios superior to those observed in hybridizations to YAC micro-colony blots. Ideally, filters should carry YAC DNA free of the endogenous yeast chromosomes; however, no practical methods are currently available for the isolation of large amounts of artificial chromosome DNA. Therefore, we used a rapid, efficient total yeast DNA preparation method devised previously for recovering plasmids from transformed yeast strains (15).

Total yeast DNA was prepared from 730 megaYACs in eight microtiter dishes (CEPH plate #'s 829-836) representing about one-quarter of a human genome (16). Preparations were diluted to a constant concentration of about 1 mg/ml nucleic acids (about 0.5 mg/ml DNA) based on A_{260} readings taken for each preparation and stored in computer files. Diluted DNAs were stored in microtiter dishes in the same well addresses as the megaYACs from which they were derived.

Pooling of YAC DNA

Sample pooling can reduce the labor required to prepare and test a large number of samples. We have used pooling both for that purpose and in order to improve the accuracy of mapping. Hybridization experiments are typically plagued by false positive and negative data (eg., apparent signals which are background noise, and absence of signals for known positives). This problem is particularly acute for high density micro-colony blots and might be as severe for high density dot blots.

Therefore, for accuracy, we adopted a pooling strategy for representing each YAC three times on the filters. In order to keep the number of dots (and thus, the filter size) from tripling, we pooled three samples in each dot. Ambiguities in deciphering addresses of true positives were reduced by following a simple pooling rule: no two YACs shared the same pool more than once. Thus, only a single YAC could possibly share two or three different pools or dots. This scheme, with a representation of three ($r=3$) and a pool size of three ($p=3$), is one of a number of such potential schemes. If signal intensity proves adequate, larger pools sizes such as $r=3$; $p=12$ would permit a very useful reduction in the number of dots (and DNA preparations) required by a factor of 4 or more.

The Biomek 1000 was programmed in the command line language SYGI (Stanford Yeast Genome Interface) to carry out the minimal pooling scheme (i.e., $r=3$; $p=3$). Due to the limited tablet size of the Biomek, one source plate was pooled into one target plate, and this was repeated eight times to accomodate all of the DNAs. Pooling of each dish required about 55 min. of Biomek time. The detailed pooling scheme is depicted in Fig. 1. The pooled microtiter dishes were sealed and stored frozen at -20°C until needed for gridding. The DNA from each individual megaYAC clone is present at 1/3 x 0.5 mg/ml after pooling, or at a final concentration of about 0.15 mg/ml.

Gridding of YAC DNA

We chose not to use the Biomek 1000 for gridding megaYAC dot blots for two reasons. First, its small tablet size and lack of precise mechanism for removing and replacing filters makes it very slow for preparation of large numbers of identical filters. Each of the eight microtiter dishes of pooled DNAs must be placed in the source position for preparation of only two filters. Then, the whole process must be repeated in order to prepare two more filters. Second, the polished, gradually tapered metal pins of the high density replicating tool exhibited very poor reproducibility in the transfer of sub-microliter quantities of megaYAC DNA.

By contrast, the 96-floating pin inoculator designed and built at Washington University in St. Louis (14) permits rapid, reproducible preparation of large numbers of dot blots because its filter frame assembly allows for the accurate removal and replacement of filters. A single microtiter dish of pooled megaYAC DNAs could be spotted onto 24 filters (each in its own frame assembly) by simply shuttling each filter into place when needed. Each spotting operation transferred about 0.4-0.5 μl with high reproducibility. We denatured the pooled megaYAC DNA with NaOH and transferred 0.5 μl, 1.0 μl, or 1.5 μl of DNA solution by simply performing the spotting operation once, twice, or three times before removing the filter. This corresponds to approximately 0.08 μg, 0.15 μg, or 0.25 μg of total yeast DNA from each megaYAC clone in each dot. By comparison, the maximum amount of total yeast DNA available from a micro-colony on a high density colony blot is about 0.05 mg. Furthermore, that DNA adheres poorly to the filter, probably due to the presence of cell debris and protein. Samples from all eight plates of pooled megaYAC DNA (730 DNAs) were placed on a single 4"x7" filter. Filters were subjected to UV irradiation. Twenty-four such filters were prepared in this manner. An example of the grid pattern is shown in Fig. 2.

292

```
A1  A2  A3  A4      A5  A6  A7  A8      A9  A10 A11 A12
B1  B2  B3  B4      B5  B6  B7  B8      B9  B10 B11 B12
C1  C2  C3  C4      C5  C6  C7  C8      C9  C10 C11 C12
D1  D2  D3  D4      D5  D6  D7  D8      D9  D10 D11 D12
E1  E2  E3  E4      E5  E6  E7  E8      E9  E10 E11 E12
F1  F2  F3  F4      F5  F6  F7  F8      F9  F10 F11 F12
G1  G2  G3  G4      G5  G6  G7  G8      G9  G10 G11 G12
H1  H2  H3  H4      H5  H6  H7  H8      H9  H10 H11 H12
```

```
(A1 A5 A9) (A2 A6 A10) (A3 A7 A11) (A4 A8 A12)   (A1 A6 A11) (A2 A7 A12) (A3 A8 B9) (A4 B5 B10)   (A1 A7 B9) (A2 A8 B10) (A3 B5 B11) (A4 B6 B12)
(B1 B5 B9) (B2 B6 B10) (B3 B7 B11) (B4 B8 B12)   (B1 B6 B11) (B2 B7 B12) (B3 B8 C9) (B4 C5 C10)   (B1 B7 C9) (B2 B8 C10) (B3 C5 C11) (B4 C6 C12)
(C1 C5 C9) [C2 C6 C10] (C3 C7 C11) (C4 C8 C12)   (C1 C6 C11) [C2 C7 C12] (C3 C8 D9) (C4 D5 D10)   (C1 C7 D9) [C2 C8 D10] (C3 D5 D11) (C4 D6 D12)
(D1 D5 D9) (D2 D6 D10) (D3 D7 D11) (D4 D8 D12)   (D1 D6 D11) (D2 D7 D12) (D3 D8 E9) (D4 E5 E10)   (D1 D7 E9) (D2 D8 E10) (D3 E5 E11) (D4 E6 E12)
(E1 E5 E9) [E2 E6 E10] (E3 E7 E11) (E4 E8 E12)   (E1 E6 E11) (E2 E7 E12) (E3 E8 F9) (E4 F5 F10)   (E1 E7 F9) (E2 E8 F10) (E3 F5 F11) (E4 F6 F12)
(F1 F5 F9) (F2 F6 F10) (F3 F7 F11) (F4 F8 F12)   (F1 F6 F11) (F2 F7 F12) (F3 F8 G9) (F4 G5 G10)   (F1 F7 G9) (F2 F8 G10) (F3 G5 G11) (F4 G6 G12)
(G1 G5 G9) (G2 G6 G10) (G3 G7 G11) (G4 G8 G12)   (G1 G6 G11) (G2 G7 G12) (G3 G8 H9) (G4 H5 H10)   (G1 G7 H9) (G2 G8 H10) (G3 H5 H11) (G4 H6 H12)
(H1 H5 H9) (H2 H6 H10) (H3 H7 H11) (H4 H8 H12)   (H1 H6 H11) (H2 H7 H12) (H3 H8 A9) (H4 A5 A10)   (H1 H7 A9) (H2 H8 A10) (H3 A5 A11) (H4 A6 A12)
```

Figure 1. A schematic representation of the pooling strategy used for these studies. The 96-well source microtiter plate carrying total yeast DNA from megaYAC clones is indicated at the top, and the 96-well target microtiter dish is shown at the bottom. Each megaYAC DNA preparation is placed in three target wells in the specified pattern (representation=3), and each target position receives three megaYAC DNAs (pool size=3). No pair of YACs is found in more than one pool. The three target positions for source positions C2, E6, and G10 are shown.

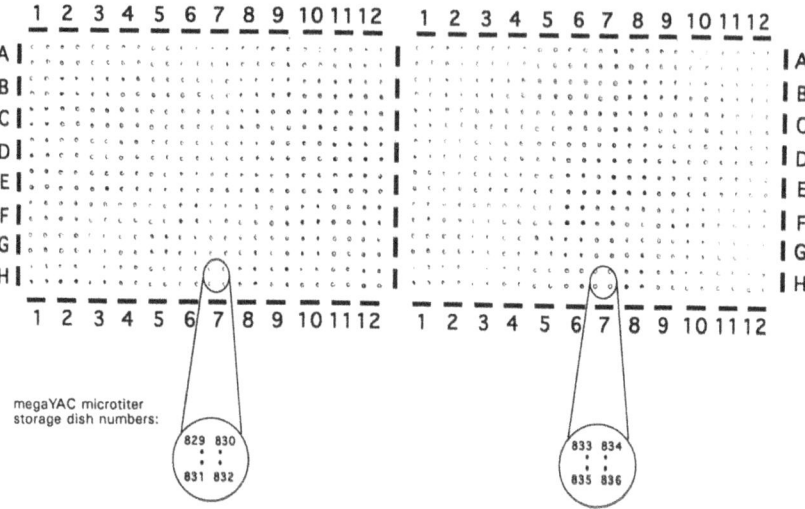

Figure 2. The grid pattern obtained from pooled megaYAC DNAs spotted in a 2x2 pattern with a 96-floating pin inoculator. All eight plates of pooled megaYAC DNA (730 DNAs) fit on a 4"x7" filter; the CEPH microtiter dish number for YACs in each corner of the 2x2 grid are shown in the enlarged circle at position H7.

Labeling cDNA Probes

The cDNAs were labeled by random priming gel-purified plasmid inserts which were either cut from isolated plasmid preparations or amplified by PCR from fresh *E. coli* colonies. Both of these methods produced excellent probes (see Figs. 3-4). For rapid, semi-automated preparation of probes, a method based solely on PCR and avoiding gel electrophoresis would be desirable. To test such a method, we used 1% of the PCR-amplified insert without any purification in a second hemi-nested PCR using primers PR1 and PR3 (5'CCCTTGCGGCCGCAGGAATT) with α-^{32}P-dCTP as the source of dCTP. The labeled PCR product was purified by spin column and used to probe megaYAC filters. Sufficient label was incorporated and the expected DNA dots did hybridize to the probe; however, several spurious dots also hybridized (data not shown). These results were encouraging and suggest that a fully nested second stage PCR might produce adequately clean probes.

Hybridization of MegaYAC Dot Blots with cDNA Probes

Inserts from a total of fifteen cDNA clones were used as probes of the quarter-genome equivalent dot blots carrying positive control YACs in known positions (see Methods). Three of the probes gave hybridization signals consistent with unique YAC addresses, for a success ratio of 20% While the number of hybridizations is small, this compares favorably with the theoretical expectation of 21.6% positives based on statistical considerations. However, the number of YACs identified for each probe was above expectations. All three successful probes identified two or more positive YACs. In two cases, HIBBN14 and HIBBN18, the probes hybridized strongly to dots corresponding to a single YAC each and weakly to dots corresponding to one or more additional YACs (Figs. 3 and 4 below). In the third case, FBL40, the probe hybridized weakly to 18 dots corresponding unambiguously to the three pools of each of six different YACs (data not shown).

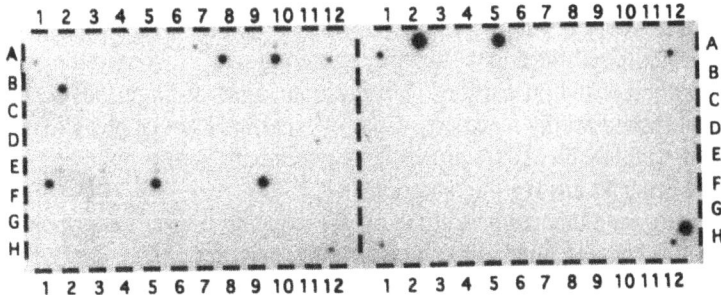

Figure 3. An x-ray film exposure of a hybridization experiment with cDNA probe HIBBN14 (EST06869; ref.1). The filter carries about 0.15 mg of DNA from each of three megaYAC clones in each dot (pooled at p=3, r=3). Triplet signals in the left panel at F1,F5,F9 (positive control added to source plate 830, position F1) and in the right panel at A2,A5,H12 (positive control added to source plate 834, position A6) correspond to control YAC DNAs added to those positions and known to carry sequences homologous to HIBBN14 cDNA. The triplet signal in the left panel at A8,A10,B2 (CEPH source plate 831, position B10) corresponds to a megaYAC which was discovered in this experiment to carry sequences homologous to HIBBN14 cDNA. Weak spots in the four corners of each half of the blot correspond to lambda DNA added for filter orientation purposes. Weak triplet signals in the left panel at C11,D3,D5 (CEPH source plate 832, position D11) and D11,E3,E5 (CEPH source plate 832, position E11) and the weak doublet A7,A10 (CEPH source plate 829, position A8) correspond to unique addresses of megaYACs which appear to carry sequences partially homologous to HIBBN14 cDNA and may represent related members of a gene family. Weak signals at E2,F3,G4 (plate 831) do not correspond to a unique megaYAC address and are seen in hybridizations with several different probes. These signals may derive from contamination in these particular wells.

Figure 4. An x-ray film exposure of a hybridization experiment with cDNA probe HIBBN18 (EST06872 and EST06873; ref.1). The filter carries about 0.15 mg of DNA from each of three megaYAC clone in each dot (pooled p=3, r=3); it was stripped once previously. No lambda DNA was included in the probe. A one day film exposure is shown. Triplet signals in the right panel at C2,C6,C10 (CEPH source plate 835, position C2) correspond to a unique megaYAC address. Weak doublet signals in the left panel at F7,F9 (CEPH source plate 829, position F9) correspond to two of three signals for a unique megaYAC address and may represent a megaYAC with partial homology to cDNA HIBBN18.

Thus, weak signals corresponding to unique megaYAC addresses were detected with all three successful probes. Tests on the purified megaYACs from these addresses confirmed that these signals are weak because of poor hybridization and not due to differences in amounts of DNA on the filters (data not shown). These results appear to represent cross-hybridization of cDNA probes to mismatched sequences, perhaps homologous gene family members. It is not surprising that some cDNA hybridization probes do not represent unique single-copy landmarks in the genome, but it is not clear what fraction of probes will recognize related genes in these mapping experiments. The addresses of YACs carrying homologous family members will be useful because it will provide chromosomal locations for the related genes. However, it remains to be determined whether the intensity of hybridization signal distinguishes accurately between the gene and related family members. Control experiments must be done on multi-genome equivalent dot blots with cDNA probes from genes with known highly homologous family members in order to answer this question.

To determine the usable lifetime for the mapping filters, we stripped and re-probed one filter five times, alternating between the two successful probes HIBBN14 and HIBBN18. Fig. 5 shows the results from the fifth probing which happened to be with probe HIBBN14. Unambiguous signals are observed for all four positive control YACs, the single megaYAC yielding a strong signal and some megaYACs yielding weak signals. Judging from these results, we expect that these filters could be used for considerably more than five hybridizations. Some of the weak signals may be undetectable on older filters, but the strong hybridization signals will be detectable.

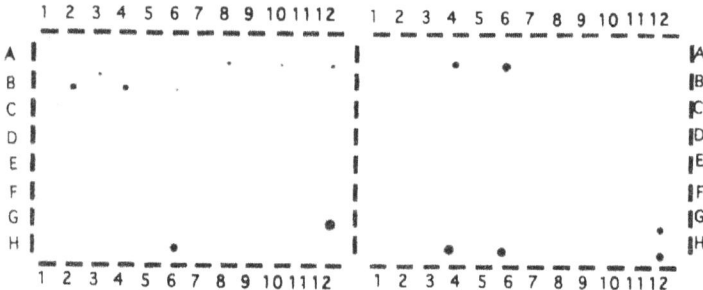

Figure 5. An x-ray film exposure of a hybridization experiment with cDNA probe HIBBN14 (as in Fig. 3). This is the fifth re-probing of this particular filter. Lambda DNA was not included in the probe. Control YACs known to hybridize with the HIBBN14 probe were added in four addresses prior to pooling. Eleven of the expected 12 signals are seen [left panel at A12,B4,B6 (positive control added to source plate 831, position B12); G12,H6 (positive control added to source plate 831, position H12); right panel at A4,A6,H12 (positive control added to source plate 835, position A12); and H4,H6,G12 (positive control added to source plate 835, position H12)]. In addition, signals corresponding to the megaYAC (identified in the experiment of Fig. 3) are also visible [left panel at A8,A10,B2 (CEPH source plate 831, position B10)]. Weak signals in the left panel at E3 and E5 represent two of the three weak signals detected for megaYAC 832E11 in the experiment of Fig. 3.

CONCLUSIONS

The data were of excellent quality. For example, most signals were triplets as expected from the pooling strategy. In addition, the signal intensity of dots was quite uniform, consistent with efforts to adjust the DNA concentrations to a constant value. Triplet signals almost always translated into unique megaYAC addresses. With only one exception (see Fig. 5), control YACs consistently yielded three signals at the appropriate locations. The one failure was apparently due to absence of the YAC DNA in that pool, because all filters made from that pooling failed to yield a control signal at that location. Application of a "two out of three" requirement to call a signal "real" appears valid since it identified all of the control YACs correctly.

The sensitivity of detection was quite adequate at 0.15 μg/dot (see Figs. 3, 4, and 5), and was even adequate at the lowest megaYAC quantity of 0.08 μg/dot (data not shown). Therefore, 50 ml megaYAC cultures will provide enough DNA for almost 500 filters. If each filter could be re-used ten times, this would permit 5,000 hybridizations. If the cDNA probes were also pooled, for example at a conservative level of r=2 and p=4, then 10,000 cDNAs could be mapped with 5,000 hybridizations.

REFERENCES

1. M.D. Adams, M.B. Soares, A.R. Kerlavage, C.F. Fields, and J.C. Venter, Rapid cDNA sequencing (expressed sequence tags) from a directionally cloned human infant brain cDNA library, *Nature Genetics* 4:373 (1993).

2. J.D. Grausz and C. Auffray, Strategies in cDNA programs, *Genomics*:17:530 (1993).

3. M.D. Adams, M. Dubnick, A.R. Kerlavage, R. Moreno, J.M. Kelley, T.R. Utterback, J.W. Nagle, C. Fields, and J.C. Venter, Sequence identification of 2,375 human brain genes, *Nature* 355:632 (1992).

4. A.S. Khan, A.S. Wilcox, M.H. Polymeropoulos, J.A. Hopkins, T.J. Stevens, M. Robinson, A.K. Orpana, and J.M. Sikela, Single pass sequencing and physical and genetic mapping of human brain cDNAs, *Nature Genetics* 2:180 (1992).

5. K. Okubo, N. Hori, R. Matoba, T. Niyama, A. Fukushima, Y. Kojima, and K. Matsubara, Large scale cDNA sequencing for analysis of quantitative and qualitative aspects of gene expression, *Nature Genetics* 2:173 (1992)

6. J. Takeda, H. Yano, S. Eng, Y. Zeng, and G.I. Bell, A molecular inventory of human pancreatic islets: sequence analysis of 1000 cDNA clones, *Human Mol. Genet.* 2:1793 (1993).

7. M.H. Polymeropoulos; H. Xiao; J.M. Sikela; M. Adams; J.C. Venter; and C.R. Merrill, Chromosomal distribution of 320 genes from a brain cDNA library, *Nature Genetics* 4:381 (1993).

8. J.R. Korenberg, X.N. Chen, K. Doege, J. Grover, P.J. Roughley, Assignment of the human aggrecan gene (AGC1) to 15q26 using fluorescence in situ hybridization analysis, *Genomics* 16:546 (1993).

9. D. Cohen, I. Chumakov, and J. Weissenbach, A first-generation physical map of the human genome, *Nature* 366:698 (1993).

10. A.J. Buckler, D.D. Chang, S.L. Graw, J.D. Brook, D.A. Haber, P.A. Sharp, and D.E. Housman, Exon amplification: a strategy to isolate mammalian genes based on RNA splicing, *Proc. Natl. Acad. Sci. USA* 88:4005 (1991).

11. M. Lovett, J. Kere, and L.M. Hinton, Direct selection: a method for the isolation of cDNAs encoded by large genomic regions, *Proc. Natl. Acad. Sci. USA* 88:9628 (1991).

12. M.S. Boguski, T.M.J. Lowe, and C.M. Tolstoshev, dbEST -- database for "expressed sequence tags", *Nature Genetics* 4:332 (1993).

13. M.T. Ross, D. Nizetic, C. Nguyen, C. Knights, R. Vatcheva, N. Burden, C. Douglas, G. Zehetner, D.C. Ward, A. Baldini, and H. Lehrach, Selection of a human chromosome 21 enriched YAC sub-library using a chromosome-specific composite probe, *Nature Genetics* 1:284 (1992).

14. B.H. Brownstein, G.A. Silverman, R.D. Little, D.T. Burke, S.J. Korsmeyer, D. Schlessinger, and M.V. Olson, Isolation of single-copy human genes from a library of yeast artificial chromosome clones, *Science* 244:1348 (1989)

15. C.S. Hoffman and F. Winston, A ten-minute DNA preparation from yeast efficiently releases autonomous plasmids for transformation of Escherichia coli, *Gene* 57:267 (1987)

16. I. Chumakov, P. Rigault, S. Guillou, P. Ougen, A. Billault, G. Gausconi, P. Gervy, I. LeGall, P. Soularue, L. Grinas, et al., Continuum of overlapping clones spanning the entire human chromosome 21q, *Nature* 359:380 (1992).

DISCUSSION

Katheleen Gardiner[1] and Ute Hochgeschwender[2]

[1]Eleanor Roosevelt Institute, 1899 Gaylord Street, Denver, CO 80206
[2]Unit on Genomics, NIMH, 9000 Rockville Pike, Bethesda, MD 20892

Many different approaches to transcribed sequence identification have been described, and all have been successful within the limits tested. What has not been clearly defined are the limitations of the various methods. For example, what fraction of isolated "Genes" are verifiable transcribed sequences, what artifacts may be generated and how can these quickly be identified or suppressed, what fraction or classes of genes may be routinely missed? The answers to these questions will vary among techniques, and generally will only be established with increased experience in an increased number of laboratories. Most procedures described in the preceding chapters require commitments in time, effort and resources to set up. Of pressing interest, therefore, to the novice transcriptional mapper is how to decide now on the most efficient approach for the project at hand. The choice of technique, or combination of techniques, will depend upon the nature of the gene identification effort. In this regard, specific disease gene isolation may be the easier problem when tissue specificity of expression is known, and in particular if genetic or deletional mapping has defined a small region of interest (< 1 Mb). Construction of transcriptional maps of whole chromosomes or large regions poses a more complicated problem. In this case, decisions must be made concerning the competing demands of efficiency versus comprehensiveness. In discussions of available techniques, three questions recurred. First, what are the theoretical limitations of the technique? For example, exon trapping will on theoretical grounds, miss genes with fewer than two introns. Second, what are the practical constraints, e.g. artefacts, biased recoveries, etc., which affect realization of the theoretical? And third, has the technique been successfully transferred to other laboratories? To address in more detail these kinds of questions and concerns, results of the numerous formal and informal discussions by the meeting participants are summarized here.

cDNA Hybridization Selection

cDNA hybrid selection, in its various forms, has been successfully established in a number of laboratories. There appear to be no significant differences in success between solution and membrane hybridization methods. Most laboratories use pools of cosmid clones or pulsed field gel purified YAC DNA (as large as 1 Mb), although in one laboratory use of total yeast DNA from a YAC strain has been successful. Several practical considerations merit discussion. First, the selection is not absolute; some fraction of the sequences recovered are not specific to the target human DNA. One example is ribosomal RNA sequences contaminating cDNA sources and being selected by yeast ribosomal DNA sequences contaminating YAC clone DNA. Another example is low and moderate frequency repeat sequences that are not efficiently blocked by prehybridization with Cot_1 DNA. In at least one case, cDNA clones homologous to a moderately repeated sequence swamped recovery of other selected cDNAs. The extent of this problem, and of the related problems posed by gene families and pseudogenes, remains to be determined. A second problem is that the hybrid selected cDNAs are frequently quite short, much less than full length. Arduous and redundant analyses of fragments from the same mRNA may be lessened if one short clone is used to isolate a full-length cDNA clone, which can in turn be used to identify all the other short, selected cDNAs derived from the same transcript. Practically, an additional problem concerns how many selected clones need to be analyzed to insure recovery of rare transcripts.

A final consideration is completeness: how many different cDNA sources need to be used in hybrid selection to insure recovery of all genes regardless of their temporal or tissue specificity of expression? One approach to reducing the magnitude of the problem is to hybrid-select simultaneously from multiple cDNA sources, each tagged with a unique adapter. This may, however, exacerbate the problem of recovery for rare cDNA clones.

Exon Trapping

Several laboratories have now had experiences with exon trapping. The consensus vector is pSPL3, an improved version of the original pSPL1, engineered to reduce recovery of vector-only sequences in the PCR and lacking a cryptic splice site that allowed retrieval of one class of repetitive sequence artifacts. Exon trapping has been applied to cosmids, pools of cosmids and even YAC clones; however, there are clear variations in success rates. Exon trapping from single cosmids has failed in some cases, but has been successful in others, even where pools of cosmids were used. In general, pooling reduces the number of trapped products, most likely because of competition in the PCR step.

The theoretical limitation of exon trapping is that the method will not detect genes with fewer than two introns. The practical limitations are threefold. First, the subcloned genomic fragments have to carry enough intronic sequences flanking the exon to insure presence of splice signals. Second, cryptic splice sites, located in both repetitive and unique sequences, can lead to trapped products of a non-exonic nature and are a source of artifacts. Third, a candidate exon must be verified as bona fide exonic. Because the detection method here is functional splicing and is not dependent on the expression of genes, there is an advantage in potentially achieving completeness, in the sense of detecting all internal exons in a stretch of genomic DNA. However, because the tissue of expression of a trapped exon is unknown, a potentially large number of tissues must be analyzed through Northern blots or cDNA library screenings. Negative results in these experiments may be due to the technical problem of hybridization with the very short DNA fragments obtained as trapped products, or due to the biological problem that the gene is expressed at a level below the detection limit of Northern blots and an average number of cDNA clones.

In summary, hybrid selection has a potential for completeness at the price of a high level of redundancy, while exon trapping has so far shown, in particular when scaled up, to be less than comprehensive. In a hybrid selection experiment, the tissue of expression may be known from the outset, but it is laborious to collect genes from all tissues. In exon trapping experiments, less work is required initially because the method is independent of the expression of the genes, but it ultimately becomes laborious to collect the information on the pattern of expression for a potential exon.

Random cDNA Sequencing and Mapping

The alternative approach, starting with genes (cDNA clones) and placing them on the physical and genetic map, has a conceptual constraint from the outset: the uneven representation of genes in mRNA populations, leading to both redundancy and incompleteness. Highly expressed genes and genes expressed ubiquitously will be analyzed repeatedly if not eliminated, while genes expressed at very low levels or in only a few cell types are missed altogether. Furthermore, unless only valid 5' or 3' ends of genes are sequenced, an individual gene may be identified many times from short sequences throughout the gene. The efficiency of the random cDNA sequencing approach is critically dependent on the quality of the starting library, and therefore, database information for ESTs need to include as much library information as possible, for instance, was cytoplasmic RNA used for library construction, or was the RNA treated with DNAse? Further, the presence of polyadenylation signals and poly A tails needs to be verified and documented for all ESTs. Another formidable problem with this approach is mapping short stretches of DNA to chromosomes at the resolution of YAC clones. These problems are major challenges for the application of random cDNA sequencing to disease gene identification and transcriptional map assembly.

Assessment of Putative Transcribed Sequences

A general problem facing all methods of gene identification is to prove the identified product is a true gene. While the isolation of putative transcribed sequences is now a less formidable task, the problem of verifying that the sequence - a trapped exon, a hybrid selected clone, a Grail-positive genomic fragment or a random sequence out of a cDNA library - indeed represents part of a bona fide mRNA remains formidable. A long open reading frame, while a presupposition unless confounded by untranslated regions, is by no means a proof - not even with moderate homologies in GenBank searches. This section summarizes 5 approaches (in roughly increasing order of difficulty) that are in use for assessment of a candidate transcribed sequence. All come with caveats, and all are subject to false positives and false negatives. Results with Grail analysis and zoo blots, in particular, are not conclusive, but can be valuable indicators (when successful) that verification may be possible.

1. Grail analysis. Frequently, submitting the putative sequence for analysis by Grail is the easiest first step in assessment. For a sequence $\geq$ 150 bp in length, a Grail score of excellent is good evidence that the sequence contains coding exon(s). Grail analysis of 100-150 bp sequences is less robust, and marginal scores are the highest obtainable (Grail does not analyze sequences < 100 bp). In any case, non-coding exons will not be scored. Despite the robustness of Grail, both false positives and false negatives occur. Thus, Grail analysis is a rapid way of sorting candidate sequences, but cannot be taken as proof, positive or negative.

2. Zoo blots. Evolutionary conservation, evidenced by cross-species hybridization, is more frequent among coding than non-coding sequences. However, the short sequences obtained from exon trapping, cDNA hybrid selection or Grail analysis make poor hybridization probes. Not only must the hybridization and wash stringencies be lowered to compensate for the short probe length, but they must be further reduced to compensate for the expected sequence divergence among related genes of different species. Under such conditions, neither the presence nor the absence of hybridization signal is considered conclusive.

While Grail and zoo blot analyses may increase confidence that a sequence is derived from a coding sequence, more direct evidence is required for proof.

3. Northern analysis. In many instances, positive results can be obtained from a Northern analysis. As a first condition, the putative transcribed sequence can be shown to be unique in the genome by Southern blot hybridization. The candidate sequence is then used as a probe against a Northern with 5-10 μg of one column poly A+ RNA from various tissues. Several tissue/cell line sources of RNA are recommended because they appear to express a complex collection of genes. These include fetal brain and the cell line HELA, with placenta and testes anecdotally joining the group. Other cell lines of potential "universal" use include HepG2 cells, embryonal carcinoma and embryonal stem cells, and teratocarcinoma cell lines. Again, a negative result is inconclusive and, for sequences expressed at low levels, at specific developmental times or only in a specific tissue, positive Northern results may be unobtainable.

4. cDNA library screening. Isolation of a cDNA clone, especially with a poly A tail, from one or more cDNA libraries, is a strong indication of the bona fides of a putative transcribed sequence. cDNA libraries, however, are not without artifacts. The possibility of repeat sequences, sequences derived hnRNA or genomic contaminations must always be borne in mind. As with Northern analysis, negative results in cDNA library screening cannot be conclusive.

5. RT-PCR. The least ambiguous, and least trivial to obtain, verification of genuine exonic materials is reverse transcription PCR across an intron. It requires a gene with at least one intron (to exclude the possibility of amplification from contaminating genomic DNA) and sufficient knowledge of the genomic structure and expression to design primers. However, when this information is available, RT-PCR is a technically rapid method for validation of a candidate exon.

Definition of a Transcriptional Map

Another general problem is the transcriptional map itself and how to define it. In disease gene searches, usually the gene is contained within 1 or a few Megabases, and enough is known from the disease phenotype to make plausible assumptions regarding tissue of expression. Because the genomic region analyzed is usually small, exhaustive analysis with a number of different techniques can be carried out, and there is a clearly defined endpoint which is reached with the identification of the disease gene, The case is different for general transcriptional map assembly. The genomic region may be large, encompassing significant fractions of, or even whole, chromosomes. This places important constraints on the choice of methods, weighing factors such as efficiency and completeness.

As yet, there is no consensus on the final form a transcriptional map should take, in particular concerning the relative importance of gene location, sequence information and expression pattern. The desired level of resolution was assigned to YAC or cosmid contigs, to facilitate anchoring to the physical map. Sequence information would ideally include the full-length coding sequence of each clone on the transcriptional map. Northern data for at least ten tissues would be desirable to give information on expression patterns which can be biologically interpreted.

These are data of obvious value. Additional information would include gene size and intron-exon boundary definition. However, as techniques for transcribed sequence identification are refined and popularized, and as detail accumulates in transcriptional maps, the most useful balance of the different types of data will define the eventual map format.

INDEX

Genes
    abundance, 3, 239
    chromosomal position
        influence on expression, 7
    clustering, 2, 7
    density, 2
    essential, 1
    expression, 3
    families, 61, 71
        globin, 2
        MHC, 2, 9, 201-203
    number, 1-3, 239
    ribosomal, *see* RNA, ribosomal
Genetic load, 1-2
Giemsa bands, 5, 37, 65, 127
GRAIL, *see* computational analysis, coding
        sequence identification

HTF islands, *see* CpG islands
Hybrid selection, *see* cDNAs, hybrid selection
Hybridization
    chromosome in situ
        analytic, 217
        preparative, 123-138
    interspecies, 81-90, 149-153
    oligonucleotide, 209, 239-251
        fingerprinting, 253-260
    $R_0 t$, 2-3
    saturation, 2

Image analysis, 24, 244-247, 258
Isochores, 2; *see* also GC content

Magnetic bead capture, 51-63, 114; *see also*
        cDNAs, hybrid selection
Map
    transcriptional, 7-8, 37, 65-79, 141-142,
        151, 194, 213-214, 225, 302
Multiplexing, 273, 275-276, 292
Mutation
    deleterious, 1-2
    lethal, 1
Mutation rate, 1-2

Normalization
    libraries, 32
    probes, 134-135

Oligonucleotide fingerprinting, *see*
        hybridization, oligonucleotide

Plasmid vectors
    pETV-SD2, 169-181
    pSPL1, 145, 184, 200, 300
    pSPL3, 184-198, 216, 300
Polymerase chain reaction, 92, 101-109, 261-
        271, 302
Positional candidate approach, 123, 135-137
Pseudogenes, 47, 61, 76

Quenching agents, 31; *see also* repetitive
        sequence elements, blocking

Recombination
    in vitro, 101-109
Redundancy, 47, 141-142, 151, 154-155, 186,
        188-189
Reference libraries
    cDNA, 24, 255-257
    genomic, 140-142, 158-159, 214, 291-292
    hybridization analysis, 23-28, 139-155, 157-
        167, 217, 239-251, 253-260, 294-296
Repetitive sequence elements, 34, 191, 204-
        205, 235
    blocking, 26, 59-60, 119
    identification, 71
    low copy, 65, 113
    removal, 134, 140-141, 142-143, 159-160
RNA
    ribosomal, 31, 32, 42, 61, 71

Sequence database
    dbEST, 55, 273
    EMBL, 2, 31
    GenBank, 2, 31, 44-45, 55
Sequencing
    by hybridization, 239-251, 253-260
    shallow shot-gun, 229-236
    single-pass, 240, 248-249, 273
Somatic cell hybrids, 6, 71, 158-159, 189-190, 274
Splicing
    cryptic, *see* exon amplification, artifacts
Subtraction
    directional tag PCR, 162-163

Telomeres, 5
Transcriptional domains, 7

UDG cloning, 56, 188

Zoo blots, *see* conserved sequences

The manufacturer's authorised representative in the EU is Springer
Nature Customer Service Centre GmbH, Europaplatz 3, 69115 Heidelberg,
Germany. If you have any concerns regarding our products, please
contact ProductSafety@springernature.com

Printed and bound by CPI Group (UK) Ltd, Croydon, CR0 4YY
23/04/2026
02095623-0015